95 Springer Series in Solid-State Sciences

Edited by Peter Fulde

Springer Series in Solid-State Sciences

Editors: M. Cardona P. Fulde K. von Klitzing H.-J. Queisser

Managing Editor: H. K. V. Lotsch

Volumes 1–89 are listed at the end of the bo

90 **Earlier and Recent Aspects of Superconductivity**
Editors: J. G. Bednorz and K. A. Müller

91 **Electronic Properties of Conjugated Polymers III**
Basic Models and Applications
Editors: H. Kuzmany, M. Mehring, and S. Roth

92 **Physics and Engineering Applications of Magnetism**
Editors: Y. Ishikawa and N. Miura

93 **Quasicrystals**
Editors: T. Fujiwara and T. Ogawa

94 **Electronic Conduction in Oxides**
By N. Tsuda, K. Nasu, A. Yanase, and K. Siratori

95 **Electronic Materials**
A New Era in Materials Science
Editors: J. R. Chelikowsky and A. Franciosi

96 **Electron Liquids**
By A. Isihara

97 **Localization and Confinement of Electrons in Semiconductors**
Editors: F. Kuchar, H. Heinrich, and G. Bauer

98 **Magnetism and the Electronic Structure of Crystals**
By V. A. Gubanov, A. I. Liechtenstein, and A.V. Postnikov

99 **Electronic Properties of High-T_c Superconductors and Related Compounds**
Editors: H. Kuzmany, M. Mehring, and J. Fink

100 **Electron Correlations in Molecules and Solids**
By P. Fulde

J. R. Chelikowsky A. Franciosi (Eds.)

Electronic Materials

A New Era in Materials Science

With 182 Figures

Springer-Verlag
Berlin Heidelberg New York
London Paris Tokyo
Hong Kong Barcelona
Budapest

Professor James R. Chelikowsky
Professor Alfonso Franciosi

Department of Chemical Engineering and Materials Science,
University of Minnesota, Minneapolis, MN 55455, USA

Series Editors:

Professor Dr., Dres. h. c. Manuel Cardona
Professor Dr., Dr. h. c. Peter Fulde
Professor Dr., Dr. h. c. Klaus von Klitzing
Professor Dr., Dr. h. c. Hans-Joachim Queisser

Max-Planck-Institut für Festkörperforschung, Heisenbergstrasse 1,
W-7000 Stuttgart 80, Fed. Rep. of Germany

Managing Editor:

Dr. Helmut K. V. Lotsch

Springer-Verlag, Tiergartenstrasse 17, W-6900 Heidelberg, Fed. Rep. of Germany

ISBN-13:978-3-642-84361-7 e-ISBN-13:978-3-642-84359-4
DOI: 10.1007/978-3-642-84359-4

Library of Congress Cataloging-in-Publication Data. Electronic materials : a new era in materials science / J. Chelikowsky, A. Franciosi (eds.). p. cm.–(Springer series in solid-state sciences : 95) Includes bibliographical references and index. ISBN-13:978-3-642-84361-7 1. Electronics–Materials. 2. Semiconductors. I. Chelikowsky, James R. II. Franciosi, A. (Alfonso), 1948– . III. Series. TK7871.E435 1991 621.381–dc20 90-23990

Softcover reprint of the hardcover 1st edition 1991

Typesetting: Springer TEX in-house system

54/3140-543210 – Printed on acid-free paper

Preface

Modern materials science is exploiting novel tools of solid-state physics and chemistry to obtain an unprecedented understanding of the structure of matter at the atomic level. The direct outcome of this understanding is the ability to design and fabricate new materials whose properties are tailored to a given device application. Although applications of materials science can range from low weight, high strength composites for the automobile and aviation industry to biocompatible polymers, in no other field has progress been more strikingly rapid than in that of electronic materials. In this area, it is now possible to predict from first principles the properties of hypothetical materials and to construct artificially structured materials with layer-by-layer control of composition and microstructure. The resulting superlattices, multiple quantum wells, and high temperature superconductors, among others, will dominate our technological future. A large fraction of the current undergraduate and graduate students in science and engineering will be directly involved in furthering the revolution in electronic materials. With this book, we want to welcome such students to electronic materials research and provide them with an introduction to this exciting and rapidly developing area of study.

A second purpose of this volume is to provide experts in other fields of solid-state physics and chemistry with an overview of contemporary research within the field of electronic materials. We hope that this publication will also benefit those segments of the industrial, governmental and academic communities whose perception of materials may not have evolved at the same pace as this discipline. This book should illustrate how quantum modeling, atomic resolution probes, and molecular and ion beam processing are now an integral part of the materials scientist's background.

In 1989, eleven distinguished scientists whose expertise is centered on the study of electronic materials were invited to present a personal perspective of their subfield in a seminar series at the University of Minnesota's Department of Chemical Engineering and Materials Science. Later they were asked to prepare a manuscript summarizing their view of the field. This volume is the result of that exercise.

We would like to thank the Department of Chemical Engineering and Materials Science, the Graduate School and the Institute of Technology at the University of Minnesota for providing the funding support for our seminar series and the production of this volume. We wish to thank the authors for their participation in our seminar series and for contributing their manuscripts. We also thank

Dr. Helmut Lotsch of Springer-Verlag for his help and advice. Finally, we thank Professor H.T. Davis and our colleagues of the Department of Chemical Engineering and Materials Science of the University of Minnesota for encouragement, innumerable stimulating conversations and their unique *esprit de corps*. Without their support, this project would not have been possible.

May 1991 *J.R. Chelikowsky*
Minneapolis, MN *A. Franciosi*

Contents

Contributors

Altarelli, Massimo
European Synchrotron Radiation Facility, B.P. 220,
F-38043 Grenoble Cedex, France
and
Max-Planck-Institut für Festkörperforschung, Hochfeld-Magnetlabor,
B.P. 166X, F-38042 Grenoble Cedex, France

Beltram, Fabio
AT&T Bell Laboratories, Murray Hill, NJ 07974, USA

Capasso, Federico
AT&T Bell Laboratories, Murray Hill, NJ 07974, USA

Chelikowsky, James R.
Department of Chemical Engineering and Materials Science,
University of Minnesota, Minneapolis, MN 55455, USA

Cohen, Marvin L.
Department of Physics, University of California,
Materials Sciences Division, Lawrence Berkeley Laboratory,
Berkeley, CA 94720, USA

Duke, Charles B.
Xerox Webster Research Center, 800 Phillips Road, 0114-38 D,
Webster, NY 14580, USA

Franciosi, Alfonso
Department of Chemical Engineering and Materials Science,
University of Minnesota, Minneapolis, MN 55455, USA

Goldman, Allen M.
School of Physics and Astronomy, University of Minnesota,
Minneapolis, MN 55455, USA

Harrison, Walter A.
Department of Applied Physics, Stanford University,
Stanford, CA 94305-4090, USA

Himpsel, Franz J.
IBM Research Division, T.J. Watson Research Center, Box 218,
Yorktown Heights, NY 10598, USA

Louie, Steven G.
Department of Physics, University of California,
Materials Sciences Division, Lawrence Berkeley Laboratory,
Berkeley, CA 94720, USA

Phillips, James C.
AT&T Bell Laboratories, Murray Hill, NJ 07974, USA

Poate, John M.
AT&T Bell Laboratories, Murray Hill, NJ 07974, USA

Sen, Susanta
AT&T Bell Laboratories, Murray Hill, NJ 07974, USA
Present address: Department of Electronics Science, University of Calcutta,
Calcutta, India

Weaver, John H.
Department of Materials Science and Chemical Engineering,
University of Minnesota, Minneapolis, MN 55455, USA

1. Introduction

James R. Chelikowsky and Alfonso Franciosi

Certain periods of history bear the name of the materials that have shaped the development of civilization: the *Stone* Age, the *Bronze* Age, the *Iron* Age. An appropriate characterization of our time would be the *Electronic Materials* Age, as semiconductor materials have truly revolutionized our lives. Consider the nature of the technological advances in recent time: from pocket radios and video cassette recorders to powerful new supercomputers, all have been made possible by our improved understanding of how to synthesize and process electronic materials. Indeed, contemporary economic competition often revolves around which country leads in the development of new electronic technologies.

Semiconductor materials have played a major role in the past 50 years, and will continue to do so in the foreseeable future, with the development of new artificially structured semiconductor materials such as superlattices and multiple quantum wells. Furthermore, attention has recently been focused on a different class of novel electronic materials: the high-temperature superconductors. These materials will certainly play a role in shaping the future of electronics, and the question which remains open is who will be the first to exploit the new superconductors in practical technological applications.

While the technological exploitation of electronic materials is a compelling economic issue, the study of these materials has been in the past, and is today, at the forefront of basic science. Consider, for example, the Nobel prizes in physics which have been directly or indirectly related to electronic materials. The start of the solid state electronic revolution is marked by the Nobel prize for the transistor in 1956. The Nobel prize in 1977 was awarded, in part, for research on disordered semiconductors. The discovery of a new quantum effect in semiconductors, the quantized Hall effect, was honored with the 1985 Nobel prize. The first major achievement of the scanning tunneling microscope (1986 Nobel prize) was real-space mapping of the surface of silicon. Research on superconducting electronic materials led to Nobel prizes in 1972, 1973 and 1987.

In electronic materials research, scientific and technological advances are often intimately connected. The technological need for ultra-pure and well-characterized electronic materials has resulted in the development of new experimental and theoretical methods. Application of these methods has tremendously improved our understanding of semiconductors so that our knowledge of materials such as silicon and gallium arsenide is more comprehensive on a microscopic level than for any other condensed matter system.

While the very development of modern solid-state physics owes much to electronic materials research, today the synthesis, processing, characterization and modeling of electronic materials is increasingly the focus of other disciplines such as chemistry, electrical engineering, and, primarily, materials science. This is a common occurrence in science. Often a field will emerge in one discipline and, as the discoveries in the field become known, spread to others. The study of the solid state follows this pattern. Today the shift of electronic materials research from the domain of solid-state physics to that of materials science is almost complete in some universities.

It is often said that the most exciting and innovative science is to be found at the boundaries between the core sciences. *Materials science* centers on the application of physics and chemistry to condensed matter systems of practical interest, and is therefore intrinsically interdisciplinary in nature. Materials science is therefore the natural recipient of the dynamic expansion of the science of electronic materials, which resides at the boundary between physics, chemistry and engineering. The new emphasis on electronic materials has been made possible by the coming of age of *modern materials science*. Once largely limited to metallurgical research, and with a cultural background made up of classical physics and chemistry concepts, materials science has seen a veritable technological explosion of new materials shake its foundations. The emergence of new polymers, composites, superalloys, ceramics, thin films, superlattices, etc. has brought about a renaissance of this discipline, and a redefinition of its cultural foundations. The modern materials scientist has to be able to handle sophisticated theoretical and experimental tools, and his/her background now encompasses methods as diverse as quantum modeling or advanced electron spectroscopy techniques.

From a theoretical perspective, the goal of modern materials science is to predict and understand the properties of real and hypothetical materials from first principles, i.e., from a knowledge of the constituent species present. From an experimental perspective, the goal is to develop characterization and processing techniques that can reveal and modify the atomic and electronic structure of real solids on the microscopic, mesoscopic, and macroscopic scale. The ultimate common goal is to synthesize novel structured materials with properties tailored to a specific technological application.

Tremendous progress has been made in characterization and processing within the general domain of materials science. On one hand, novel techniques such as synchrotron-radiation-based spectroscopies and microscopies (see Chaps. 4 and 8 in this volume), and a number of new electron microscopies (atomic resolution transmission electron microscopy, scanning tunneling microscopy, etc.) are providing an unprecedented degree of characterization at the microscopic level. One can now reveal the atomic and electronic structure of solids on the scale of the individual atoms. On the other hand, new synthesis techniques such as molecular beam epitaxy and chemical vapor deposition allow materials scientists to build artificially structured materials with almost any composition (see Chap. 10). Local processing methods include ion beam, electron beam, and laser-assisted techniques, which have extended the domain of accessible materials parameters well beyond the limits of thermodynamic equilibrium (see Chap. 12).

Progress from a theoretical perspective has been almost as rapid. Ten years ago, it was virtually impossible to predict from *first principles* the band gap, crystal structure or surface structure of the simplest elemental solids. Now such predictions are routine (see, for example, Chap. 3). In fact, one may be able to predict the high pressure phases of solids better than they can be experimentally determined (see Chap. 5). At the other end of the theoretical spectrum, new *empirical* methods are being developed which have predicted new materials (see Chap. 11).

Given the importance of these new developments, a review which tries to present both the theoretical and the experimental perspective seemed timely. We chose our title to emphasize that materials scientists have initiated a *New Era in Materials Science*. We believe this era will shape our technological future, and that the interdisciplinary approach of materials science is the one which will show the way to this future.

At the University of Minnesota, the Department of Chemical Engineering and Materials Science has always been at the forefront of interdisciplinary research in materials. From its very inception, the department called upon the expertise of mathematicians and physical chemists to start a new era of chemical engineering. Department heads such as Neal Amundson, Rutherford Aris, Kenneth Keller, and Ted Davis consistently expanded the program with total disregard for labels and traditional discipline boundaries, seeking to add the expertise and the methods best suited to address the current issues in the science and technology of materials. Over the years, the department has evolved into what it is today, i.e., an eclectic mixture of scientists from different disciplines, systematically crossing the traditional boundaries to apply the methods of physics and chemistry to materials research.

Following an original suggestion by Ted Davis, we decided to illustrate and celebrate the birth of the electronic materials era by inviting to Minneapolis in the Spring of 1989 eleven distinguished scientists whose expertise is centered on the study of electronic materials to present an overview of their subfield. We also requested that the lecturers prepare a manuscript summarizing their personal perspective of the field. This volume is the result of their effort.

In view of the breadth attained by modern electronic materials research, no review can attempt to present a truly comprehensive survey. In choosing which subfields to include, we were driven by our own tastes and we were forced to omit entire areas of great importance. We apologize in advance for these omissions. Because of the editors' own expertise, our review emphasizes electronic properties, methods, and applications, and shortchanges somewhat the field of structural characterization and synthesis methods. Many subfields in these two areas, however, have been the subject of recent reviews, including molecular beam epitaxy [1.1–3] and chemical vapor deposition [1.4–6], synthesis techniques, electron microscopies [1.7–9], extended fine structure techniques [1.10–12] and scanning tunneling microscopy [1.13–15]. From a theoretical perspective we mention recent reviews on impurities and localized states [1.16–18], heterojunctions [1.19–21], and molecular dynamics methods [1.22]. We address interested readers to these excellent surveys already available in the literature.

The first contribution, from Walter Harrison of Stanford University, focuses on "simple" methods for characterizing the electronic structure of solids. He notes that many first attempts in this area were centered in treating the solid as a "huge molecule". While such an idea was appealing, its implementation had never been very successful. Professor Harrison reversed this situation by combining two seemingly disparate views of a semiconducting solid. One view is that a semiconductor is a covalent solid with bonding orbitals that are relatively localized, and therefore necessarily far from free-electron-like. Another view is that the observed energy bands of a semiconductor are not that far removed from a free-electron-like energy band description. By determining the coupling between localized orbitals in a semiconductor required to produce free-electron-like bands, one can obtain coupling constants transferable to molecules or other solids. In addition, utilizing simple approximations based upon the nature of the electron states in ionic, covalent and metallic systems, it becomes possible to describe a wide range of dielectric and electronic properties of solids in terms of universal electronic structures.

While the "simple" approach of Professor Harrison is very valuable for obtaining a qualitative understanding of the solid-state chemical bond, it cannot be used to make quantitative predictions from first principles. In Chap. 3, Steven Louie of the University of California at Berkeley reviews methods to explain and predict many material properties solely from a knowledge of the constituent atomic species. As Professor Louie explains, most of our knowledge about the electronic structure of solids is derived from spectroscopic studies. Examples include band gaps, optical transition energies, and photoemission spectra. The methods reviewed in Chap. 3 have for the first time successfully predicted spectroscopic energies of semiconductors and other materials, and will have a far-reaching impact on our understanding of condensed matter.

Some of the experimental techniques that are providing a direct description of the electronic structure of solids and a critical test of the theoretical models are described in Chap. 4 by Franz Himpsel of the IBM T.J. Watson Research Center. While most of the early spectroscopy work on solids involved optical absorption studies, the availability of ultrahigh vacuum techniques and advanced vacuum ultraviolet and X-ray sources, including synchrotron radiation, and of novel electron beam sources and detectors has dramatically shifted the emphasis to electron spectroscopy techniques. These methods have greatly enhanced our ability to examine both surface and bulk electronic structures and stimulated new theoretical work as discussed in Chap. 3. Dr. Himpsel has elected to review primarily angle resolved photoemission and inverse photoemission spectroscopies, and gives some special attention in the last section of the chapter to recent promising developments in the area of spin-resolved, spatially resolved and time-resolved techniques. Throughout the chapter, the examples selected emphasize materials problems of industrial interest.

The next contribution is by Marvin L. Cohen from the University of California at Berkeley. Professor Cohen emphasizes that one of the breakthroughs that marked the emergence of a "new era of materials science" is the newly acquired ability of theorists to explain and predict a variety of structural, vibrational, electronic, thermodynamic and even superconductive properties of materials using quantum theory. The group of Professor Cohen, for example, was the first to produce first principles calculations able to reproduce the equations of state for a *real* solid other than for an inert gas solid. Indeed, calculations of this kind have almost single-handedly revolutionized our ability to predict the properties of solids. The phase diagram calculations for silicon polytypes resulted in correct predictions for the phase behavior os silicon under pressure. Moreover, on the basis of these theoretical studies, a new *superconducting* high pressure of silicon was predicted. This predicition was later confirmed by experiment. Given the paucity of "correct" superconducting predicitions, this achievement must be considered one of the landmarks in the science of materials.

In Chap. 6 Allen Goldman of the University of Minnesota addresses another landmark of modern materials science: the discovery of *high-temperature* superconductors, one of the great surprises in materials research. Professor Goldman reviews this rapidly developing field, summarizing our current experimental understanding of these materials, and notes that despite the publication of literally thousands of papers in the past few years, the details of the mechanism of high temperature superconductivity remain obscure. Also, it is not known at this time whether the promise of enhanced performance of high T_c superconductors will be realized. Professor Goldman points to the source of both of these problems: the complexity of the new materials and the "subtle" differences between the low- and high-temperature superconductors. For example, a key issue with respect to the mechanism is whether the superconductors are strongly or weakly coupled. If the superconducting gap could be measured accurately, this key issue would be resolved. However, materials problems and characterization issues have resulted in measured gaps which span the weak to the strong limits. Professor Goldman discusses the prospects for defining and executing the critical experiments required to resolve this and other issues.

One of the areas of electronic materials research which has seen the most rapid expansion is surface science. In Chap. 7 Charles Duke from Xerox Webster Research Center notes that during the past decade a comprehensive description of surface geometries, chemical bonding and electronic structure has been achieved for the cleavage faces of both zincblende and wurtzite structure semiconductors. Dr. Duke presents in his contribution the theoretical approach which is mostly responsible for such impressive achievements. The approach is an intermediate one, between the simple theory of chemical bonds in solids presented by Professor Harrison in Chap. 2, and the first principles approach described by Professor Cohen in Chap. 5. More recently, theoretical predictions of the surface atomic vibrational spectra of semiconductors have begun to appear in the literature, and Dr. Duke reviews the major concepts and results within this area. A brief, stimulating paragraph discusses some of the technological aspects of metal overlayers on semiconductors, and results for Sb and Al overlayers on GaAs, GaP and InP surfaces are examined as examples.

Metal-semiconductor interfaces are considered from an experimental perspective by John H. Weaver of the University of Minnesota in Chap. 8. Since electronic devices inherently involve metal–semiconductor contacts, the study of such contacts has always been an important part of electronic materials research. Professor Weaver focuses on microscopic studies of metal–semiconductor interfaces, and describes some of the experimental methods that are allowing *local* investigations with high spatial and energy resolution. Such techniques have greatly improved our understanding of surface processes such as intermixing, chemical reactions and Schottky barrier formation. The first goal of this review is to examine composition profiles and chemical changes associated with the evolution of interfaces between metals and III-V semiconductors, as a function of metal coverage, metal deposition method, and substrate temperature. A second goal is to examine Schottky barrier formation in the context of surface chemical reaction, the process of overlayer growth, temperature and other system parameters.

As we mentioned earlier, one of the most exciting aspects of modern materials science is our ability to synthesize artificial solids with specific properties. One class of artificial materials which is revolutionizing the design of solid-state electronic devices are semiconductor superlattices. Massimo Altarelli of the European Synchrotron Radiation Facility in Grenoble reviews in Chap. 9 the current theoretical understanding of the properties of semiconductor superlattices. Dr. Altarelli centers his review on how the energy bands of these solids differ from those predicted by simple particle-in-a-box-models, and examines some key examples drawn from the spectroscopy of heterostructures. Specific topics addressed include interband and excitonic optical properties, and resonant tunneling of holes in double-barrier heterostructures. Also discussed is the electronic structure of superlattices in a magnetic field parallel to the interface, or in a perpendicular electric field. Dr. Altarelli points out that in all cases the three-dimensional superlattice band structure can accurately describe such configurations in terms of the semiclassical dynamics of band electrons.

Some of the most exciting device applications of superlattices and other artificially structured semiconductor materials are reviewed in Chap. 10 by Federico Capasso of AT&T Bell Laboratories, together with two of his collaborators, Fabio Beltram and Susanta Sen. As they note, the dramatic progress in the past decades toward faster and smaller devices and circuits has continued and even accelerated in some cases in the past few years. The progress in device performance has so far defied the many predictions of what constitutes the "ultimate device performance". Dr. Capasso and coauthors emphasize in their chapter the new degrees of freedom in device design which derive from our ability to tailor semiconductor composition and electronic properties via epitaxial growth techniques. Their review focuses both on new devices, which depend on novel applications of quantum effects, e.g., resonant tunneling, and on more traditional devices in which the growth parameters are altered to optimize performance.

As mentioned earlier, one of the goals of materials science is to predict routinely new materials with specific properties. The technological implications are obvious, but what is considerably less obvious is whether there is any hope of realizing such a goal. Many scientists have felt this goal is unachievable; a notable exception is James C. Phillips of AT& T Bell Laboratories. His review in Chap. 11 centers on the connection between quantum ideas derived from pseudopotential theory, see Chap. 5, and crystal chemistry and metallurgy. The ideas here are subtly simple. If the atomic constituents of one crystal are similar to those of another crystal, then we expect the properties to be *similar*. For example, if the atomic species of a hypothetical crystal are similar to the species of a known superconducting crystal, then one might expect the hypothetical crystal to be a superconductor. The central issue is what constitutes *similar*. In traditional chemistry, the ideas of Pauling come to mind. If two atomic species have similar electronegativities and atomic radii, they may form similar structures with similar properties. The electronegativities and atomic radii form *chemical scales* by which one may attempt to quantify similarities. Given the success of pseudopotential theory in describing the electronic and structural properties of solids, one might hope to construct some modern, i.e., quantum mechanical, scales whose accuracy should exceed that of the traditional scales. Dr. Phillips reviews new quantum scales that have generated dramatically successful results for some 10^5 binary and ternary solid-state systems.

In the last chapter, we abandon modeling and characterization of electronic materials in thermodynamic equilibrium, and examine some of the frontiers of nonequilibrium processing of electronic materials. John Poate of AT&T Bell Laboratories focuses in Chap. 12 on ion beam and laser processing of semiconductors. These methods have tremendous technological implications, as the evolution of ion implantation and laser annealing has shown in recent years. Such techniques are routinely used to synthesize materials with impurity concentrations far exceeding those attainable by conventional crystal growth and even epitaxial thin film growth methods. Moreover, these processes have provided us with a much improved fundamental understanding of crystal growth and phase transitions in electronic materials. The field of ion- and laser-assisted processing is today far too extended to allow a comprehensive survey within a single chapter, and Dr. Poate has elected to emphasize recrystallization in silicon, and segregation and diffusion phenomena at the amorphous–crystal interface and in amorphous silicon. Dr. Poate discusses in some detail laser heating techniques, which have been employed to explore novel regimes of solidification. The thermodynamic parameters of amorphous silicon are discussed and contrasted with those of crystalline silicon, highlighting important implications for our understanding of phase transitions in silicon.

In summary, we make no pretension to have proposed in this volume an exhaustive survey of the field of electronic materials, but we do hope to have been able to communicate to the reader some of the excitement that pervades this dynamic area of condensed matter research. A new era of materials science has indeed begun, and electronic materials research is at the very core of it. If this book has captured an image of this rapidly changing field, while challenging a few young minds to become the materials scientists of the future, it will have served its purpose.

References

1.1 A very useful review can be found in: E.H.C. Parker (ed.): *The Technology and Physics of Molcular Beam Epitaxy* (Plenum, New York 1985)

1.2 See, for example, L.L. Chang, K. Ploog (eds.): *Molecular Beam Epitaxy and Heterostructures*, NATO ASI Series, Series E, No. 387 (Martinus Nijhoff, Dordrecht, The Netherlands 1985)

1.3 A recent review for silicon is: E. Kapser, J.C. Beam (eds.): *Silicon Molecular Beam Epitaxy* (CRC, Boca Raton, FL 1988)

1.4 P. Sroeve (ed.): *Integrated Circuits: Chemical and Physical Processing*, Am. Chem. Soc. Symp. Series, Vol. 290 (American Chemical Society, Washington, DC 1984)

1.5 T.F. Kuech: Mater. Sci. Rep. **2**, 1 (1987)

1.6 D.W. Hess, K.F. Jensen, T.J. Anderson: Rev. Chem. Eng. **3**, 97 (1985)

1.7 E. Ruska: Rev. Mod. Phys. **59**, 627 (1987)

1.8 J.C.H. Spence: *Experimental High Resolution Electron Microscopy*, 2nd ed. (Oxford University Press, New York 1988)

1.9 W. Krakow, F.A. Ponce, D.J. Smith (eds.): *High Resolution Microscopy of Materials*, MRS Proc., Vol. 139 (Materials Research Society, Pittsburgh, PA 1989)

1.10 B.K. Teo, D.C. Joy (eds.): *EXAFS Spectroscopy Techniques and Applications* (Plenum, New York 1981)

1.11 E.A. Stern, S.M. Heald: In *Handbook of Synchrotron Radiation*, Vol. 1b, ed. by E.E. Koch (North-Holland, Amsterdam 1983) p.995

1.12 P.A. Lee, P.H. Citrin, P. Eisenberger, B.M. Kincaid: Rev. Mod. Phys. **53**, 769 (1981)

1.13 G. Binnig, H. Rohrer: Rev. Mod. Phys. **59**, 615 (1987)

1.14 G. Binnig, H. Rohrer: Surf. Sci. **152/153**, 17 (1985)

1.15 J.E. Demuth: In *Physics in a Technological World*, ed. by A.P. French (American Institute of Physics, New York 1988) p.141

1.16 S.T. Pantelides: In Proc. 19th Int. Conf. on the Physics of Semiconductors, ed. by W. Zawadzki (IOP, Warsaw, Poland 1988) p.29, and references therein

1.17 See, for example, J. Bourgoin, M. Lannoo (eds.): *Point Defects in Semiconductors II*, Springer Ser. Solid-State Sci., Vol. 35 (Springer, Berlin, Heidelberg 1983)

1.18 M. Stavola, S.J. Pearton, G. Davies (eds.): *Defects in Electronic Materials*, Mater. Res. Soc. Proc., Vol. 104 (Materials Research Society, Boston, MA 1988)

1.19 H. Kroemer: In [1.2]

1.20 See, for example, F. Capasso, G. Margaritondo (eds.): *Heterojunction Band Discontinuities: Physics and Device Applications* (Elsevier, Amsterdam 1987)

1.21 G. Margaritondo, A. Franciosi: Annu. Rev. Mater. Sci. **14**, 67 (1984)

1.22 A recent review is given in G. Ciccotti, D. Frenkel, I.R. McDonald (eds.): *Simulation of Liquids and Solids* (North-Holland, Amsterdam 1987)

2. The Simplest Ab Initio Theory of Electronic Structure

Walter A. Harrison

With 2 Figures

It has long been thought that electronic states in molecules and solids could be viewed as combinations of free-atom states. This idea even appeared in the thesis of *Bloch* [2.1] in 1928 but was thought to be so crude as to only be useful for illustrative purposes. *Hückel* [2.2] used this concept for molecular orbitals in 1931, noting that for estimating electron energies one needed only a few parameters, which could be calculated or guessed, with many taken equal to zero. This theory is called *tight-binding theory*. By 1963 *Hoffmann* [2.3] had proposed the *extended Hückel theory*, in which he reintroduced many of the discarded parameters and found approximate ways to compute them. There was a flexibility to these theories in the sense that, once they were programed and used to calculate a property, it was easy to modify the program to improve the result and the next application used the new scheme, with a new name. See [2.4] for a comprehensive discussion of such approaches.

Once computers became more powerful, and density-functional methods became available for ab initio theories of electronic structure, approximate schemes had little to offer except some saving in computing costs, and they sacrificed very much in the way of accuracy and reliability. The simple schemes have tended to fall into disuse, at least in the theory of solids. At the same time an alternative approach arose (see [2.5] for a complete account), in which further approximations were made to allow analytic determination of electronic-state energies and properties without the aid of a computer. The accuracy was comparable to the extended Hückel theory and the resulting formulas clearly displayed all of the trends in properties from system to system. As with extended Hückel theory, the results were considerably less accurate than the full computations, and of course much less accurate than semiempirical schemes which essentially interpolated experimental properties among similar systems. However, for many purposes the simple theory was at least as useful, was much easier to understand and use, and was more easily extended over the entire range of properties and solids. Furthermore, it had a stability in that most attempts to improve the accuracy quickly led to machine computations and the loss of the analytic simplicity which was one of the theory's major virtues. It is the analytic theory which we wish to discuss here.

In order to develop this approximate theory it was necessary to call upon a vast amount of quantitative information which was available from the more complete theories [2.6]. However, an interesting feature of the development was that once the understanding was achieved, a means for obtaining all needed parameters was also found and there was no cause to continue to use the findings of the computational theory. In that sense the theory became independent of the more quantitative theory and became an approximate ab initio theory. This can give the misleading impression that the development of the simple theory was independent of the full analysis; on the contrary, it owes its origin to the full theory.

2.1 Tight-Binding Theory

The essential idea of tight-binding theory is that the electronic states $|\psi_k\rangle$ in a molecule or a solid can be written as a linear combination of atomic states $|\psi_i\rangle$ of the constituent atoms

$$|\psi_k\rangle = \sum_i u_i|\psi_i\rangle \ . \tag{2.1}$$

This becomes a simplification if we also assume that only a *minimal basis set* of such atomic states is needed, only those states from shells which are partially occupied in the free atom, and that is the approximation we make. For most systems this consists of the valence s- and p-states appropriate to the row in the periodic table from which the atom comes. The electronic eigenstates and eigenvalues ε_k are obtained by minimizing the expectation value of the Hamiltonian for the state $|\psi_k\rangle$ with respect to the coefficients u_i. If, when these states are occupied by the number of electrons present, the atoms remain essentially neutral, the sum of the eigenvalues of the occupied states

$$E_{\text{tot}} = \sum_{\text{occ}} \varepsilon_k \tag{2.2}$$

can be regarded as the total electronic energy of the system in the sense that the *change* in energy of the system as the atoms are rearranged (including separating them to free atoms) is equal to the change in this sum of eigenvalues. This provides a simple theory of virtually the entire range of properties of solids, since most properties (cohesion, elasticity, dielectric constants, etc.) can be written in terms of the energy.

2.2 Universal Parameters

In order to proceed we need values for the tight-binding parameters, which mathematically are the matrix elements of the Hamiltonian matrix in the basis of the atomic states. These parameters include the atomic term values ε_i (the diagonal matrix elements), which we take as Hartree-Fock free-atom term values from [2.7], which are listed in Table 2.1. They also include the coupling V_{ij} between electronic states on neighboring atoms (the off-diagonal matrix elements). It has

Table 2.1. Hartree-Fock atomic term values from [2.7]. The first entry is ε_s, the second is ε_p (values in parentheses are highest core level). All values are in eV

I	II	III	IV	V	VI	VII	VIII	IA	IIA
							He	Li	
							−24.97	−5.34	
	Be	B	C	N	O	F	Ne	Na	
	−8.41	−13.46	−19.37	−26.22	−34.02	−42.78	−52.52	−4.95	
	−5.79*	−8.43	−11.07	−13.84	−16.76	−19.86	−23.14	(−41.30)	
	Mg	Al	Si	P	S	Cl	Ar	K	Ca
	−6.88	−10.70	−14.79	−19.22	−24.01	−29.19	−34.75	−4.01	−5.32
−3.84*	−5.71	−7.58	−9.54	−11.60	−13.78	−16.08	(−25.96)		(−36.47)
Cu	Zn	Ga	Ge	As	Se	Br	Kr	Rb	Sr
−6.49	−7.96	−11.55	−15.15	−18.91	−22.86	−27.01	−31.37	−3.75	−4.85
−2.35*	−4.01*	−5.67	−7.33	−8.98	−10.68	−12.43	−14.26	(−22.04)	(−29.88)
Ag	Cd	In	Sn	Sb	Te	I	Xe	Cs	Ba
−5.98	−7.21	−10.14	−13.04	−16.02	−19.12	−22.34	−25.69	−3.36	−4.29
−2.59*	−3.98*	−5.37	−6.76	−8.14	−9.54	−10.97	−12.44	(−18.59)	(24.59)
Au	Hg	Tl	Pb	Bi	Po	At	Rn	Fr	Ra
−6.01	−7.10	−9.82	−12.48	−15.18	−17.96	−20.82	−23.78	−3.21	−4.05
−2.65*	−3.94*	−5.23	−6.52	−7.79	−9.05	−10.33	−11.64	(−17.10)	(−22.30)

* Values extrapolated from neighboring values.

turned out that to a good approximation those can be neglected except for nearest-neighbor atoms and that these nearest-neighbor couplings can be taken to have the universal values

$$\begin{aligned} V_{ss\sigma} &= -1.32\hbar^2/md^2\,, \quad & V_{pp\sigma} &= 2.22\hbar^2/md^2\,, \\ V_{sp\sigma} &= 1.42\hbar^2/md^2\,, \quad & V_{pp\pi} &= -0.63\hbar^2/md^2\,. \end{aligned} \tag{2.3}$$

The first two subscripts indicate the angular-momentum quantum number of the atomic state and the third indicates the angular momentum of the two states around the internuclear axis (σ for zero and π for one unit of $\hbar$).

Where do these couplings come from? This question has an interesting answer which also gives us some idea how much confidence we can have in the universal values. Initially these forms came from fitting an analytic form to values obtained from careful band calculations on semiconductors [2.8]. It was then realized by *Froyen* and *Harrison* [2.9] that these forms followed from the fact that semiconductor band structures could be well described by tight-binding theory but at the same time were very free-electron-like.

This is most easily seen for the simplest case: that of a one-dimensional chain of atoms, each containing an atomic s-state. With nearest-neighbor coupling it is easily seen that the energy bands are given by $\varepsilon_k = \varepsilon_s + 2V_{ss\sigma}\cos kd$, where d is the spacing between atoms and k varies from $-\pi/d$ to $+\pi/d$. If the spacing were such that the electrons also behaved as free electrons, their energy could also be written $\varepsilon_k = \varepsilon_0 + \hbar^2k^2/2m$. The total band width, as k varies from 0 to π/d, is given in tight-binding theory by $-4V_{ss\sigma}$, and for free electrons by $\hbar^2(\pi/d)^2/2m$.

If these are to be consistent we conclude that $V_{ss\sigma} = -(\pi^2/8)\hbar^2/md^2$, of the same form as that given in (2.3). In fact, the coefficient $-\pi^2/8$ is -1.23. For the three-dimensional diamond lattice an analogous treatment gave the same form for all four interatomic matrix elements listed in (2.3) and for $V_{ss\sigma}$ the value $-9\pi^2/64 = -1.39$; the value -1.32 given in (2.3) came from fitting the known band structure of germanium. The differences are not so important and it might have been more satisfying to use the geometric values for the coefficients, but we shall use the customary values given in (2.3). It is important, however, that these values cannot be very far off at the equilibrium spacing because in real semiconductors the band structures *are* quite free-electron-like. It has also turned out that these interatomic matrix elements remain quite the same when the atoms are rearranged, as in an ionic structure, so that it is reasonable to take them as universal.

We now have in Table 2.1 and Eq. (2.3) all the parameters needed for studying the electronic structure, and therefore the properties, of simple covalent (ordinarily semiconducting) and ionic solids.

We should point out, however, what has turned out to be the most serious approximation given above: the assumption that each atom remains neutral in the solid. Frequently within the tight-binding context the atoms become charged (a sodium atom in rocksalt is found to have a net charge near 0.8), and this must be taken into account to obtain reasonably accurate predictions of some properties. When this has been done, with a self-consistent determination of the charges, agreement between theory and experiment improves significantly [2.10]. The corrections are not as large as one might at first think, since the shift in energy levels on one atom resulting from its charge tends to be cancelled (usually to within 10%) by the shift due to its neighboring atoms, which have the opposite charge. We need not worry about these Coulomb shifts in the present discussion as we review the theory of a variety of systems.

2.3 A Diatomic Molecule, N_2

We begin by applying the above approach to a simple diatomic molecule, nitrogen, where the calculations are completely trivial, but at the same time show exactly how we may proceed in the solid. We take the term values for nitrogen from Table 2.1, $\varepsilon_s = -26.22\,\text{eV}$ and $\varepsilon_p = -13.84\,\text{eV}$. Here we take the internuclear distance from experiment, $d = 1.09\,\text{Å}$, although it is possible to extend the theory to estimate these values [2.10, 11]. We immediately obtain the couplings from (2.3), giving for example $V_{pp\pi} = -4.04\,\text{eV}$ and molecular π-states at energies $\varepsilon_p = \varepsilon_p \pm V_{pp\pi} = -17.88\,\text{eV}$ and $-9.80\,\text{eV}$. In N_2 only the lower state is occupied. The four σ-states are also immediately obtained by evaluating parameters and solving a quadratic equation.

In spite of the fact that they are reduced to trivial calculations, these results for the occupied states are in rather good accord with more complete theories requiring extensive computer calculations. The predicted value $-17.88\,\text{eV}$ for the bonding π-state may be compared with the results of a full calculation by *Ransil* [2.12], who obtained -16.7 eV. For the three occupied σ-states we obtain -41.1, -21.7, and -21.5 eV, compared to *Ransil*'s -38.6, -20.3, and -15.1 eV. As in many cases in solids, the empty states here are not at all well given, but they do not directly enter the ground-state properties. It is necessary to extend the theory if one is interested in the excited states [2.5].

2.4 A Simplification Using Hybrids

Before proceeding, it will be interesting to simplify the electronic structure, a necessity for solids though completely unimportant for the simple diatomic molecule. We introduce the familiar concept of *hybrid states*. In the calculation of the σ-states in N_2 we may use as a basis hybrids in the form $(|s\rangle \pm |p\rangle)/\sqrt{2}$ which are oriented on each atom either toward or away from their neighbors, as illustrated in Fig. 2.1. If we again solve the appropriate quadratic equation we obtain precisely the values obtained above. However, we may instead make the approximation of neglecting the coupling of an outward-directed hybrid with any states on the other atom, also indicated in Fig. 2.1. Then these two states become nonbonding at an energy $\varepsilon_{\text{nb}} = (\varepsilon_s + \varepsilon_p)/2 = -20.0$ eV. Further, the remaining two states are written immediately as $\varepsilon_{\text{nb}} \pm (V_{ss\sigma} - 2V_{sp\sigma} - V_{pp\sigma})/2$ with the lower having the value -40.4 eV. We have not lost any appreciable accuracy by this simplification, but all we have saved is the solution of a quadratic equation, so it is of no importance. Such a simplification in a solid, on the other hand, replaces a major computational task by a trivial estimate such as that in the diatomic molecule. It will allow us to calculate the energy bond by bond, rather than all at once for the entire crystal, and enable us to treat complicated atomic arrangements which without this simplification would be a formidable task even for a computer.

The magnitude of the coupling between hybrids is called a *covalent energy*, in this case given by

$$V_2 = \tfrac{1}{2}|V_{ss\sigma} - 2V_{sp\sigma} - V_{pp\sigma}| = 3.19\hbar^2/md^2 \,. \tag{2.4}$$

An analogous expression will apply to covalent solids.

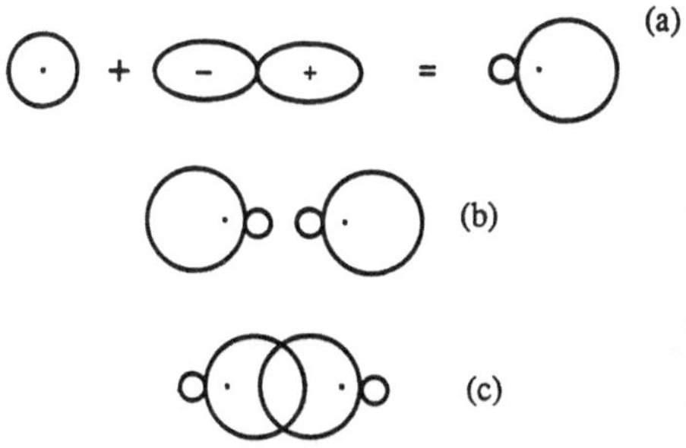

Fig. 2.1. **(a)** An s-state and a p-state can be combined to make a hybrid state which is directed to the right. **(b)** In a diatomic molecule the outward-directed hybrids are called nonbonding states and their coupling to other states can be neglected. **(c)** The inward-directed hybrids are coupled by the covalent energy V_2, given in this case by (2.4), and form bonding and antibonding states

2.5 Cohesion of N_2

We can immediately evaluate properties for this diatomic molecule. We need, however, to add a repulsion between atoms, which arises from the overlap of atomic states on adjacent atoms and a corresponding increase in the kinetic energy of the electrons occupying these states. Without it, the bond energies proportional to $1/d^2$ would cause the two nitrogen atoms to fall together indefinitely. This is called an *overlap repulsion* and we may guess, on the basis of the virial theorem [2.11], that it will vary as twice the power of the attraction, or as a constant divided by d^4. (Then this kinetic energy will end up half as large as the potential energy arising from the interatomic coupling, as appropriate to the virial theorem.) Combining this with the energy of the occupied (and nonbonding) states we obtain a total distance-dependent energy of $-2V_2+4V_{pp\pi}+A/d^4$. (We have used hybrids for this evaluation.) This must be minimum at the observed spacing and that condition gives the value of A for the nitrogen molecule. We may now substitute that value back and subtract the sum of energies of the occupied states in the molecule from that in the two free atoms to obtain a cohesive energy of $|\varepsilon_p-\varepsilon_s-V_2+2V_{pp\pi}| = 16.2\,\mathrm{eV}$. (Note that the term A/d^4 has cancelled half of the terms $-2V_2+4V_{pp\pi}$.) This may be disappointing compared to the experimental cohesive energy of 8.7 eV, but this factor of two may also be familiar. Extended Hückel theory [2.3], which we discussed at the beginning, also has been found to overestimate cohesive energies by a factor of two. Indeed, this turns out to be a feature of carbon-row systems. We will find that this factor-of-two discrepancy disappears for heavier systems. It is not too much of a problem, since we know to expect just this discrepancy for carbon-row systems. I do not believe that the origin of the discrepancy is known, but our comparison with more accurate energy levels for nitrogen would suggest that it is not an error in the one-electron eigenvalues but in the use of the sum of those eigenvalues to represent the total energy, i.e., the accuracy of (2.2).

We could also obtain the interatomic force constant as the second derivative of this total energy with respect to the internuclear distance and thus obtain the vibrational frequencies, or from the full curve we could obtain the full set of frequencies, including anharmonicity.

2.6 Polarizability of N_2

This description of the electronic structure also allows an analysis of the dielectric properties. The effect of the field $\boldsymbol{E}$, in tight-binding theory, is simply to shift the term values on the two atoms, separated from each other by $\boldsymbol{d}$, with respect to each other by $-e\boldsymbol{E}\cdot\boldsymbol{d}$. This makes the bond asymmetric, shifting the charge to one side. The effect is obtained by solving a quadratic equation and leads to a polarizability for fields along the molecular axis of

$$\alpha = e^2 d^2 \left(\frac{1}{V_2} + \frac{2}{|V_{pp\pi}|} \right) . \tag{2.5}$$

We see that the entire range of properties of the system follows once we have a representation of the electronic structure. The same is true in solids. We begin with semiconductors, covalent solids which have bonds closely analogous to those of N_2.

2.7 Tetrahedral Semiconductor Bonds

Silicon, gallium arsenide, and other tetrahedral semiconductors all have four valence electrons per atom. It is possible to make just four orthogonal hybrids with the four valence orbitals (sp^3) per atom, and thus four independent bonds. Four electrons per atom are just enough to fill the bonds, if each atom has four neighbors. If we construct these four hybrids [each of the form $(|s\rangle + \sqrt{3}|p\rangle)/2$, with the p-state oriented along an internuclear distance] and neglect their coupling with any but the opposite hybrid directed into the same bond, in exact analogy with our treatment of hybrids in N_2, we may treat the system as consisting of independent bonds. The analysis then becomes as simple as it was for N_2. The new covalent energy is

$$V_2 = \tfrac{1}{4} |V_{ss\sigma} - 2\sqrt{3} V_{sp\sigma} - 3V_{pp\sigma}| = 3.22\hbar^2/md^2 . \tag{2.6}$$

This covalent energy for sp^3-hybrids is slightly larger than the covalent energy for sp-hybrids ($3.19\hbar^2/md^2$) given in (2.4). In a polar semiconductor such as gallium arsenide the two hybrids making up the bond have different energy, but we still use (2.6) for the covalent energy and find the bond energy by solving a quadratic equation, just as we did for N_2 when a field was applied. We obtain

$$\varepsilon_b = \frac{\varepsilon_h^c + \varepsilon_h^a}{2} - \sqrt{\left(\frac{\varepsilon_h^c - \varepsilon_h^a}{2} \right)^2 + V_2^2} , \tag{2.7}$$

where the hybrid energies are denoted as coming from the cation or anion. Half the difference of the hybrid energies arises so often that it is convenient to define it as the *polar energy*,

$$V_3 = \tfrac{1}{2} \left(\varepsilon_h^c - \varepsilon_h^a \right) . \tag{2.8}$$

2.8 Semiconductor Energy Bands

A third energy which enters many properties is called the *metallic energy*,

$$V_1 = \tfrac{1}{4}(\varepsilon_p - \varepsilon_s) . \tag{2.9}$$

In a polar semiconductor there are two metallic energies, one for the anion and one for the cation. The coupling between two hybrids sharing the same atom in a tetrahedral solid can easily be seen to be equal to the negative of this metallic energy, $-V_1$. Thus adjacent bonds are coupled due to this V_1 (in a nonpolar semiconductor the coupling between bonds is easily seen to be $-V_1/2$). It is V_1 which broadens the bonding levels into bands just as the coupling $V_{ss\sigma}$ broadened the s-levels in the atomic chain discussed above into a band of width $4V_{ss\sigma}$.

In a simplest approximation the width of the bonding band, or valence band, in a tetrahedral semiconductor is given by $4V_1$. In that approximation the band gap between the occupied valence band and empty conduction band is

$$E_0 = 2V_2 - 4V_1 . \tag{2.10}$$

This turns out to be a good estimate of the direct gap even in a more complete calculation of the bands. If the metallic energy exceeds half the covalent energy, the gap goes to zero and the system becomes metallic; this occurs in tetrahedral tin. It is the increase in the ratio $2V_1/V_2$, called the *metallicity*, from diamond to silicon to germanium to tin (principally due to the decrease in V_2 from the increased interatomic distances) which provides the main trend in the properties of these systems.

2.9 Cohesion in Semiconductors

The calculation of the total energy in terms of the two-center bonds described above is directly parallel to that for molecular N_2. It gives $V_2 - 2V_1$ per bond for nonpolar ($V_3 = 0$) semiconductors in direct analogy with the $|\varepsilon_p - \varepsilon_s - V_2 + 2V_{pp\pi}|$ which we found for N_2. Interestingly enough, this goes to zero, as did the gap (2.10), when the metallicity approaches 1. However, when the metallicity grows, the effects on the energy of the coupling between each bond and the neighboring antibonds become important and must be included to get meaningful results. This is quite easily done in perturbation theory giving corrections (called *metallization*) to the cohesive energy [2.11] proportional to V_1^2/V_2. The resulting cohesion, for polar as well as nonpolar semiconductors, is good to approximately 20% [2.11], except for carbon-row systems in which we overestimate the cohesion by a factor of two, as previously indicated. It may be noted that for tin the cohesion arises entirely from metallization, since $V_2 - 2V_1 = 0$; the bonding energy ($-V_2$) is just cancelled by promotion energy ($2V_1$).

2.10 The Dielectric Properties

We may also directly calculate the polarizability of these two-center bonds for fields along the bond, as in N_2, and average over angles to obtain the dielectric susceptibility of a semiconductor, which is given by [2.5]

$$\chi = \frac{\sqrt{3}e^2V_2^2}{8d(V_2^2+V_3^2)^{3/2}} \ . \tag{2.11}$$

This is not one of the most accurate predictions. It tends to be too small by a factor of order one half for the homopolar semiconductors, but it gives quite accurate trends. In particular, the ratio of the susceptibilities of the polar semiconductors is quite close. For example, the ratios χ(GaAs)/χ(Ge), χ(ZnSe)/χ(Ge), and χ(CuBr)/χ(Ge) are very close to experiment.

In judging the predictions, we should not forget that they have been made without computation, with all of the dependences upon bond length and composition directly as displayed in (2.11). Furthermore, it is the same three fundamental parameters, V_1, V_2, and V_3, which determine both the dielectric and the bonding properties (though this was not assumed in the earliest treatment [2.5]). This is of course possible because the theory is based upon the fundamental electronic structure, crudely described as it is.

2.11 Ionic Crystals

We turn next to compounds such as rock salt. If we considered a molecule of NaCl rather than a crystal we could of course proceed with the two-center bond as in N_2. However, in the crystal each atom has six neighbors and it is not possible to make six independent hybrids from the four orbitals per atom. Nor does it make any sense to try to add higher-energy orbitals such as the d-states in order to do this, since in reality the coupling of these d-states to the valence states of the sodium and chlorine pushes them still higher and leaves them unoccupied. We need a new view of the electronic structure, but fortunately can proceed with the same parameters. Again using the Hartree-Fock term values of Table 2.1 we see that the lowest sodium valence state is far above the highest chlorine valence state,

$$\varepsilon_s(\mathrm{Na}) - \varepsilon_p(\mathrm{Cl}) = 8.8\,\mathrm{eV} \ . \tag{2.12}$$

The coupling between them is quite small, at $d = 2.82$ Å we obtain from (2.3) the value $V_{sp\sigma} = 1.36$ eV. This is characteristic of the ionic solid. Electrons drop from the cation states to the anion states and are little disturbed by the residual coupling.

As a first approximation we can neglect the coupling altogether and treat the system as composed of independent ions. Then the energy gap between occupied and full states is given by the 8.8 eV of (2.12), in agreement with the experimental value of 8.5 eV. It is at first astounding that the answer is even near correct since we have used an energy near the ionization energy of chlorine for an added electron, while the electron affinity, which might seem more appropriate for an added electron, is smaller by 10 eV. However, as we indicated at the outset, the Madelung shift very nearly cancels the intra-atomic Coulomb repulsion giving the difference between electron affinity and ionization energy. The extent of the agreement is perhaps fortuitous. These Coulomb potentials do not quite cancel and there are corrections to the one-electron theory of opposite sign which remove the resulting discrepancy [2.10].

We also obtain the cohesion immediately from the differences in one-electron energies. As the solid is formed we gain $\varepsilon_s(\text{Na}) - \varepsilon_p(\text{Cl})$ per atom pair from the transfer of the electron. Thus the cohesion is predicted to be just the 8.8 eV of (2.12), again in agreement with the experimental 8.0 eV. In fact, this theory has produced an empirical rule, which we believe to be new: that the cohesion of an alkali halide is equal to the gap. This same, almost trivial theory gives quite good trends both for the gap and the cohesion for all of the alkali halides, and similar agreement for the divalent alkaline-earth chalcogenides.

We turn finally to the dielectric susceptibility, which we find equal to zero without coupling. On the scale of the semiconductor susceptibilities this is not a bad estimate. However, if we wish to obtain values we must incorporate the coupling between levels. We examine this more closely in the next section.

2.12 Covalency in Ionic Compounds

We first perform a simple band calculation, analogous to that for the atomic chain, and find that at $\boldsymbol{k} = 0$ the symmetry is high enough that there is no coupling of the s-states to the p-states. Our prediction of the gap is unchanged by the inclusion of coupling and remains in good accord with experiment. At other wavenumbers the bands push each other apart.

For other properties it is most appropriate to include this weak covalent coupling in perturbation theory. The coupling of each occupied chlorine p-state to each of two neighbors gives a shift of $-V_{sp\sigma^2}/[\varepsilon_s(\text{Na})-\varepsilon_p(\text{Cl})]$. Thus the six p-electrons per chlorine have their energy shifted by $-12V_{sp\sigma^2}/[\varepsilon_s(\text{Na})-\varepsilon_p(\text{Cl})] = -2.5\,\text{eV}$, which is a small contribution to the cohesion. Half of this is cancelled by the overlap repulsion, which must also be included as it was for the covalent bonds. The independent-ion theory is not importantly affected.

The susceptibility may also be directly calculated in perturbation theory [2.5]. The coupling of each p-state electron to a neighbor transfers a charge $V_{sp\sigma^2}/[\varepsilon_s(\text{Na}) - \varepsilon_p(\text{Cl})]^2$ to that atom. When a field is applied that charge changes, due to the change in $\varepsilon_s(\text{Na}) - \varepsilon_p(\text{Cl})$. The corresponding dipole leads directly to a susceptibility of

$$\chi = \frac{4e^2 V_{sp\sigma^2}}{[\varepsilon_s(\text{Na}) - \varepsilon_p(\text{Cl})]^3 d} = 0.06 \tag{2.13}$$

for NaCl, again nearly a factor of two smaller than the observed 0.10. Again it is of the right order, and the formula applies to all of the other alkali halides and alkaline-earth chalcogenides. As before, the same theory and the same parameters give us the entire range of properties of these ionic solids.

2.13 Transition-Metal Compounds

When transition-metal atoms are present, we must include their d-states among the valence states. These states have atomic term values near those of the atomic s-state, but the d-states are much more strongly localized near the nucleus; thus their coupling with neighbors tends to be weaker. For our purposes here we shall again neglect Coulomb shifts in the d-states, though frequently they are far from negligible [2.10].

Usually the lowest-energy empty states and highest-energy occupied states (in the sense of an ionic crystal) are the d-states from the metallic atom and the p-states from the nonmetallic atom. The coupling between these can be obtained [2.5] by a theory analogous to, though somewhat more intricate than, that given above for coupling between s- and p-states. (The d-state theory actually predated the sp-theory and suggested that the latter was a possibility.) The largest coupling is given by

$$V_{pd\sigma} = -2.95\hbar^2 r_d^{3/2} md^{7/2} . \tag{2.14}$$

The parameter r_d is known for each transition metal, being for example 0.80 Å for iron and 0.67 Å for copper [2.5]. Again we have all the parameters needed to calculate the electronic structure and properties of the compounds.

Frequently, the coupling $V_{pd\sigma}$ is not sufficiently small compared to $\varepsilon_d(M) - \varepsilon_p(X)$ that we can use perturbation theory, e.g., in TiC and in the cuprates. We need a new view. A particularly convenient one is to use the method of moments to estimate the width $\langle(\varepsilon - \varepsilon_{\mathrm{av}})^2\rangle$ of the pd-bands that are formed. It can be shown that the five atomic d-states, coupled to three atomic p-states, produce two nonbonding p-bands, three bonding bands and three antibonding bands (just as, in fact, the single sodium s-state coupled to three chlorine p-states gave two nonbonding p-bands, one bonding and one antibonding band). We can approximate the density of states as illustrated in Fig. 2.2 and fit the subband widths to agree with our estimated second moment.

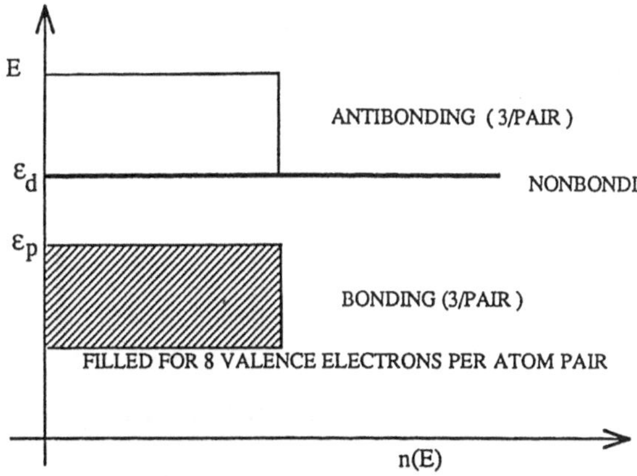

Fig. 2.2. Simplified pd density of states for transition-metal compounds

The average energy of the bonding band, obtained from the second moment, is easily found to be

$$\langle \varepsilon_b \rangle = \frac{1}{2}(\varepsilon_d + \varepsilon_p) - (V_2^2 + V_3^2)^{1/2} \tag{2.15}$$

with $V_3 = (\varepsilon_d - \varepsilon_p)/2$ and $V_2 = \sqrt{n}(\frac{1}{3}V_{pd\sigma}^2 + \frac{2}{3}V_{pd\pi}^2)^{1/2}$ from the second-moment calculation, with n the number of neighbors, six in the rock-salt structure. Using these forms we can directly proceed with the bonding and dielectric properties of these systems.

It is interesting to view the energy from (2.15) as arising from *resonant bonds* [2.13]. There are three electron pairs forming a bond with a covalent energy and a polar energy, but shared for each atom with six neighbors. Of particular interest is the enhancement of the covalent energy by the factor $\sqrt{n}$; the energy of a resonating bond is enhanced by the square root of the number of bond sites among which it resonates. This resonance energy comes from the theory directly without additional parameters. Similarly the coupling in rock salt added an energy $-12V_{sp\sigma^2}/[\varepsilon_s(\text{Na}) - \varepsilon_p(\text{Cl})]$ for the two electrons in the resonant bond, with the coupling enhanced by a factor of $\sqrt{6}$. Similarly also, for benzene and graphite, the resonant π-bond energy is enhanced by $\sqrt{2}$ and $\sqrt{3}$, respectively, in comparison to that in ethane.

2.14 Summary

We see that tight-binding theory and universal parameters make it possible to estimate most of the properties of molecules and of covalent and ionic solids. The estimates become simple if we take an appropriate outlook, that outlook being different for different systems. Indeed, the outlook in each case is quite close to that traditionally taken for these systems. For full numerical solution of the electronic structure and properties of these systems, it would not be necessary to understand them, but understanding is essential if we wish to simplify the theory. It is sometimes difficult to differentiate between the inessential and the essential features in a full numerical solution containing both.

References

2.1 F. Bloch: Z. Phys. **52**, 555 (1928)

2.2 E. Hückel: Z. Phys. **70**, 204 (1931)

2.3 R. Hoffmann: J. Chem. Phys. **39**, 1397 (1963)

2.4 C.A. Coulson: In *Physical Chemistry, An Advanced Treatise*, Vol. V/Valency, ed. by H. Eyring (Academic, New York 1970) pp.288ff., 370 ff.

2.5 W.A. Harrison: *Electronic Structure and the Properties of Solids* (W.H. Freeman, New York 1980), (Dover, New York, 1989)

2.6 See, for example, J.R. Chelikowsky, M.L. Cohen: Phys. Rev. B **14**, 556 (1976)

2.7 J.B. Mann: *Atomic Structure Calculations, 1: Hartree-Fock Energy Results for Elements Hydrogen to Lawrencium* (Clearinghouse for Technical Information, Springfield, VA 1967)
2.8 Calculated values from D.J. Chadi, M.L. Cohen: Phys. Status Solidi (b) **68**, 405 (1975) were fit by W.A. Harrison: In *Festkörperprobleme*, Vol. 17, ed. by J. Treusch (Vieweg, Braunschweig 1977) p. 135
2.9 S. Froyen, W.A. Harrison: Phys. Rev. B **20**, 2420 (1979)
2.10 W.A. Harrison: Phys. Rev. B **31**, 2121 (1985);
G.K. Straub, W.A. Harrison: Phys. Rev. B **39**, 10325 (1989)
2.11 W.A. Harrison: Phys. Rev. B **27**, 3592 (1983), especially the appendix
2.12 B.J. Ransil: Rev. Mod. Phys. **32**, 245 (1960)
2.13 L. Pauling: *The Nature of the Chemical Bond* (Cornell University Press, Ithaca, NY 1960)

3. Theory of Electronic Excitations in Solids

Steven G. Louie

With 8 Figures

One of the long-sought-after goals of condensed matter theorists has been to explain and predict the properties of real materials from first principles using quantum theory. Several recent advances and the availability of modern computers are making this goal approachable ([3.1–3] and Chap. 5). The application of first principles methods has given fundamental understanding of the behavior of a host of materials and materials systems, such as surfaces and interfaces. As in atoms and molecules, the physical and chemical properties of materials are ultimately governed by their electronic structure. In particular, the ability to predict the electronic excitation spectra is important in the understanding and application of the electronic materials.

Materials properties may be generally divided into two groups, those related to the electronic ground state of the system, and those associated with exciting the system electronically into a higher energy state. Ground-state properties include the structure, the binding energy, various static structural parameters, and the vibrational properties. These quantities can be determined from knowing the ground-state total energy as a function of the atomic coordinates ([3.1, 2] and Chap. 5). Electronic excited-state properties, on the other hand, involve the excitation of an electron away from the ground state and require knowing the quantum energy levels of the electrons or electronic structure [3.3]. They are measured in optical, direct and inverse photoemission, and in various spectroscopic experiments. For example, the excited-state properties of a surface can be quite sensitively dependent on the existence and energy of the localized surface states.

The band structure theory of solids has given a conceptual framework for understanding electron excitations. In fact, empirical schemes [3.4] based on the single-particle picture have often been very successful in interpreting experimental spectroscopic data. However, because of the complexity of the strong electron–electron interactions in solids, predictive calculations based on first principles [3.5] have only been possible in the past few years. In this chapter, we present an overview of a recent theoretical development in this area of first-principles calculation of the excited-state properties of real materials. The theory [3.5, 6] is shown to be very general and has been applied successfully to the study of semiconductors and metals as well as surfaces, interfaces, and small metal clusters. Selected examples from these studies are presented to illustrate the capabilities and predictive power of the approach.

3.1 Quasiparticle Theory of Electron Excitations

Solid-state theory has made great progress since the 1960s in first-principles calculation of the properties of real materials. These advances made it possible in the early 1980s to compute with good accuracy the structural energies and related ground-state properties (such as lattice constant, bulk modulus, vibrational frequencies, structural stability, surface structural parameters, etc.) of many solids. However, these ab initio total energy methods, which are mostly based on the local density functional (LDA) formalism [3.7, 8] in treating electron correlations, do not directly give the excitation energies. Indeed, in most cases, the practice of comparing LDA electronic band structure energies to spectroscopic data has led to rather severe discrepancies. For example, the best calculation using the LDA predicted Ge to be a metal and gave for Si a band gap of only 0.5 eV instead of the experimental value of 1.17 eV. In general, the band gaps of semiconductors and insulators are underestimated in the LDA by 50% or more. Other self-consistent field methods such as the Hartree-Fock (HF) approach have given even worse results (Table 3.1). This is known as the "band gap" problem. This problem exists not only for the band gap of insulators, but extends to all excitation spectra of solid-state systems.

The source of the problem is traced to an inadequacy of previous first-principles calculations in treating many-electron interaction effects for excited-state properties. A quantitative understanding of spectroscopic data requires the concept of quasiparticles [3.6], the particle-like excitations in an interacting many-electron system. Owing to electron–electron interactions arising from the Pauli exclusion principle and Coulomb repulsion, the excitation energies (or energy band structure) for the electrons are modified. The excited electron or quasiparticle is dressed with an electronic polarization cloud giving rise to a renormalized energy, an effective mass, and a finite lifetime. Thus, an accurate treatment of the dynamical correlations seen by an electron in a solid is crucial in calculating the excitation energies. Much effort has been devoted to developing theoretical methods [3.5, 6, 9–12], either semiempirically or from first principles, for calculating the quasiparticle properties going beyond the one-particle picture.

One successful and fruitful development has been a first-principles self-energy approach [3.5], in which the electron's energy is determined by directly calculating the contribution of the dynamical polarization of the surrounding electrons. There are no adjustable parameters in the calculation. The only inputs are the atomic numbers of the constituent elements and the crystal or surface symmetry. In this approach, the quasiparticle energies and wavefunctions are determined by solving a Schrödinger-like equation:

$$(T + V_{\text{ext}} + V_H)\psi_{n\boldsymbol{k}}(\boldsymbol{r}) + \int d\boldsymbol{r}'\, \Sigma(\boldsymbol{r}, \boldsymbol{r}'; E_{n\boldsymbol{k}})\psi_{n\boldsymbol{k}}(\boldsymbol{r}') = E_{n\boldsymbol{k}}\psi_{n\boldsymbol{k}}(\boldsymbol{r}) , \qquad (3.1)$$

where T is the kinetic energy operator for the electron, V_{ext} is the external potential due to the ions, V_H is the average (Coulomb) potential. The many-electron exchange and correlation effects are included in the so-called self-energy operator Σ. The self-energy operator describes an effective potential on the quasiparticle resulting from the response of the other electrons in the system. In general, Σ is nonlocal, energy-dependent, and non-Hermitian, with the imaginary part giving the lifetime of the quasiparticles. The excited electron's energy E_{nk} (with n and $\boldsymbol{k}$ denoting the band and $\boldsymbol{k}$-vector quantum indices) is usually written as a sum of a single-particle term, E^0_{nk}, plus a self-energy Σ_{nk} due to exchange-correlation effects.

The self-energy operator Σ is taken to be the first-order term in a series expansion in terms of the screened Coulomb interaction W and the dressed Green function G of the electron:

$$\Sigma(\boldsymbol{r},\boldsymbol{r}';E) = \mathrm{i}\int \frac{d\omega}{2\pi}\mathrm{e}^{-\mathrm{i}\delta\omega}G(\boldsymbol{r},\boldsymbol{r}';E-\omega)W(\boldsymbol{r},\boldsymbol{r}';\omega)\,, \tag{3.2}$$

where δ is a positive infinitesimal. The two major components of the theory are the interacting Green function

$$G(\boldsymbol{r},\boldsymbol{r}';E) = \sum_{nk}\frac{\psi_{nk}(\boldsymbol{r})\psi^*_{nk}(\boldsymbol{r}')}{E-E_{nk}-\mathrm{i}\delta_{nk}}\,, \tag{3.3}$$

and the dynamically screened Coulomb interaction

$$W(\boldsymbol{r},\boldsymbol{r}';\omega) = \Omega^{-1}\int d\boldsymbol{r}''\varepsilon^{-1}(\boldsymbol{r},\boldsymbol{r}'';\omega)V_{\mathrm{c}}(\boldsymbol{r}''-\boldsymbol{r}')\,, \tag{3.4}$$

where ε is the time-ordered dielectric response function of the system and V_{c} is the usual bare Coulomb interaction.

The screened Coulomb interaction incorporates the dynamical many-electron interaction effects between an electron and the other electrons in the solid. The dielectric response function ε is the key in determining the electron self-energy in the present approach. Unlike the case of the electron gas, $\varepsilon(\boldsymbol{r},\boldsymbol{r}';\omega)$ is dependent separately on the spatial variables $\boldsymbol{r}$ and $\boldsymbol{r}'$, because of electron charge density inhomogeneity in a solid. In Fourier space, $\varepsilon_{GG'}(\boldsymbol{q},\omega)$ for a crystal is a matrix in the reciprocal space lattice vectors whose off-diagonal elements describe the so-called local fields (variations in electronic polarizability at different points in the crystal). These local fields are physically very important [2.13] and represent a major difficulty in evaluating the self-energy operator for a real material such as a semiconductor. Note that neglect of screening in determining Σ (i.e., using the bare Coulomb interaction instead of the screened one) would reduce the quasiparticle equation (3.1) to the usual Hartree-Fock equation. The screening potentials of a static point charge at two different locations in the Si crystal are depicted in Fig. 3.1. The significant differences in the two potentials illustrate the importance of local field effects.

The quasiparticle energies together with Σ and G must be obtained in a self-consistent fashion. In practical applications, the electron Green function may be constructed initially using the LDA Kohn-Sham eigenfunctions and eigenvalues,

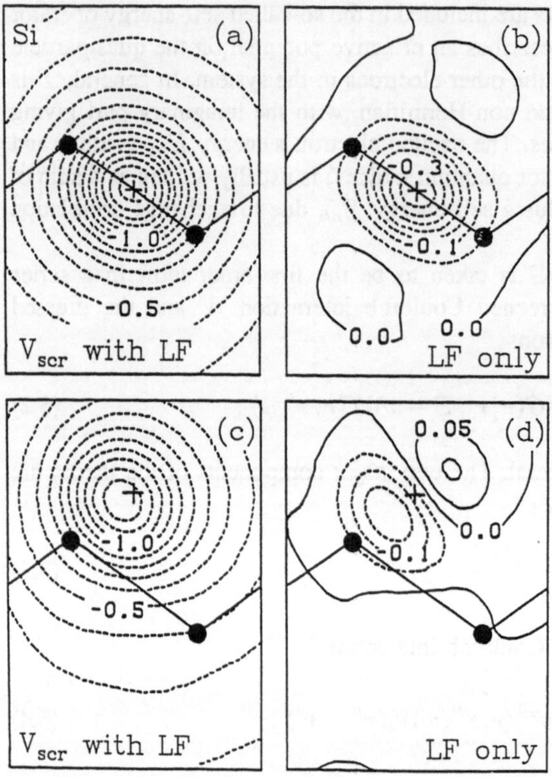

Fig. 3.1. (a,c) The screening potential in response to a single electron at r' (indicated by +) displayed in the $(1\bar{1}0)$ plane of Si in units of Ry. **(b,d)** The contribution from local fields only

and subsequently updated with the quasiparticle spectrum from (3.1). The dynamical response function is obtained in two steps. First, the static dielectric matrix ε is calculated [3.13] as a ground state property from the LDA. Second, the dielectric matrix is extended to finite frequency using a generalized plasmon pole model [3.5], which represents each component $\boldsymbol{q}$, $\boldsymbol{G}$, and $\boldsymbol{G}'$ of the imaginary part of ε^{-1} by a delta function with its strength and position determined by exact sum rules. There are no adjustable parameters. Comparisons with experiment and with subsequent calculations using alternative methods [3.12] for calculating the frequency dependence of ε^{-1} showed that the generalized plasmon pole scheme is highly accurate.

This approach has been used to compute the excited-state properties of a variety of solid-state systems, including semiconductors and metals as well as surfaces, interfaces, superlattices, and small metal clusters. It has been shown that the use of the crystalline Green function and inclusion of local fields (the full dielectric matrix) and dynamical screening effects are all important factors in a full physical description of the quasiparticle properties.

3.2 Band Gaps and Excitation Spectra of Bulk Crystals

The first application and major success of the quasiparticle self-energy approach was the resolution of the band gap problem in semiconductors and insulators [3.5, 14]. Table 3.1 compares the calculated minimum gaps for several selected crystals with experimental values [3.15, 16]. As seen from the table, with the excitation energies properly interpreted as transitions between quasiparticle states, the calculated energy gaps are in excellent agreement with experiment. For semiconductors, the quasiparticle gaps open up dramatically as compared to the LDA results. In general, the gaps are within approximately 0.1 eV of the experimental values. As discussed in the previous section, there are no adjustable parameters in the calculations. The only inputs are the atomic numbers and the crystal structure. Similarly, accurate results have been obtained for other semiconductors and insulators, including ionic compounds [3.12, 17] and alloy systems [3.18]. This level of accuracy is achievable only when both local fields and dynamical (frequency-dependent) screening effects are included in the evaluation of the electron self-energy operator.

Table 3.1. Comparison of calculated band gaps E_g (in eV) with experiment

	HF	LDA	Present theory	Expt. [Ref.]
Diamond	13.6	3.9	5.6	5.48 [3.15]
Si	6.4	0.5	1.29	1.17 [3.15]
Ge	4.9	< 0	0.75	0.74 [3.15]
LiCl	16.9	6.0	9.1	9.4 [3.16]

The optical properties of solids may also be analyzed in terms of the quasiparticle energies. In general, the agreement between theory and experiment is at the same level as for the minimum gaps. A comparison between experimental optical transitions and the calculated transition energies is presented in Table 3.2 for the crystals diamond, Si, and Ge. The experimental values are from high precision electroreflectance and wavelength modulation spectroscopy measurements [3.15, 19–22]. The theoretical results are typically within 0.1–0.2 eV of the experimental values for all transitions except for the very high energy ones in diamond, where the experimental uncertainties are large. This kind of agreement is quite a dramatic improvement over previous theories [3.2, 9–11]. In fact, the level of agreement with experiment using the quasiparticle theory is comparable to that of the highly successful "empirical pseudopotential method" [3.4], in which the band structure is obtained by fitting to optical data using several adjustable parameters. Furthermore, the first-principles nature of the quasiparticle results has been of predictive value in interpreting optical data.

In addition to the minimum band gaps and optical transition energies, the theory yields excellent band dispersions. Figure 3.2 depicts the calculated Ge quasi-

Table 3.2. Comparison between theory and experiment for optical transitions (in eV) in Ge, Si, and diamond

	LDA	Present work	Expt. [Ref.]
Ge			
$\Gamma_{7v} \rightarrow \Gamma_{8v}$	0.30	0.30	0.297 [3.19]
$\Gamma_{8v} \rightarrow \Gamma_{7c}$	−0.07	0.71	0.887 [3.19]
$\Gamma_{8v} \rightarrow \Gamma_{6c}$	2.34	3.04	3.006 [3.19]
$\Gamma_{8v} \rightarrow \Gamma_{8c}$	2.56	3.26	3.206 [3.19]
$X_{5v} \rightarrow X_{5c}$	3.76	4.45	4.501 [3.19]
Si			
$\Gamma_{25'v} \rightarrow \Gamma_{15c}$	2.57	3.35	3.4 [3.20]
$\Gamma_{25'v} \rightarrow \Gamma_{2'c}$	3.26	4.08	4.2 [3.15]
$L_{3'v} \rightarrow L_{1c}$	2.72	3.54	3.45 [3.20]
$L_{3'v} \rightarrow L_{3c}$	4.58	5.51	5.50 [3.20]
Diamond			
$\Gamma_{25'v} \rightarrow \Gamma_{15c}$	5.5	7.5	7.3 [3.21]
$\Gamma_{25'v} \rightarrow \Gamma_{2'c}$	13.1	14.8	15.3 ±5 [3.22]
$X_{4v} \rightarrow X_{1c}$	10.8	12.9	12.5 [3.15]

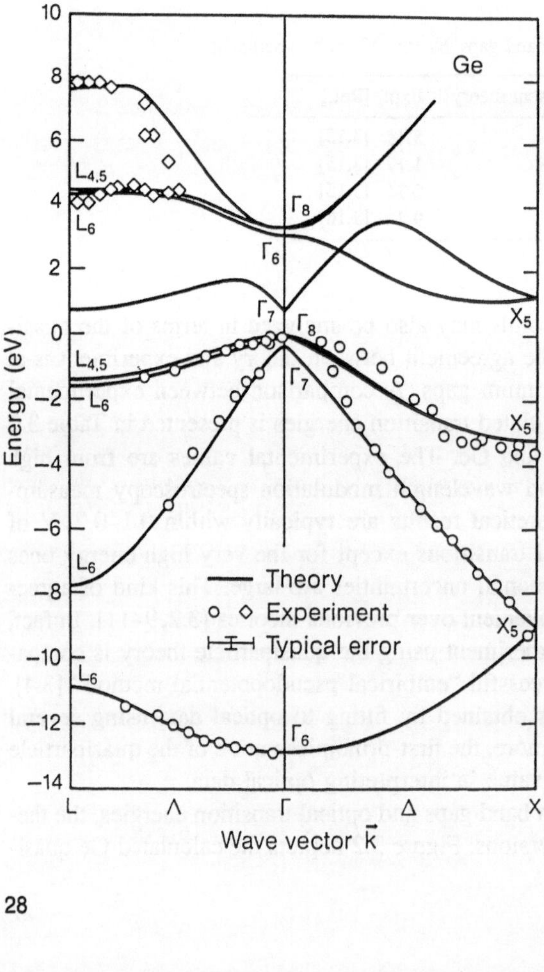

Fig. 3.2. Direct ○ [3.23] and inverse photoemission ◊ ([3.24] and Chap. 4) data versus theoretical quasiparticle energies of Ge

Table 3.3. Comparison of the calculated conduction band critical point energies (in eV) at L relative to the valence band edge to the results of angle-resolved inverse photoemission experiments for Si and Ge

	Theory	Expt. [3.24]
Si		
$\Gamma_{25'v} \to L_{1c}$	2.27	2.4 ± 0.15
$\Gamma_{25'v} \to L_{3c}$	4.25	4.15 ± 0.1
Ge		
$\Gamma_{8v} \to L_{6c}$	0.75	0.8
$\Gamma_{8v} \to L_{6c}$	4.33	
$\Gamma_{8v} \to L_{4,5c}$	4.43	4.2 ± 0.1
$\Gamma_{8v} \to L_{6c}$	7.61	7.8 ± 0.1

Table 3.4. The results (in eV) of the present theory for LiCl are compared with experiment for band gap E_g, Cl $3p$ band width W_{3p}, and the separation between the Cl $3s$ and $3p$ bands $E_{3p} - E_{3s}$

LiCl	Present theory	Expt. [Ref.]
E_g	9.1	9.4 [3.16]
W_{3p}	3.8	4.0 ± 0.2 [3.25]
$E_{3p} - E_{3s}$	11.6	11.6 ± 0.5 [3.26], 11.0 ± 0.6 [3.27]

particle band structure. The valence band structure is in excellent agreement with results from angle-resolved photoemission experiments [3.23]. The predicted conduction band energies have been quantitatively verified by subsequent data from angle-resolved inverse photoemission studies ([3.24] and Chap. 4). The theoretical results are, in general, well within the experimental errors. A more detailed comparison for the conduction band states at L with inverse photoemission [3.24] is given in Table 3.3. A similar level of agreement with experiment has been obtained for other materials. Some results together with experimental data [3.16, 25–27] for the very ionic insulator LiCl are given in Table 3.4.

A particularly interesting case is that of the band width of the alkali metals. Although conceptually these are the simplest of the metals, their measured band widths [3.28, 29] are in substantial disagreement with the corresponding free-electron values or results from traditional band structure calculations using LDA or Hartree-Fock methods. A comparison of the calculated quasiparticle band structure [3.30] with photoemission data for Na is given in Fig. 3.3. The surprisingly large observed band-width reduction is explained by the self-energy effects, although the origin of the dispersionless feature near the Fermi energy is not yet understood. The calculated quasiparticle energies for Na are also consistent with X-ray absorption edge measurements [3.31], which are sensitive to the position of the empty density of states features deriving from the gap at the zone face. Similar narrowing of the occupied states has been calculated for other simple metals [3.32, 33]. The inclusion of exchange-correlation effects in the dielectric screening has been found to be important in evaluation of the self-energy in this case.

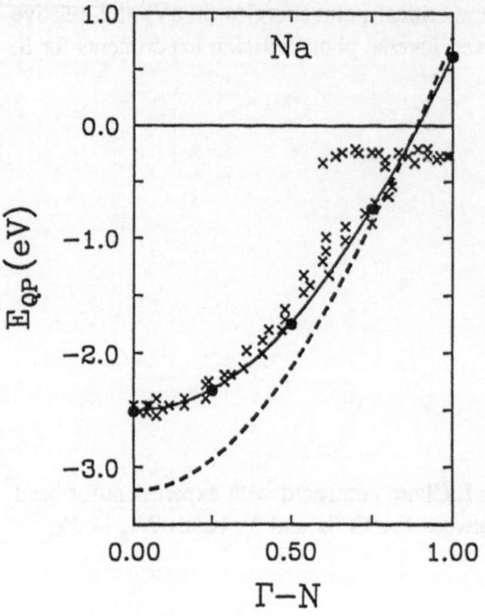

Fig. 3.3. Quasiparticle energies for Na: (×) experimental data from [3.29]; (- -) LDA eigenvalues; and (•) calculated quasiparticle energies

Since most modern electronic structure calculations for solids are performed with the LDA and since the LDA eigenvalues have been widely used in interpreting experimental spectra, it is of theoretical interest to compare the calculated quasiparticle energies with the LDA eigenvalues. The difference between the electron energies from the two theories is plotted as a function of the quasiparticle energy in Fig. 3.4 for four materials: diamond, Si, Ge, and LiCl. The plotted values correspond to the corrections needed to be added to the LDA results to obtain the quasiparticle energies. The required correction is dominated by a large jump at the energy gap and has a rather smooth energy dependence away from the gap region. This discontinuous jump at the gap reflects a change in the character of the semiconductor wavefunction from bonding to antibonding across the gap. The self-energy operator, being a nonlocal operator and extending over a range of one bond length [3.5], is much more sensitive to the wavefunction nodal structure than the simplified local exchange-correlation potential $V_{xc}(\boldsymbol{r})$ used in local density functional theory. In addition, the correction for the conduction states shows a different energy dependence from that for the valence states. This energy dependence can be quite large for insulators such as diamond. Thus, the needed correction is generally not a rigid shift of the bands. Furthermore, the distribution of the correction between the conduction and valence states is materials dependent. As will be discussed below, this has important implications for theories on heterojunction band offsets or Schottky barriers. An accurate determination of the line up of quasiparticle band edges is needed in both cases.

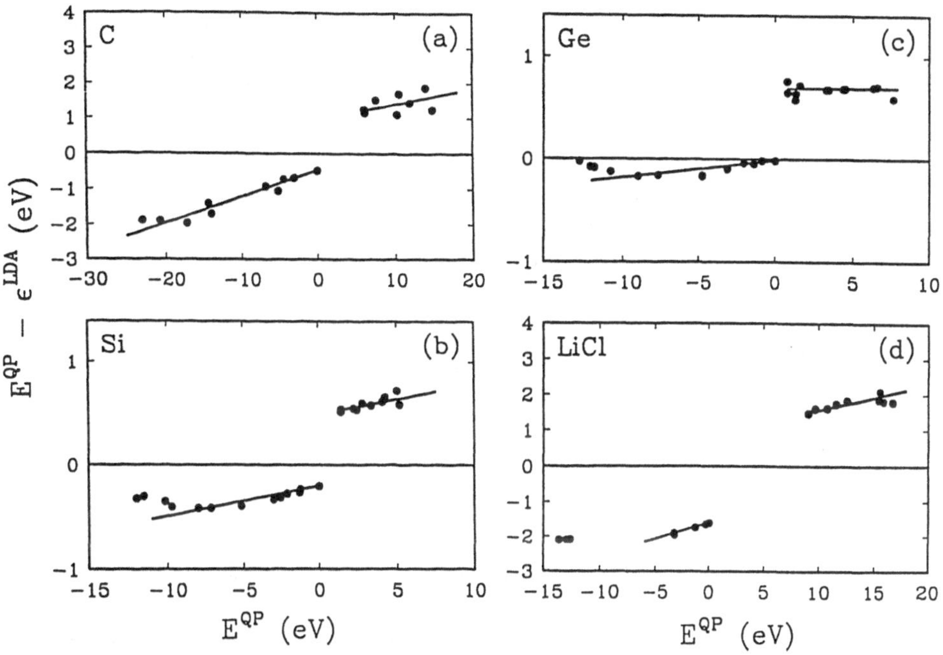

Fig. 3.4. Difference between calculated quasiparticle energies and LDA eigenvalues for (a) diamond, (b) Si, (c) Ge, and (d) LiCl

3.3 Surfaces, Interfaces, Superlattices, and Clusters

The successes of the quasiparticle self-energy calculations for bulk materials have led to refinement and extension of the approach to more complex systems, including various surfaces, interfaces, superlattices, and clusters. The power of a first-principles theory is of particular importance for these new and often experimentally less well-characterized systems for which empirical methods are much less effective. This section focuses on several prototypical examples: the As chemisorbed Ge(111) and Si(111) surfaces [3.34, 35], the GaAs(110) surface [3.36], the GaAs/AlAs heterojunction [3.37], and the alkali metal clusters [3.38].

For surfaces and interfaces, the goal is to provide a theoretical framework for predicting not only the structure but the excitation spectra, in particular the surface-state energies and band offsets at interfaces. As in the bulk band gap problem, use of LDA eigenvalues as surface-state energies has significantly underestimated the gap between the occupied and empty surface states. The many-body calculations involve two processes: first, the determination of the atomic coordinates from an ab initio pseudopotential local density functional total energy calculation and then the calculation of the quasiparticle energy for both bulk and surface states. This allows a well-founded comparison of theory to surface spectroscopic measurements.

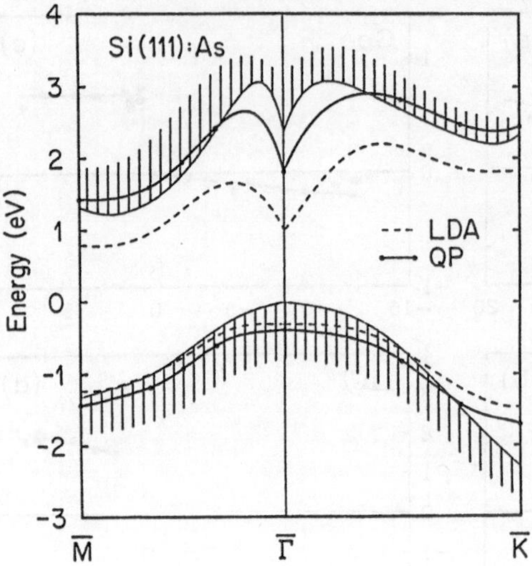

Fig. 3.5. Calculated quasiparticle surface bands for the As-capped Si(111) surface plotted against the bulk projected bands in the surface Brillouin zone. LDA surface band energies are also shown (*dashed lines*)

Experimentally, it is observed that at saturation coverage the chemisorbed As atoms substitute for the outermost-layer atoms on the Si(111) and Ge(111) surfaces [3.39, 40]. The surfaces become chemically passivated and stable against reconstruction showing a 1×1 periodicity. Currently there is much activity on these systems because of the interest in the growth of GaAs on covalent semiconductors. They are good prototypes for the many-body calculations because of their geometric simplicity and the availability of detailed experimental data. The calculations [3.34] were carried out using a repeated slab geometry with a 12-layer thick slab, and the As adatom position was determined by total energy minimization.

Figure 3.5 depicts the calculated quasiparticle energies for the As-capped Si(111) system. The quasiparticle surface-state bands, together with the LDA surface-state bands, are plotted against the projected quasiparticle band structure. Very similar results were obtained for the As/Ge(111) surface. The fully occupied surface-state band corresponds to the lone-pair states on the As adatoms. These lone-pair surface states have been carefully studied using angle-resolved photoemission measurements [3.39, 40]. The theory also predicted an empty surface-state band in the gap. The empty states correspond to localized surface states splitting off from the Si conduction band continuum.

In comparison with the LDA results, the occupied quasiparticle surface-state band is lower in energy and has an enhanced dispersion. Both are needed to bring theory into better agreement with experiment. Figure 3.6 compares the calculated lone-pair surface-state energies with angle-resolved photoemission data. For both the Si and Ge surfaces, the agreement is excellent in both the placement and the width of the band and is well within the estimated uncertainties of ± 0.1 eV with experiment and theory. Since there are no adjustable parameters in the theory, this lends strong support to the conclusion that both the electronic and geometric structures determined theoretically are correct.

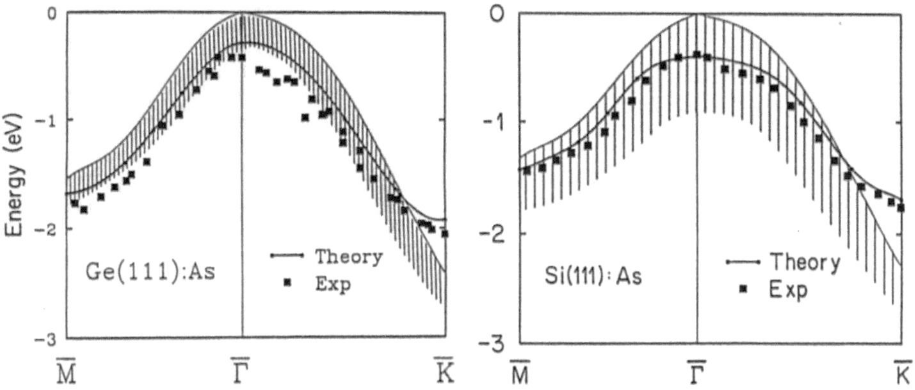

Fig. 3.6. The calculated occupied surface band is compared to photoemission results [3.39, 40] for the As-capped Ge(111) surface (*left panel*) and the As-capped Si(111) surface (*right panel*)

The effect of the many-body correction on the empty surface states is quite dramatic, as seen in Fig. 3.5. The empty states are substantially shifted upwards, opening up the gap between the empty and filled surface states by nearly an extra 1 eV at some $\boldsymbol{k}$-points in the surface Brillouin zone. The empty surface states are accessible to experimental observation using probes such as angle-resolved inverse photoemission, surface optical transition techniques, or scanning tunneling spectroscopy. Recent scanning tunneling [3.35] and inverse photoemission [3.41] experiments have given quantitative confirmation to the theoretical predictions in Fig. 3.5.

Another good example of surface calculation is a recent study [3.36] on the optical and photoemission properties of the clean GaAs(110) surface. The GaAs(110) relaxed 1×1 surface is one of the most studied semiconductor surfaces. The relaxation of the surface atoms is that of a buckling type with the surface As atoms moving outward to the vacuum and the Ga atoms moving inward. The geometric structure of this surface is now quite well determined from both theoretical and experimental analyses. However, its excited-state properties are far from well understood. In particular, the position of the empty surface band has been a matter of controversy both theoretically and experimentally.

Figure 3.7 shows the calculated quasiparticle surface-state energies near the band gap region for the relaxed GaAs(110) surface. Also plotted in the figure are four sets of experimental data [3.42–45]. For the occupied surface state band, the theory agrees very well with the data from angle-resolved photoemission experiments [3.42]. The calculated empty surface-state energies are also in good agreement with one set of inverse photoemission data [3.43] and with results from a laser excite and probe (2-step) photoemission measurement [3.45]. But they are in disagreement with the interpretation of data from a second inverse photoemission experiment [3.44]. First-principles calculations of this kind are very helpful in sorting out the physics of a situation with conflicting experimental results. For the clean GaAs(110) surface, taking theory and all the experiments

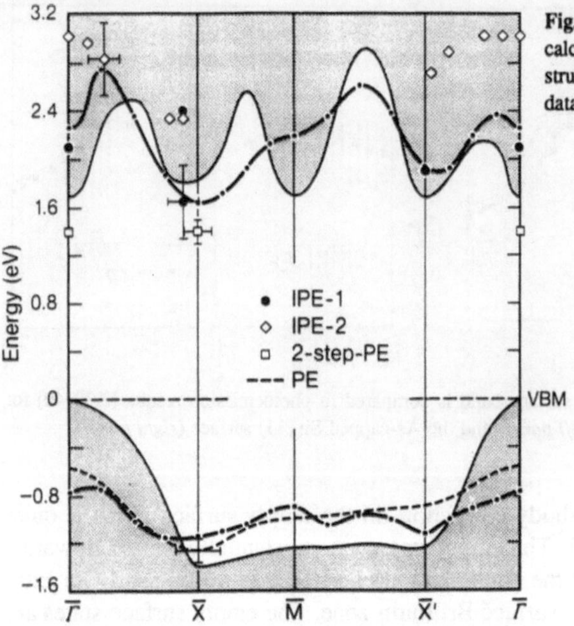

Fig. 3.7. Comparison of the GaAs(110) calculated quasiparticle surface band structure with various experimental data (see text)

together, a clear picture emerges favoring a low-lying band of surface states with a band minimum in the gap.

In comparison to the LDA results, the quasiparticle calculations yield the correct surface excitation energies by substantially opening up the gap between the empty and filled surface states, in ways very similar to the bulk case. However, analysis [3.34] of the calculated results shows that the size of the many-body corrections depends on the detailed character of the surface states, which can have a substantially different character than the bulk states. This means that an accurate determination of the excitation spectra of a surface will generally require a full quasiparticle calculation and would not be obtainable from a ground-state LDA study and knowledge of the bulk spectra.

Artificial structures such as semiconductor heterojunctions, superlattices, and quantum wells are becoming increasingly important in modern materials science. A crucial parameter in determining the properties of these systems is the band offset or band discontinuity at the semiconductor interface. These band offsets are simply the differences in the quasiparticle energy E_{qp} across the junction for the band edge states. A prototypical calculation is one on the band offsets at the GaAs-AlAs(001) heterojunction [3.37], a technologically important system with a nearly perfect lattice-matched interface.

The band offsets calculation is simplified by noting that, from bulk and surface calculations, the quasiparticle wavefunctions are found to be virtually identical [3.5] to the LDA wavefunctions. E_{qp} near an interface may therefore be written as

$$E^{\mathrm{qp}}_{nk} = \varepsilon^{\mathrm{LDA}}_{nk} + \Sigma_{nk} - V^{\mathrm{xc}}_{nk}\,, \tag{3.5}$$

where Σ_{nk} and V^{xc}_{nk} are, respectively, the expectation value of the self-energy operator and the LDA exchange-correlation potential for a given state. Then, the valence band offsets ΔE_v becomes

$$\Delta E_v = \Delta E_v^{LDA} + \delta_{vbm} , \tag{3.6}$$

where ΔE_v^{LDA} is the LDA calculated valence band offset and δ_{vbm} is a many-body correction given by

$$\delta_{vbm} = (\Sigma - V_{xc})^{vbm}_{GaAs} - (\Sigma - V_{xc})^{vbm}_{AlAs} . \tag{3.7}$$

Since E_{qp} should be evaluated at a distance away from the interface in determining the band offsets and both Σ and V_{xc} are short range operators, $(\Sigma - V_{xc})^{vbm}$ can be replaced by bulk values. In obtaining the LDA band offsets, a fully converged 12-layer superlattice calculation was performed.

For the GaAs-AlAs(001) interface, it was found that the many-body correction is quite significant; that is, δ_{vbm} = 0.12 eV. It is about 30% of the LDA result for the valence band offset, which is ΔE_v^{LDA} = 0.41 eV. Combining the two gives a calculated value for the valence band offset of ΔE_v = 0.53 eV. This result is in good agreement with the most recent experimental values [3.46–48] of 0.50–0.56 eV (Table 3.5). The many-body correction can be understood in terms of a more localized valence band wavefunction for AlAs, which leads to a more negative self-energy for AlAs as compared to GaAs and, thus, a positive δ_{vbm}. The many-body correction is expected to play an even more important role for cases where the junctions are made of materials with lesser chemical similarities than GaAs and AlAs.

Quasiparticle calculations have also been carried out for short period superlattices including the Si/Ge strained layer superlattices [3.49] and the GaAs/AlAs superlattices [3.50]. Superlattice periods of up to several layers were studied. As in the case of surface studies, the calculations have played an important role in the interpretation of the excited-state properties of these systems, in particular the various band orderings at different periodicity and the optical response functions. The effects of disorders and strains have also been studied in some detail.

Another area of rapid growth, bordering physics, chemistry, and materials science, is the study of small clusters [3.51]. The recent intensive activities in this field stem from a fundamental interest in reduced size and dimensions and from technological interest in possible applications of clusters to miniaturized devices. Among the simplest of the small clusters is the alkali metal cluster, which showed striking size-dependent behavior [3.52]. Many of the ground-state

Table 3.5. Valence band offset of GaAs/AlAs(001) (in eV)

Experiment	*Menendez* et al. (1986) [3.46]	0.49 ± 0.5
	Dawson et al. (1987) [3.47]	0.535 ± 0.015
	Wolford (1987) [3.48]	0.56 ± 0.03
Theory	Present work	0.53 ± 0.05

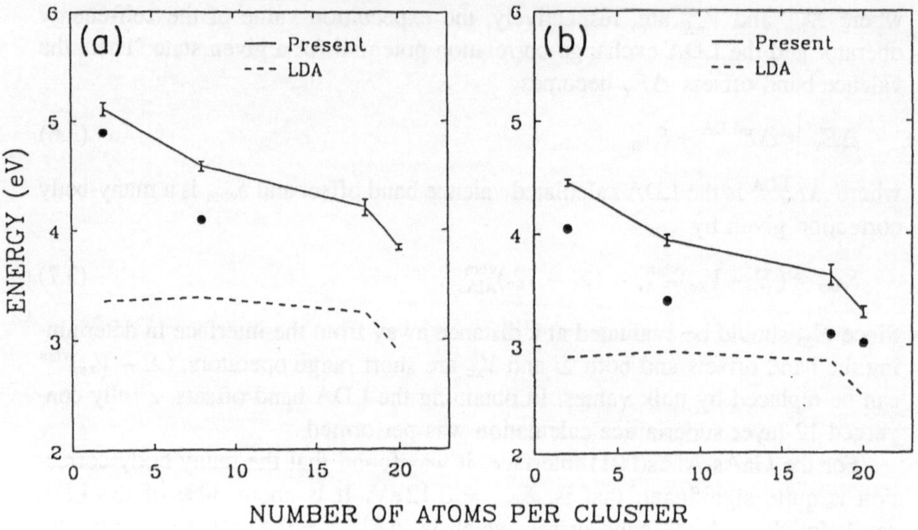

Fig. 3.8a,b. Absolute values of the highest-occupied quasiparticle energies of alkali-metal clusters in the GW approximation (present calculation) and in the LDA with the jellium-sphere-background model. (a) Na_n and (b) K_n ($n = 2, 8, 18$ and 20). (•): Experimental ionization potentials [3.53]

properties have been successfully explained using LDA-type calculations using a simple model to replace the ionic potentials with that of a positive jellium background charge. For example, the mass abundance of these clusters has been explained in terms of the energy shell structure of the electrons in a spherical or distorted spherical jellium potential [3.52, 53]. The quantitative understanding of the excited-state properties is, however, more difficult and again requires knowledge of the electron excitation energies.

A quasiparticle calculation for the alkali metal clusters within the jellium model required some refinement [3.38] in the self-energy approach. Since one is dealing with a localized system consisting equivalently of several to several tens of atoms, a real-space formalism for dielectric screening including local field effects was developed. The theory yielded ionization potentials in good agreement with experiment for various size clusters. Figure 3.8 depicts the absolute values of the highest-occupied quasiparticle energies for Na_n and K_n clusters with $n = 2, 8, 18$ and 20. These quasiparticle energies, which correspond to the energy needed to remove an electron from the cluster, are compared with available experimental data [3.53] and with LDA results. Both the trend and magnitude of the experimental data are well reproduced by the theory[1]. The optical transition energies for these clusters have been predicted and await experimental verification. Table 3.6 lists the calculated quasiparticle energies of a jellium sphere

[1] The experimental ionization potentials for metal clusters depend on the temperature, and hotter clusters are expected to give smaller ionization potentials [3.53]. The calculations were carried out for $T = 0$.

Table 3.6. Calculated quasiparticle energies of the jellium sphere corresponding to Na_{20} (present calculation) and in the LDA (eV)

	Present calculation	LDA
E_{1s}	-5.8 ± 0.4	-5.1
E_{1p}	-5.2 ± 0.4	-4.4
E_{1d}	-4.4 ± 0.1	-3.5
E_{2s}	-3.8 ± 0.05	-2.9
E_{1f}	-1.7 ± 0.1	-2.4

corresponding to a cluster of 20 Na atoms. For occupied states ($1s, 1p, 1d$, and $2s$), the quasiparticle energies are much lower than the LDA eigenvalues. But the unoccupied quasiparticle energy (E_{1f}) becomes higher, giving rise to a larger energy gap as in the case of bulk semiconductors.

3.4 Model Dielectric Matrix

The dielectric response matrix is at the heart of the quasiparticle calculations. For semiconductors and insulators, the static ε^{-1} is obtained within the random phase approximation by directly calculating the electronic polarizability [3.13], and the local fields (off-diagonal elements of ε^{-1} in the Fourier space representation) are found to be central for quantitative results. Corrections to ε^{-1} due to exchange-correlation effects may be introduced with the LDA. These were found to be negligible for the band gaps of the semiconductors but, as mentioned above, these effects are more important in the case of simple metals.

The calculation of the local fields from first principles, however, is computationally very demanding. It is therefore desirable to develop simpler schemes to obtain the screened Coulomb interaction without sacrificing accuracy. Some progress has been made recently in finding a model [3.54] for the screening which takes into account the essential role of local fields. The model consists of taking the screening potential around an added electron at $\boldsymbol{r}'$ to be the same as for a homogeneous medium of the local density at $\boldsymbol{r}'$. For semiconductors, the model for the homogeneous medium is given the finite dielectric constant of the semiconductor. This model dielectric matrix approach has produced results [3.54, 55] for bulk and surface systems in excellent agreement with the full calculations but requiring significantly less computer time. This approach would be most useful in studying systems with complex structure involving many atoms.

3.5 Summary and Conclusions

In this chapter, an overview of a first-principles self-energy approach for calculating the electronic excitation (quasiparticle) energies in solids has been presented. The focus has been on use of the method for understanding and predicting excited-state properties of real materials. The approach has proved to be well-founded as well as being practically applicable in computing the spectral features seen in a wide range of systems. Applications to bulk crystals, surfaces, interfaces, superlattices, and metallic clusters are discussed. Results in excellent agreement with experiments have been obtained for band gaps, optical transition energies, direct and inverse photoemission spectra, and interfacial band offset parameters. In many cases, the theory has been predictive. Although the method described here is a rather recent development, its successes so far have been impressive and wide ranging. Combined with total energy methods ([3.1, 2, 56] and Chap. 5) for structural determination, the quasiparticle self-energy approach should provide a powerful new avenue for first-principles study of materials properties.

Acknowledgements. This work was supported by National Science Foundation Grant DMR-8818404 and by the Director, Office of Energy Research, Office of Basic Energy Sciences, Materials Sciences Division of the U.S. Department of Energy under Contract No. DE-AC03-76SF00098. The support of a John S. Guggenheim Fellowship is also gratefully acknowledged.

References

3.1 S. Lundqvist, N.H. March (eds.): *Theory of Inhomogeneous Electron Gas* (Plenum, New York 1983) and references therein
3.2 S.G. Louie: In *Electronic Structure, Dynamics and Quantum Structural Properties of Condensed Matter*, ed. by J. Devreese, P. van Camp (Plenum, New York 1985) p. 335
3.3 M.S. Hybertsen, S.G. Louie: Comments Cond. Mat. Phys. **13**, 223 (1987)
3.4 M.L. Cohen, J.R. Chelikowsky: *Electronic Structure and Optical Properties of Semiconductors*, 2nd ed., Springer Ser. Solid-State Sci., Vol. 75 (Springer, Berlin, Heidelberg 1989)
3.5 M.S. Hybertsen, S.G. Louie: Phys. Rev. Lett. **55**, 1418 (1985); Phys. Rev. B **34**, 5390 (1986)
3.6 L. Hedin, S. Lundqvist: Solid State Phys. **23**, 1 (1969)
3.7 P. Hohenberg, W. Kohn: Phys. Rev. **136**, B864 (1964)
3.8 W. Kohn, L.J. Sham: Phys. Rev. **140**, A1133 (1965)
3.9 C.S. Wang, W.E. Pickett: Phys. Rev. Lett. **51**, 597 (1983)
3.10 C. Strinati, H.J. Mahausch, W. Hanke: Solid State Commun. **51**, 23 (1984)
3.11 S. Horsch, P. Horsch, P. Fulde: Phys. Rev. B **29**, 1870 (1984)
3.12 R.W. Godby, M. Schluter, L.J. Sham: Phys. Rev. B **35**, 4170 (1987)
3.13 M.S. Hybertsen, S.G. Louie: Phys. Rev. B **35**, 5585 (1987); ibid. **35**, 5602 (1987)
3.14 M.S. Hybertsen, S.G. Louie: Phys. Rev. B **32**, 7005 (1985)
3.15 *Landolt-Börnstein: Numerical Data and Functional Relationships in Science and Technology*, Group 3, Vol. 17a, *Physics of Group IV Elements and III-V Compounds* (Springer, Berlin, Heidelberg 1982)
3.16 G. Baldini, B. Bosacchi: Phys. Status Solidi **38**, 325 (1970)
3.17 S.B. Zhang, D. Tomanek, M.L. Cohen, S.G. Louie, M.S. Hybertsen: Phys. Rev. B **40**, 3162 (1989)
3.18 X. Zhu, S.G. Louie: To be published
3.19 D.E. Aspnes: Phys. Rev. B **12**, 2797 (1975)
3.20 R.R.L. Zucca, Y.R. Shen: Phys. Rev. B **1**, 2668 (1970)
3.21 R.A. Roberts, W.C. Walker: Phys. Rev. **161**, 730 (1967)
3.22 F.J. Himpsel, J.F. van der Veen, D.E. Eastman: Phys. Rev. B **22**, 1967 (1980)

3.23 A.L. Wachs, T. Miller, T.C. Hsieh, A.P. Shapiro, T.C. Chiang: Phys. Rev. B **32**, 2326 (1985)
3.24 D. Straub, L. Ley, F.J. Himpsel: Phys. Rev. B **33**, 2607 (1986)
3.25 R.T. Poole, J.G. Jenkin, R.C.G. Leckey, J. Liesegong: Chem. Phys. Lett. **22**, 101 (1973)
3.26 L.I. Johanson, S.B.M. Hagstrom: Phys. Scr. **14**, 55 (1976)
3.27 R. Hulthen, N.G. Nilsson: Solid State Commun. **18**, 1341 (1976)
3.28 E.W. Plummer: Surf. Sci. **152/153**, 162 (1985)
3.29 E. Jensen, E.W. Plummer: Phys. Rev. Lett. **55**, 1912 (1985)
3.30 J.E. Northrup, M.S. Hybertsen, S.G. Louie: Phys. Rev. Lett. **59**, 819 (1987)
3.31 P.H. Citrin, G.K. Wertheim, T. Hashizume, F. Sette, A.A. MacDowell, F. Comin: Phys. Rev. Lett. **61**, 1021 (1988)
3.32 M. Surh, J.E. Northrup, S.G. Louie: Phys. Rev. B **38**, 5976 (1988)
3.33 J.E. Northrup, M.S. Hybertsen, S.G. Louie: Phys. Rev. B **39**, 8198 (1989)
3.34 M.S. Hybertsen, S.G. Louie: Phys. Rev. Lett. **58**, 1551 (1987); Phys. Rev. B **38**, 4033 (1988)
3.35 R.S. Becker, B.S. Swartzentuber, J.S. Vickers, M.S. Hybertsen, S.G. Louie: Phys. Rev. Lett. **60**, 116 (1988)
3.36 X.-J. Zhu, S.B. Zhang, S.G. Louie, M.L. Cohen: Phys. Rev. Lett. **63**, 2112 (1989)
3.37 S.B. Zhang, D. Tomanek, S.G. Louie, M.L. Cohen, M.S. Hybertsen: Solid State Commun. **66**, 585 (1988)
3.38 S. Saito, S.B. Zhang, S.G. Louie, M.L. Cohen: Phys. Rev. **40**, 3643 (1989)
3.39 R.D. Bringans, R.I.G. Uhrberg, R.Z. Bachrach, J.E. Northrup: Phys. Rev. Lett. **55**, 533 (1985)
3.40 R.I.G. Uhrberg, R.D. Bringans, M.A. Olmstead, R.Z. Bachrach, J.E. Northrup: Phys. Rev. B **35**, 3945 (1987)
3.41 W. Drube, R. Ludeke, F.J. Himpsel: In Proc. 19th Int. Conf. on the Phyiscs of Semiconductors, ed. by W. Zawadzki (Institute of Physics, Polish Academy of Sciences, Warsaw 1988), p. 637
3.42 A. Huijser, J. van Laar, T.L. van Rooy: Phys. Lett. **65A**, 337 (1978)
3.43 D. Straub, M. Skibowski, F.J. Himpsel: Phys. Rev. B **32**, 5237 (1985)
3.44 B. Reihl, T. Riesterer, M. Tschudy, P. Perfetti: Phys. Rev. B **38**, 13456 (1988)
3.45 R. Haight, J.A. Silberman: Phys. Rev. Lett. **62**, 815 (1989)
3.46 J. Menendez, A. Zinczuk, D.J. Werder, A.C. Gossard, J.H. English: Phys. Rev. B **33**, 8863 (1986)
3.47 P. Dawson, K.J. Moore, C.T. Foxon: In *Quantum Well and Superlattice Physics* (Bay Point), Proc. SPIE, Vol. 792, ed. by G.H. Dohler, J.N. Schulman (SPIE, Washington, DC 1987) p. 208
3.48 D.J. Wolford: Proc. 18th Int. Conf. on the Physics of Semiconductors, ed. by O. Engstrom (World Scientific, Singapore 1987) p. 1115 and private communication
3.49 M.S. Hybertsen, M. Schluter: Phys. Rev. B **36**, 9683 (1987)
3.50 S.B. Zhang, M.S. Hybertsen, M.L. Cohen, S.G. Louie, D. Tomanek: Phys. Rev. Lett. **63**, 1495 (1989)
3.51 R.P. Andres et al.: J. Mater. Res. **4**, 704 (1989)
3.52 W.A. de Herr, W.D. Knight, M. Y. Chou, M.L. Cohen: In *Solid State Physics*, Vol. 40, ed. by H. Ehrenreich, D. Turnbull (Academic, New York 1987) p. 94
3.53 W.D. Knight, K. Clemenger, W.A. de Heer, W.A. Saunders, M.Y. Chou, M.L. Cohen: Phys. Rev. Lett. **52**, 2141 (1984)
3.54 M.S. Hybertsen, S.G. Louie: Phys. Rev. B **37**, 2733 (1988)
3.55 X.-J. Zhu, S.G. Louie: To be published
3.56 S. Fahy, X.W. Wang, S.G. Louie: Phys. Rev. Lett. **61**, 1631 (1988)

4. Determination of the Electronic Structure of Solids

Franz J. Himpsel

With 8 Figures

Materials science is undergoing a transformation from an art to a science. While many novel classes of materials [4.1] have been developed without a systematic search, we are presently at a stage where the perfection of materials plays a crucial role. For example, the optical absorption by impurities in fiber optics materials has been reduced to a level that makes fiber optics a viable technology for information transfer. Likewise, the high purity and crystalline perfection of semiconductor materials like Si and GaAs is crucial for the functioning of electronic devices. An important factor is also the quality of interfaces between different device materials. Here, the perfection does not match that of bulk materials yet. The relatively low density of charged interface defects at the SiO_2/Si interface (typically 10^{10} defects/cm^2 or 10^{-5} monolayer) is one of the main reasons for today's dominance of silicon technology. When it comes to systematically improving materials, it is of prime interest to know the electronic structure in great detail. Examples are III-V compounds, where the band structure allows for higher mobilities, ergo faster speed, than in Si. The character of the band gap (direct vs indirect), the effective masses (curvatures of the valence and conduction bands), and the spacing of higher-lying indirect band minima (which limit the mobility at high velocities by intervalley scattering) all play a role in the electrical properties of devices. III-V and II-VI materials with optimal properties can be engineered by producing ternary (and higher order) compounds, such as GaInAs, GaAlAs, and HgCdTe. By adding magnetic materials, such as in MnCdTe, one is able to produce semiconductors whose properties can be tuned by a magnetic field. In a class by themselves are superlattices, i.e., layers of alternating, lattice-matched materials. By low-temperature growth techniques (molecular beam epitaxy, chemical vapor deposition) these artificial crystals can be produced with sharp interfaces and periods of atomic dimensions. The properties of these metastable materials are not limited by thermodynamic constraints. With all these opportunities, one can easily get lost in a forest of materials combinations unless one systematically studies the electronic structure with respect to the various parameters and optimizes desired properties.

The electronic structure of a solid is given by the quantum numbers of the electrons. For a crystalline material there are energy E, momentum $\hbar\boldsymbol{k}$, point group symmetry (i.e., angular symmetry), and spin. This information can be summarized by plotting $E(\boldsymbol{k})$ energy band dispersions with the appropriate labels for point group symmetry and spin. Disordered solids can be characterized by the average values of these quantities, while the disorder introduces characteristic

Table 4.1. Various techniques for probing electronic states. Photoemission and inverse photoemission have the advantage of probing both energy and momentum of an electronic state

INCOMING PARTICLE ↓ / OUTGOING PARTICLE →	PHOTON	ELECTRON POSITRON	ION ATOM
PHOTON	Optical Luminescence Raman De Haas V. Alphen Cyclotron Resonance	Photoemission Auger	Photon Stimulated Desorption
ELECTRON POSITRON	Inverse Photoemission Appearance Potential Positron Annihilation	Electron Energy Loss Auger Appearance Potential Tunneling	Electron Stimulated Desorption
ION ATOM	Chemoluminescence	Ion Neutralization Penning Ionization	Secondary Ion

broadenings. An ideal technique for probing the electronic structure of a solid would be able to determine all these quantum numbers. Searching for such a technique, one may consider a variety of probing particles and draw a table like Table 4.1. It is helpful to concentrate on a spectroscopy that involves only electronic transitions and not atomic motion, so that the Franck–Condon principle is valid. With atoms or ions in the incoming or outgoing channel, this is difficult to achieve. Among the remaining photon- and electron-based techniques, photoemission (Fig. 4.1a) and inverse photoemission (Fig. 4.1b) distinguish themselves through the following features: (i) they are able to probe energy and momentum of electrons simultaneously, and (ii) they measure absolute electron energies, not energy differences. In inverse photoemission an electron impinges onto a surface and a photon is emitted, i.e., just the reverse of photoemission. The two techniques complement each other since they probe occupied and unoccupied states, respectively. A detailed discussion of photoemission and inverse photoemission techniques can be found in [4.2] and [4.3], respectively. Sometimes, it is advantageous to employ optical techniques which achieve higher energy resolution. This

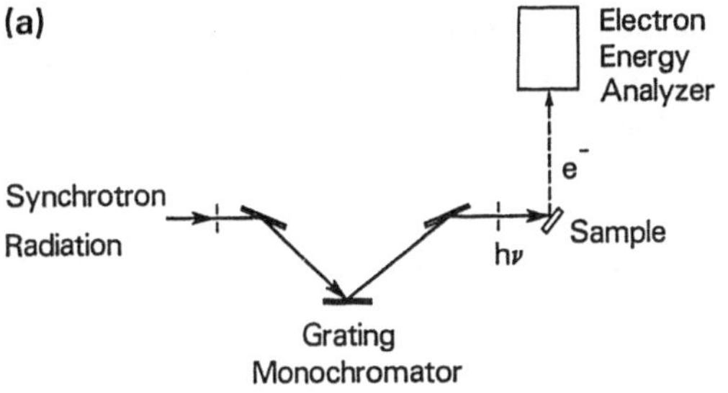

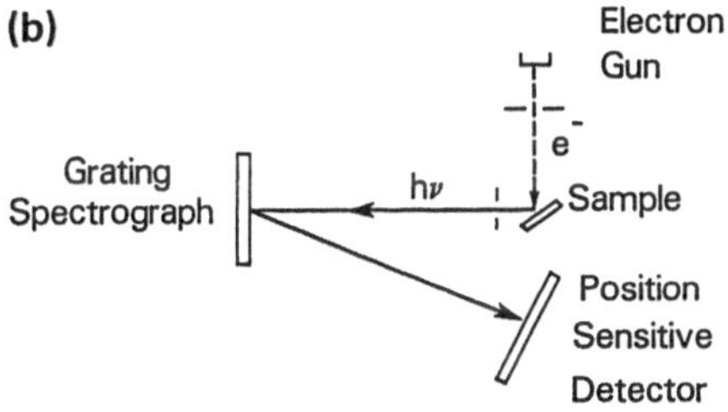

Fig. 4.1. Schematic state of the art setups for (**a**) photoemission and (**b**) inverse photoemission. Photons are labelled $h\nu$, electrons e^-

is useful for mapping sharp, long-lived states near the band edges in semiconductors or near the Fermi level in metals. Momentum resolution can be achieved in this case by methods like cyclotron resonance and the de Haas–van Alphen effect. The field of band structure determination has progressed to a stage where the results obtained from a large variety of elements and compounds are making their entry into reference books [4.4]. Knowledge of the $E(\boldsymbol{k})$ band dispersion characterizes the electronic properties of a solid rather completely. For example, one can parametrize the band dispersion by an empirical pseudopotential [4.5] or by a tight binding scheme [4.6], and obtain wavefunctions and their matrix elements. From these one can derive optical constants, magnetic moments, and other electronic properties of a solid. This process has already been carried out for a variety of semiconductors [4.5] and metals [4.6].

4.1 Band Mapping with Photoemission and Inverse Photoemission

How are energy band dispersions determined? A first look at the task reveals that photoemission (and inverse photoemission) provides just the right number of independent measurable variables to establish a unique correspondence to the quantum numbers of an electron in a solid. The energy E is measured from the kinetic energy E_{kin} of the electron. The two momentum components parallel to the surface $\boldsymbol{k}^{\parallel}$ are measured from the polar and azimuthal angles of the electron. The third momentum component $k^{\perp}$ is determined by tuning the photon energy $h\nu$, thus requiring a tunable photon source (or a tunable photon detector in inverse photoemission) for a complete band structure determination. Finally, the spin-polarization of the electron provides the spin quantum number, and the polarization of the photon provides the point group symmetry.

For two-dimensional states (e.g., in layered crystals and at surfaces) the determination of energy bands is almost trivial since only E and $\boldsymbol{k}^{\|}$ – the momentum parallel to the surface – have to be determined. These quantities obey the conservation laws

$$E_1 = E_u - h\nu \,, \tag{4.1}$$

$$\boldsymbol{k}_1^{\|} = \boldsymbol{k}_u^{\|} - \boldsymbol{g}^{\|} \,, \tag{4.2}$$

where $\boldsymbol{g}^{\|}$ is a vector of the reciprocal surface lattice, u denotes the upper state and l the lower state. These conservation laws can be derived from the invariance of the electron states relative to time and any translation parallel to the surface by a lattice vector. For the photon, only its energy $h\nu$ appears in the balance because the momentum of an ultraviolet photon is negligible compared to the momentum of the crystal electrons. The subtraction of a reciprocal lattice vector $\boldsymbol{g}^{\|}$ simply corresponds to plotting energy bands in a reduced Brillouin zone, i.e., within the unit cell in $\boldsymbol{k}^{\|}$ space.

In the three-dimensional case, the momentum perpendicular to the surface is not conserved because of momentum transfer at the surface potential step. Several methods have been conceived to overcome this difficulty. As an extra parameter to determine $k^{\perp}$ one can use either tunable photon energy or data from two different crystallographic surfaces. Before working out rigorous band mapping methods in three dimensions it is helpful to have an example like the energy bands of Si shown in Fig. 4.2 [4.7]. The momentum parallel to the surface has been fixed at $\boldsymbol{k}_u^{\|} = 0$ by using the conservation law in Eq. 4.2. All possible initial and final states are located along the line $\Gamma\Lambda L$ in $\boldsymbol{k}$ space which is obtained by varying $k^{\perp}$. The corresponding band structure is rather complex for the upper bands, but it can be simplified to just one free-electron-like band, shifted by an inner potential. This can be seen from the fact that only a single, intense transition appears in the data at each photon energy, not multiple ones as expected from the calculated bands. The physics behind this simplification is that the electron in the upper band has high enough kinetic energy to feel only a weak, average influence of the crystal potential. Transitions between bulk states are vertical in the $E(k^{\perp})$ diagram of Fig. 4.2, since for states with three-dimensional periodicity not only E and $\boldsymbol{k}^{\|}$ are conserved but also $k^{\perp}$. (Roughly speaking, the $k^{\perp}$ transfer happens "after" the excitation, when the photoelectron penetrates the surface potential step.) It becomes clear from this picture that a knowledge of the upper band provides the information on the perpendicular momentum component that is missing from the conservation laws (4.1, 2). The kinetic energy of the electron can be converted directly into the momentum $k^{\perp}$ using the $E(k^{\perp})$ relation of the upper band. By tuning the photon energy one is able to move the transition to various $k^{\perp}$ values along the upper band.

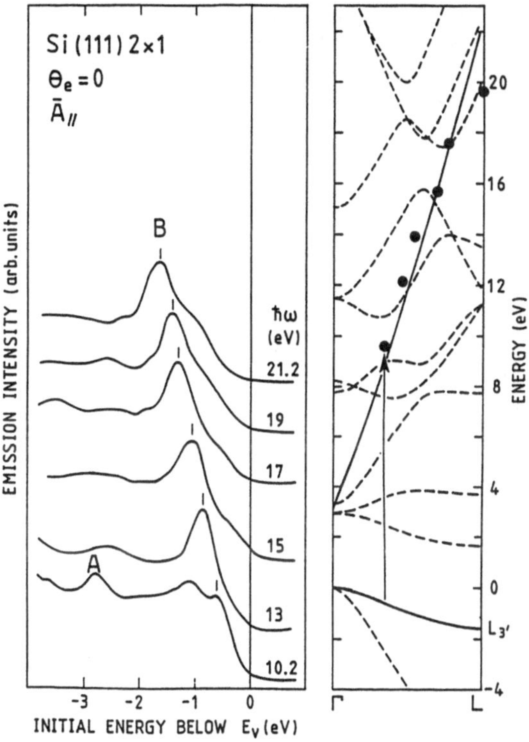

Fig. 4.2. Mapping of the $E(\boldsymbol{k})$ bulk band dispersion in Si by photoemission. The momentum parallel to the surface is fixed at $\hbar \boldsymbol{k}^{\parallel} = 0$ by choosing photoelectrons emitted normal to the surface, and the momentum perpendicular to the surface $\hbar \boldsymbol{k}^{\perp}$ is varied by tuning the photon energy by $h\nu$. The maze of calculated upper bands can be approximated by a parabolic free-electron-like band for practical purposes. (From [4.7])

The method just described, which uses a single, free-electron-like band sketched in Fig. 4.2 as the final state, is the simplest possible approach. The next level of sophistication is shown in Fig. 4.3 [4.8, 9], again for normal emission ($\boldsymbol{k}^{\parallel} = 0$). In this case, several sets of upper bands are required to explain all peaks occurring in photoemission and inverse photoemission. They are all derived from the same free-electron-like parabola, but using different reciprocal bulk lattice vectors $\boldsymbol{g}$. Such $\boldsymbol{g}$ vectors have to be added to the momentum in order to fold it back into the first Brillouin zone. In the simplest approximation, only the $\boldsymbol{g}$ vector closest to the escape direction of the photoelectron is used (the so-called "primary cone", see [4.10]). This is the case in Fig. 4.2. In Fig. 4.3, there is at least one "secondary cone" (dashed line) contributing to the spectra, in addition to the primary cone (solid line). This can be inferred from the appearance of a second set of transitions in the spectra, which disperse in a direction opposite to the primary transitions when the photon energy is varied.

It is surprising to see that the complex array of calculated upper bands behaves effectively like a parabolic band for generating the (inverse) photoemission spectra. This can be verified quantitatively by calculating the emission intensity using the proper transition matrix elements, including the lifetime broadening of the bands [4.11]. Essentially, only the band closest to the free-electron-like parabola has the proper Fourier components [4.12] to couple to the outgoing photoelectron, or to the incoming electron in inverse photoemission. The discontinuities

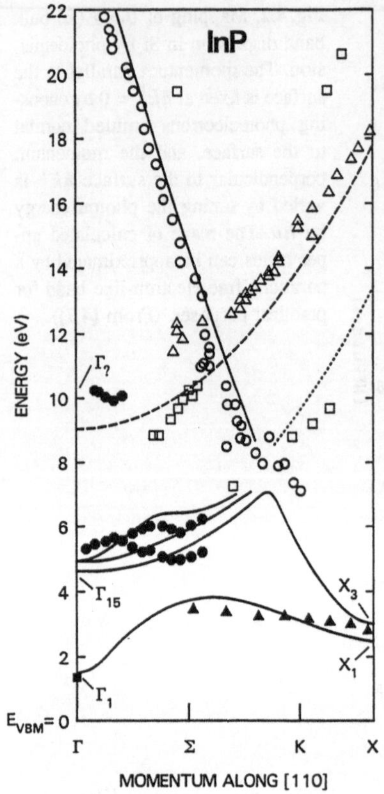

Fig. 4.3. Experimental band structure of the conduction bands of InP exhibiting two sets of upper bands ("primary and secondary cones"). Inverse photoemission results [4.8] are shown by *full symbols*, photoemission results [4.9] by *open symbols*

between the bands which account for the parabola are smoothed by lifetime broadening. While most semiconductors (and insulators) have complicated upper bands (often due to unoccupied d-states), most metals are characterized by a much simpler band structure. Consequently, they are amenable to more rigorous band mapping methods, which do not rely on a free-electron-like upper band.

The "exact" band mapping methods can be classified according to the scheme shown in Fig. 4.4. By fixing $\boldsymbol{k}^{\parallel}$ (e.g., $\boldsymbol{k}^{\parallel} = 0$ as in Fig. 4.2) one scans $k^{\perp}$ by changing the photon energy. When a symmetry line is crossed (e.g., at $k^{\perp} = 0$), the sign of the band dispersion reverses. Thus, one has nailed down one point in $\boldsymbol{k}$-space (point 1 in Fig. 4.4). The same is true at the boundary of the Brillouin zone (point 3 in Fig. 4.4). In this manner, the critical points of the $E(k^{\perp})$ band dispersion are obtained exactly, independent of any approximation for the upper bands. Between the critical points one may interpolate the upper band rather accurately by adding the next Fourier component to the free-electron-like parabola, which opens up band gaps at critical points. There are two ways which avoid any interpolation altogether. The first consists of repeating the $E(k^{\perp})$ scan for different $\boldsymbol{k}^{\parallel}$ values (point 2 in Fig. 4.4), thereby obtaining the critical points along a symmetry line. The second method is triangulation from two different crystal surfaces. One tries to find the same transition on both surfaces,

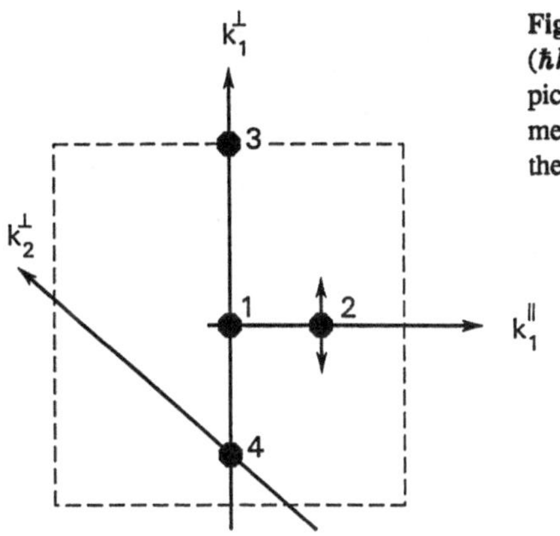

Fig. 4.4. Various methods fo determine the momentum ($\hbar k^\parallel$, $\hbar k^\perp$) of bulk electronic states. The *dashed line* depicts the Brillouin zone boundary. $k^\parallel$ is given by momentum conservation (4.2), $k^\perp$ is obtained by varying the photon energy or by using two crystal faces

identified by identical photon and electron energies. This is accomplished by varying $k_1^\perp$ and $k_2^\perp$ for the two crystal surfaces, respectively. The corresponding lines intersect in a well-defined point (point 4 in Fig. 4.4). For general-purpose band mapping, the approximate interpolation method is the most convenient. The absolute methods are more tedious since a whole series of spectra is required for each $E(\boldsymbol{k})$ point. They are useful for refining a band structure that is already known qualitatively.

4.2 Understanding Semiconductors from First Principles

After determining energy band dispersions experimentally, it is interesting to see if they can be understood theoretically. There are two opposite approaches. The empirical method attempts to fit the experimental bands, and then to calculate the interesting properties of a solid, such as optical constants, magnetic moment, and superconductivity [4.5, 6, 13]. The first-principles approach tries to obtain the band dispersion with just the atomic number as input, finding everything else by total energy minimization, see Chaps. 3 and 5). The energy bands appear in these calculations as the eigenvalues of the Schrödinger equation at various $\boldsymbol{k}$ points. However, most of the published energy bands refer to the neutral ground state, which describes neither the photoemission final state (involving a positive ionic state) nor the inverse photoemission final state (involving a negative ionic state). In fact, one cannot conceive an experiment for measuring energy band dispersions in the ground state. To determine the binding energy of an electron, one has to remove it, thereby creating an excited state. Correspondingly, the energy eigenvalues calculated for the ground state depend on the calculation scheme (e.g., local density versus Hartree–Fock, the first of which yields calculated semiconductor band gaps which are too small, the second one too large). Only properties like the charge density or the total energy can be measured without disturbing the ground state. The energy difference between the excited state and the ground state is the self-energy (see Chap. 3 for details). It can be

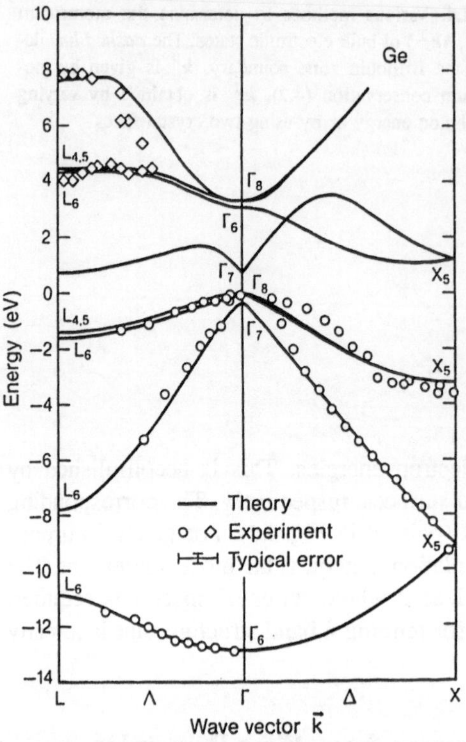

Fig. 4.5. Comparison of experimental critical points of the Ge band structure with state-of-the-art self-energy calculations. The calculations are from [4.15], the photoemission data for the valence band from [4.17], and the inverse photoemission data for the conduction band from [4.18]

calculated by using Dyson's equation for the electron and photon propagators [4.14–16]. The largest self-energy effect in semiconductors is an increase of the band gap relative to the ground state as given by local density theory. The self-energy has an overall step-function behavior across the band gap, reflecting the opening of the gap, and the conduction bands experience a much larger shift than the valence bands. However, the behavior of individual critical points may deviate from this trend (e.g. L_2' in the conduction band). Generally, the agreement between such self-energy calculations and photoemission (or inverse photoemission) experiments is within the experimental and theoretical accuracy (see Fig. 4.5 and [4.15–18]).

The self-energy has not only a real part, describing energy shifts, but also an imaginary part, describing lifetime broadening in the excited state. The broadening increases away from the Fermi level due to the increasing phase space available for electron–hole pair creation [4.2]. There is a threshold for pair production at one gap energy from the band extrema. Closer to the band edges, only electron–phonon scattering contributes to the lifetime. In this regime, the measurement accuracy is often not limited by the intrinsic lifetime broadening, but by the experimental energy and momentum resolutions. With current light sources one obtains typical resolutions of 0.1 eV and 0.1 Å^{-1} for energy and momentum, respectively, in photoemission. Ten times better resolutions can easily be achieved by available monochromators and spectrometers, but the count rates become very low. With new undulator-based synchrotron radiation sources [4.19] this restriction will be removed.

For inverse photoemission [4.3] the energy resolution is more limited (0.3 eV), while the momentum resolution is comparable to that of photoemission spectroscopy. In order to cope with the extremely low yield in inverse photoemission (typically 10^{-8} photons per electron) one cannot afford to monochromatize electrons. A case for high energy resolution can be made for all systems which exhibit electronically driven phase transitions e.g., charge density, waves, metal–insulator transitions, superconductors, heavy Fermions, Kondo systems. The relevant energy scale is determined by the transition temperature T_c, with $k_B T_c$ on the order of meV. In recent studies of high temperature superconductors [4.20], one was able to detect a band gap of $\Delta \approx 25$ meV opening up below the Fermi level when the sample was cooled below the transition temperature.

4.3 Magnetic Storage and Thin Film Magnetism

The magnetic properties of thin films may deviate strongly from the bulk properties. Generally, one would expect magnetically dead layers near the surfaces of a magnetic film. After all, magnetism is a cooperative phenomenon, and the number of magnetic neighbor atoms decreases at the surface or at the interface to a nonmagnetic material. However, the opposite can happen for materials with a nearly half-filled shell (e.g., V, Mn, Cr, Gd). In this case, the magnetic moment of an isolated atom is large, because Hund's rule aligns the spins. Diluting the atoms increases the magnetic moment, while it is suppressed in bulk V, Mn, Cr, etc. by band structure effects. All of these phenomena have been observed [4.21]. Such effects can be either detrimental or a bonus for magnetic storage applications. They affect not only the storage medium, but also the read/write heads, which are often in thin film form for optimum miniaturization.

For measuring the magnetic band structure one has to analyze the spin-polarization of the photoelectrons or produce spin-polarized electrons for inverse photoemission. Spin analysis costs a factor of 10^4 in intensity. It is accomplished utilizing the left-right asymmetry observed when a spin-up electron backscatters off a high-Z target. In conventional Mott scattering, the energy-analyzed photoemitted electron is accelerated to about 100 keV and experiences the Coulomb field of the nucleus of a gold target. In low-energy electron scattering (about 100 eV), the Coulomb field is due to the electron cloud of the gold target. Targets with high atomic number, such as Au and W, produce a large effect due to their large nuclear and electronic charge. Low-energy electrons can be backscattered from either a polycrystalline material [4.22], or from a single crystal [4.23], for which the intensity of the diffraction spots is monitored. Mott detectors have the advantage of an absolute calibration and high stability. However, they are bulky and complex. Although more compact Mott detectors have recently been developed, the trend is towards using low-energy spin detectors, which are so small that they can be attached to mobile angle-resolving spectrometers. In order to combine spin detection with a tunable and polarized light source, such as synchrotron radiation, one has to compensate for the four orders of magnitude in signal that are typically lost during spin analysis. Such instruments are usually coupled to special, high flux synchrotron radiation beam lines, accepting either a large solid angle of photons from a bending magnet, or using undulator-type insertion devices [4.19].

Spin-polarized inverse photoemission is relatively easy compared to spin-resolved photoemission. Compact spin-polarized electron sources have been available for several years [4.24]. They are based on the emission of spin-polarized photoelectrons from negative affinity GaAs photocathodes by circularly polarized laser light. As in Mott detectors, the spin–orbit interaction plays an essential role in creating spin-polarization. One of the future applications of spin-polarized inverse photoemission will be in the area of surface and thin film magnetism. The feasibility of such experiments has been demonstrated [4.25], but the required sophisticated materials growth techniques have yet to be established.

4.4 Optoelectronics and Excited State Spectroscopy

One of the most important, and most difficult, areas in optoelectronics is the generation of light in semiconductor devices. In most cases such devices have to operate at excitation densities approaching the destruction limit. Therefore, the electronic properties of materials in highly excited states are of particular interest. Also, the lifetime of various excited states is an important parameter for the emitted light flux. The lifetime is affected by phonons, surfaces, and interfaces. Time-resolved versions of various spectroscopies can now be extended down to sampling times of less than a picosecond. Examples are two-photon photoemission and time-resolved absorption or luminescence spectroscopy. In two-photon photoemission one excites the solid with the first photon and probes the excited state with a second photon (Fig. 4.6, [4.26]). The excitation typically takes place near the band extrema, where long-lived states can be probed. Away from the band edges, the lifetime becomes rapidly too short to build up a measurable excited state population. The high photon flux required to excite a significant fraction of the electrons in the solid is usually provided by a laser. The second, probe photon needs to have higher photon energy to overcome the work function. It can be produced by multiplying the frequency of the pump photon in a nonlinear medium, or by creating soft X-rays in a laser-induced plasma. Only a relatively low flux is required for the probe photons, since at higher fluxes one may create a space charge which distorts the energy distribution of the photoelectrons. Synchrotron radiation is also a candidate for producing probe photons, but it is not easy to match the high repetition rate of a storage ring to that of a pump laser.

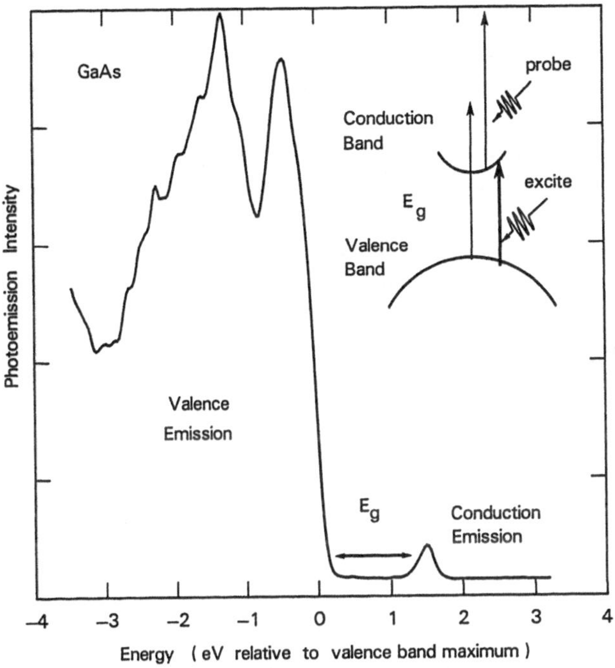

Fig. 4.6. Two-photon photoemission measurement of a long-lived surface state (after [4.26]). An intense pump laser pulse partially fills conduction band states. The probe laser ionizes them. By delaying the probe pulse one can determine the time dependence of the hot electron states

4.5 Spatial Resolution

Spatially resolved spectroscopy is also becoming increasingly sophisticated. Two levels of spatial resolution will be considered: the atomic level, and a much coarser level that is relevant to electronic devices (typically 0.1 μm). At an atomic level there are fascinating possibilities today to create atom by atom new materials with metastable structures. Manipulation along a surface may be achieved by a scanning tunneling microscope tip. So far, only relatively crude "scratches" or "dimples" have been made with typical widths of 20 Å [4.27] or larger. However, as far as spatially resolved probes of the electronic structure are concerned, the use of scanning tunneling spectroscopy makes it already possible to routinely determine the local electronic structure at surfaces with nearly atomic resolution. A momentum-integrated density of states can be obtained by measuring the current vs voltage characteristics, more specifially $(dI/dV)/(I/V)$. For a state that is spatially localized within atomic dimensions, e.g., a defect state, the momentum cannot be resolved due to the Heisenberg uncertainty relation. For an extended state, one would need to know the momentum in addition to the energy for a complete band structure determination. The real space information obtained through scanning tunneling spectroscopy can be transformed into momentum space by fitting wavefunctions to the tunneling spectra [4.28]. Likewise, the momentum space information from photoemission and inverse photoemission can be transformed into real space information by using the wavefunction of empirical band calculations.

Atomic resolution in the direction perpendicular to a surface (interface) can be achieved by mapping two-dimensional surface (interface) states with the band mapping methods described above. The $k^{\perp}$ momentum information is lost due to the uncertainty relation. Such two-dimensional states can be found in parts of $\boldsymbol{k}$-space where no three-dimensional states exist. In semiconductors with localized bands these two-dimensional states often persist as resonances, even when bulk states are present. The surface (interface) band structure can be very different from that of a bulk semiconductor, opening up a whole new field in which new materials are engineered in the form of atomic layer superlattices. For example, the band gap of Si shrinks to nearly zero at the Si(111) 7×7 surface, and the band gap of the CaF_2/Si(111) interface [4.29] is twice as large as that of Si and 5 times smaller than that of CaF_2.

Spatial resolution at a much coarser level, typically 0.1 μm, is relevant for the analysis of electronic device structures. Using spatially resolved photoelectron spectroscopy [4.30–34], one could not only carry out an elemental analysis, like in scanning Auger microscopy, but also distinguish different chemical bonding states of the same element via the core level shifts (see Sect. 4.6 and Chap. 8). Another big advantage would be the near absence of radiation damage, which severely restricts any kind of electron microscopy. With electron excitation, most of the energy goes into production of damaging secondary electrons, whereas with X-ray excitation, the photon energy can be tuned just above a core level threshold, where the cross section for core excitation is highest. Actually, with a tunable X-ray source, like synchrotron radiation, one could tune the photon energy to just below and just above the core level threshold for a given element and obtain a map of the distribution of the element by taking the ratio between the two pictures. Thus the nearly constant background from other core levels and from the valence band would be eliminated. Note that the absorption coefficient depends exponentially on the cross section, so that the *ratio* and not the difference has to be taken.

The development of photoelectron microscopes has grown rapidly in the last few years and is expected to become a major factor with the new undulator-based synchrotron radiation sources that are under construction [4.19]. An undulator increases the spectral brilliance (photons per energy interval, source area and solid angle) by four orders of magnitude relative to an ordinary synchrotron light source. This gain can be converted directly into usable flux in a microscopy experiment. As shown in Fig. 4.7, there are two types of microscopes that use the photoelectrons emitted from the sample for producing a magnified image. There are also microprobes of the scanning type, but they are much slower than imaging microscopes, which process all image elements in parallel. One design [4.32, 33] accelerates the photoelectrons to about 20 keV right in front of the sample with a special immersion lens, so that they can be subsequently

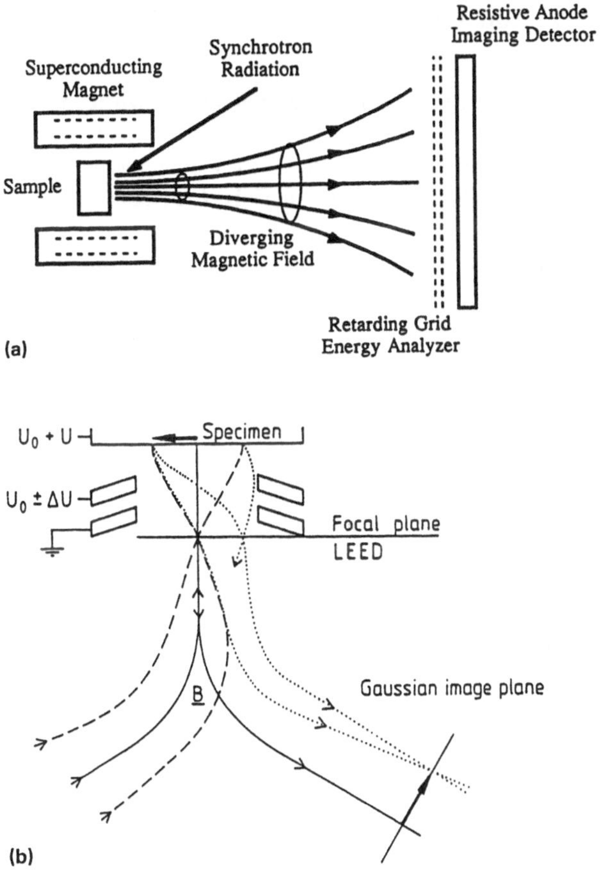

Fig. 4.7. Two types of photoelectron microscopes, (a) using magnification along diverging magnetic field lines [4.30, 31], and (b) using an accelerating lens followed by standard electron microscope optics [4.32, 33]. In the second case, the microscope can also be used for LEED, with a magnetic field $\boldsymbol{B}$ perpendicular to the drawing separating the incident from the reflected beam

imaged in a conventional electron microscope. The other design [4.30, 31] lets the photoelectrons spiral out along diverging magnetic field lines. Both types of microscopes have achieved submicron resolution. The energy analysis of the electrons is still somewhat of a sore point since it degrades the resolution. Best spatial resolution is achieved for electrons with nearly zero kinetic energy (which is obvious for the magnetic design where the diameter of the spirals limits the resolution). Therefore, current microscopes often operate near the photoelectric threshold and use the contrast induced by a locally varying work function. To select a given core level would require electron energy analysis with a resolution of about 0.3 eV. One possible solution of the energy analysis problem is to image secondary electrons without energy selection and to select the different core level contributions (and differently shifted components of one core level) by tuning the photon energy to the corresponding thresholds, as outlined above.

4.6 Packaging, Polymers, and Core Levels

When it comes to packaging, i.e., mounting semiconductor chips, wiring them and cooling them, the range of materials in use broadens still further. Often ceramics or polymers are used as insulating materials, in which the level of perfection is far from that of semiconductor device materials. Consequently, the band-mapping techniques desribed in Sect. 4.1 do not apply anymore. The useful information about the electronic structure consists mainly of nearest-neighbor bonding. For such problems, it is appropriate to switch to a more local probe, i.e. core level spectroscopy. An energy shift is induced on atomic core levels by a charge transfer to or from its neighbors. This chemical shift scales with the number of neighbors and with the electronegativity difference. The method is sensitive enough to distinguish inequivalent carbon atoms in polymers, as shown in Fig. 4.8 [4.35]. With this selectivity, one can follow the reaction of a polymer film with other materials (e.g., metals in polymer/metal wiring sandwiches) on an atom-specific basis. A more detailed discussion of the applications of core level spectroscopy to interfaces can be found in Chap. 8 of this book.

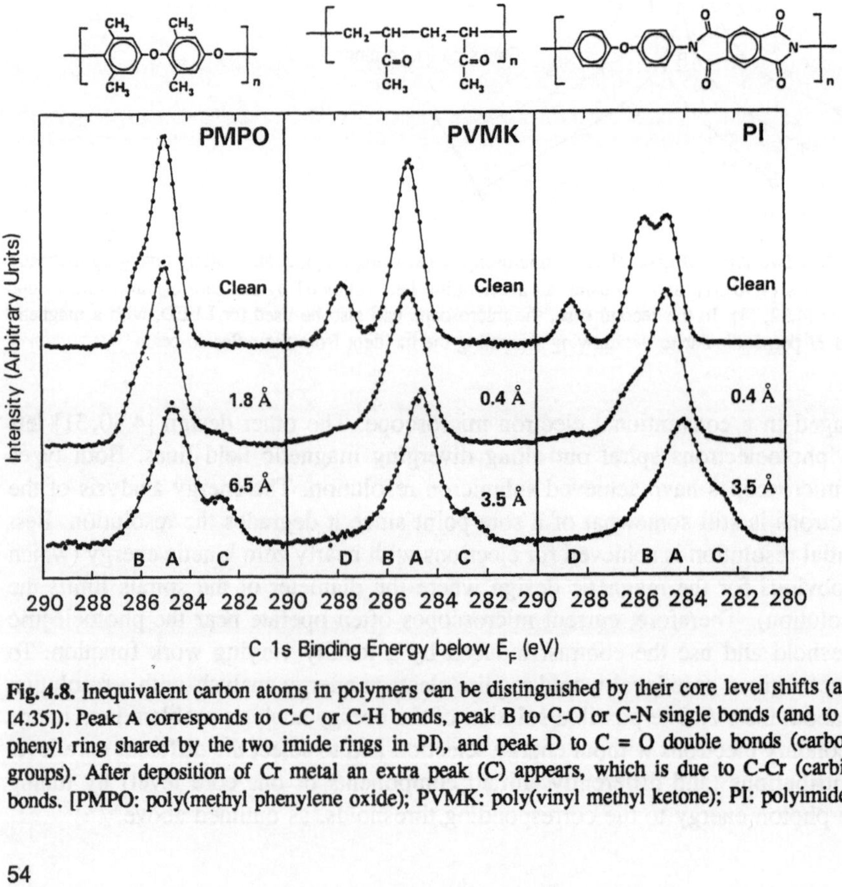

Fig. 4.8. Inequivalent carbon atoms in polymers can be distinguished by their core level shifts (after [4.35]). Peak A corresponds to C-C or C-H bonds, peak B to C-O or C-N single bonds (and to the phenyl ring shared by the two imide rings in PI), and peak D to C = O double bonds (carbonyl groups). After deposition of Cr metal an extra peak (C) appears, which is due to C-Cr (carbide) bonds. [PMPO: poly(methyl phenylene oxide); PVMK: poly(vinyl methyl ketone); PI: polyimide]

4.7 Summary

The purpose of this overview has been to demonstrate the capabilities of various techniques in determining the electronic structure of solids. Photoemission and inverse photoemission come out as the most versatile methods, since they provide the complete set of quantum numbers that characterize electrons in a crystalline solid, i.e., energy, momentum, spin and angular (point group) symmetry. These measurements can be condensed into a set of a few parameters by empirical band calculations, or can be predicted accurately by first-principles band calculations, as long as excited state effects are included. New experimental avenues are beginning to open up, such as time- and spatially-resolved measurements. They promise to push the field of materials science into new territory, where excited states and interfaces drastically modify the properties of a solid.

References

4.1 A. Psaras, H.D. Langford (eds.): *Advancing Materials Research* (National Academy, Washington, DC 1987). For a compilation of novel materials see the article by F.J. Di Salvo on p. 161

4.2 E.W. Plummer, W. Eberhardt: Adv. Chem. Phys. **49**, 533 (1982);
F.J. Himpsel: Adv. Phys. **32**, 1 (1983)

4.3 V. Dose: Surf. Sci. Rep. **5**, 337 (1985);
N.V. Smith: Rep. Prog. Phys. **51**, 1227 (1988);
F.J. Himpsel: Comments Cond. Mat. Phys. **12**, 199 (1986); Surf. Sci. Reports **12**, 1 (1990)

4.4 *Landolt-Börnstein, Numerical Data and Functional Relationships in Science and Technology*, Group 3, Vol. 23a, *Electronic Structure of Solids: Photoemission Spectra and Related Data*, (Springer, Berlin, Heidelberg 1989)

4.5 J.R. Chelikowsky, M.L. Cohen: Phys. Rev. B **14**, 556 (1976);
M.L. Cohen, J.R. Chelikowsky: *Electronic Structure and Optical Properties of Semiconductors*, Springer Ser. Solid-State Sci. (Springer, Berlin, Heidelberg 1988)

4.6 N.V. Smith, R.L. Benbow, Z. Hurych: Phys. Rev. B **21**, 4331 (1980);
R. Lässer, N.V. Smith, R.L. Benbow: Phys. Rev. B **24**, 1895 (1981);
N.V. Smith, R. Lässer, S. Chiang: Phys. Rev. B **25**, 793 (1982)

4.7 R.I.G. Uhrberg, G.V. Hansson, U.O. Karlsson, J.M. Nicholls, P.E.S. Persson, S.A. Flodström, R. Engelhardt, E.-E. Koch: Phys. Rev. Lett. **52**, 2265 (1984)

4.8 W. Drube, D. Straub, F.J. Himpsel: Phys. Rev. B **35**, 5563 (1987)

4.9 G.P. Williams, F. Cerrina, G.J. Lapeyre, J.R. Anderson, R.J. Smith, J. Hermanson: Phys. Rev. B **34**, 5548 (1986)

4.10 G.D. Mahan: Phys. Rev. B **2**, 4334 (1970); Phys. Rev. Lett. **24**, 1068 (1970)

4.11 D.J. Spanjaard, D.W. Jepsen, P.M. Marcus: Phys. Rev. B **15**, 1728 (1977);
J.F.L. Hopkinson, J.B. Pendry, D.J. Titterington: Comput. Phys. Commun. **19**, 69 (1980);
D.W. Jepsen, F.J. Himpsel, D.E. Eastman: Phys. Rev. B **26**, 4039 (1982);
D.W. Jepsen, Th. Fauster, F.J. Himpsel: Phys. Rev. B **29**, 1078 (1984)

4.12 F.J. Himpsel, W. Eberhardt: Solid State Commun. **31**, 747 (1979);
T.-C. Chiang, J.A. Knapp, M. Aono, D.E. Eastman: Phys. Rev. B **21**, 3513 (1980)

4.13 T. Schneider, H. De Raedt, M. Frick: Z. Phys. B **76**, 3 (1989)

4.14 L. Hedin: Phys. Rev. **139**, A796 (1965);
For a review see L. Hedin, S. Lundqvist: In *Solid State Physics: Advances in Research and Applications*, Vol. 23, ed. by F. Seitz, D. Turnbull, H. Ehrenreich (Academic, New York 1969) p. 1
4.15 M.S. Hybertsen, S.G. Louie: Phys. Rev. B **34**, 5390 (1986);
S.G. Louie: Private communication
4.16 W. von der Linden, P. Horsch: Phys. Rev. B **37**, 8351 (1988)
4.17 A.L. Wachs, T. Miller, T.C. Hsieh, A.P. Shapiro, T.-C. Chiang: Phys. Rev. B **32**, 2326 (1985)
4.18 D. Straub, L. Ley, F.J. Himpsel: Phys. Rev. B **33**, 2607 (1986)
4.19 E.E. Koch (ed.): Handbook on Synchrotron Radiation, Vols. 1a, 1b (North-Holland, Amsterdam 1983)
4.20 J.-M. Imer, F. Patthey, B. Dardel, W.-D. Schneider, Y. Petroff, A. Zettl: Phys. Rev. Lett. **62**, 336 (1989);
C.G. Olson, R. Liu, A.-B. Yang, D.W. Lynch, A.J. Arko, R.S. List, B.W. Veal, Y.C. Chang, P.Z. Jiang, A.P. Paulikas: Science 731 (Aug. 1989); Phys. Rev. B **42**, 381 (1990)
4.21 D. Weller, S.F. Alvarado, W. Gudat, K. Schroeder, M. Campagna: Phys. Rev. Lett. **54**, 1555 (1985);
C. Rau, C. Liu, A. Schmalzbauer, G. Xing: Phys. Rev. Lett. **57**, 2311 (1986)
4.22 D.T. Pierce, R.J. Celotta, R.J. Kelley, J. Unguris: Nucl. Instrum. Methods A **266**, 550 (1988)
4.23 J. Kirschner, R. Feder: Phys. Rev. Lett. **42**, 1008 (1979);
J. Kirschner: *Polarized Electrons at Surfaces*, Springer Tracts in Modern Physics, Vol. 106 (Springer, Berlin, Heidelberg 1985) p. 62
4.24 C.S. Feigerle, D.T. Pierce, A. Seiler, R.J. Celotta: Appl. Phys. Lett. **44**, 866 (1984)
4.25 J. Unguris, A. Seiler, R.J. Celotta, D.T. Pierce, P.D. Johnson, N.V. Smith: Phys. Rev. Lett. **49**, 1047 (1982);
J. Kirschner, M. Glöbl, V. Dose, H. Scheidt: Phys. Rev. Lett. **53**, 612 (1984);
M. Donath: Appl. Phys. A **49**, 353 (1989)
4.26 R. Haight, J.A. Silberman: Phys. Rev. Lett. **62**, 815 (1989)
4.27 P. Avouris, R. Wolkow: Mater. Res. Soc. Symp. Proc. **131**, 157 (1989);
E.J. van Loenen, D. Dijkamp, A.J. Hoeven, J.M. Lenssinck, J. Dieleman: Appl. Phys. Lett. **55**, 1312 (1989);
This field is advancing rapidly, see: D.M. Eigler, E.K. Schweizer: Nature **344**, 524 (1990)
4.28 J.A. Stroscio, R.M. Feenstra, A.P. Fein: Phys. Rev. Lett. **57**, 2579 (1986); J. Vac. Sci. Technol. A **5**, 838 (1987)
4.29 T.F. Heinz, F.J. Himpsel, E. Palange, E. Burstein: Phys. Rev. Lett. **63**, 644 (1989);
A.B. McLean, F.J. Himpsel: Phys. Rev. **39**, 1457 (1989)
4.30 G. Beamson, H.Q. Porter, D.W. Turner: Nature **290**, 556 (1981);
I.R. Plummer, H.Q. Porter, D.W. Turner, A.J. Dixon, K. Gehring, M. Keenlyside: Nature **303**, 599 (1983)
4.31 P. Pianetta, P.L. King, A. Borg, C. Kim, I. Lindau, G. Knapp, M. Keenlyside, R. Browning: J. Electron. Spectrosc. Relat. Phenom. **52**, 797 (1990)
4.32 W. Telieps, E. Bauer: Surf. Sci. **162**, 163 (1985); Ultramicroscopy **17**, 57 (1985);
E. Bauer: Ultramicroscopy **17**, 51 (1985)
4.33 B.P. Tonner, G.R. Harp: Rev. Sci. Instrum. **59**, 853 (1988)
4.34 F. Cerrina et al.: Nucl. Instrum. Methods A **266**, 303 (1988)
4.35 J.L. Jordan, P.N. Sanda, J.F. Morar, C.A. Kovac, F.J. Himpsel, R.A. Pollak: J. Vac. Sci. Technol. A **4**, 1046 (1986);
J.L. Jordan, C.A. Kovac, J.F. Morar, R.A. Pollak: Phys. Rev. B **36**, 1369 (1987)

5. Predicting the Properties of Solids, Clusters and Superconductors

Marvin L. Cohen

With 9 Figures

In the past, theoretical physicists characteristically focused on properties of matter that could be examined using highly idealized models. However, in recent years a large part of the research on the theory of condensed matter [5.1, 2] has been concentrated in areas formerly considered the domain of materials science. This trend has been aided by the development of methods capable of addressing problems related to the properties of real materials. These methods are an outgrowth of the quantum mechanics invented 50 or more years ago. It took this long period of practice with model calculations along with detailed studies of atoms, refinements in quantum theory and solid state spectroscopy, and the advent of modern computers to reach the present state of the art. Among these, a major ingredient was the unraveling of optical spectra of solids [5.3] and their interpretation in terms of electronic energy bands. This collaborative research between experimentalists and theorists served a similar role for solids to the one that atomic spectroscopy had for the development of quantum mechanics itself. It provided a data base of measured energy levels which could be used to test theoretical proposals. At first empirical theories were developed and tested using the experimental data. These evolved and now the theories are *ab initio* and require little or no experimental information about the solid.

This review traces some of the evolution of the theoretical development and concentrates on applications to materials science. Some emphasis is placed on the predictive nature of the theoretical schemes to illustrate their robustness and the usefulness of the approaches. Although the development of methods to compute the properties of solids took different paths, the schemes in use today have much in common; they allow accurate determination of electronic energy levels and wavefunctions. The focus here is on the use of pseudopotentials, which have been shown to produce useful results starting from first principles.

5.1 Background

In much of the research on materials, the goal is to compute the electronic energy levels and wavefunctions for the solid and to use these to explore electronic as well as structural and vibrational properties. The pseudopotential approach starts with a model composed of valence electrons moving in a periodic array of ionic cores. It is assumed in this model that the cores remain unchanged in going from the atom to the solid. For example, in the case of Si each core has a +4 charge and is composed of a Si nucleus together with 10 tightly held core electrons. Valence electrons move freely through this array of periodic cores. For Si, the four valence electrons per atom arrange themselves to form the covalent bonds which stabilize the structure.

The first pseudopotential applications used experimental data. In the Empirical Pseudopotential Method (EPM), the valence electrons are assumed to move in an average potential caused by electron–core and electron–electron interactions. Because the Pauli principle prevents valence electrons from occupying core states, there is a "Pauli force" which pushes electrons away from the core. This repulsive potential tends to cancel the attractive Coulomb potential in the core region and produces a net pseudopotential which is quite weak. The EPM assumes this model, and in this scheme, the potential is fit using approximately three parameters for each element. Dozens of solids were examined in the 1960s and 1970s [5.3] and in addition to providing interpretations of optical spectra, the EPM yielded energy band structures, wavefunctions, and electronic density plots.

The focus for much of the EPM was on the group IV, III–V, and II–VI octet tetrahedrally bonded semiconductors. The early calculations [5.3–5] concentrated on the electronic energy bands of this group, and then calculations of optical spectra [5.6] and the electron density were done [5.7]. In the mid-1970s [5.8], it was possible to provide detailed results for energy bands, electronic charge density, optical spectra, and density-of-states spectra. The scheme was extended to a large number of systems, and these calculations are still considered to be the most accurate available. There was agreement between theory and experiment for optical spectra, X-ray studies of electron density, photoemission electron energy spectra, and angle-resolved photoemission determination of electronic energy bands. Figure 5.1 illustrates a comparison of the EPM band structure of GaAs with the results of angle-resolved photoemission measurements. In Fig. 5.2 a comparison between theory and experiment is shown for the valence electronic charge density of Si. Figure 5.3 illustrates the results for calculations of the modulated reflectivity. Almost all of the structure in these spectra was interpreted in terms of transitions between specific energy bands.

The results of these studies are well-documented in the literature and in texts [5.3, 6] and serve as the foundation for our understanding of the electronic structure of the tetrahedral semiconductors. The results and methods also form a basis for understanding a much larger class of materials. It was discovered [5.12] early in studies using the EPM that the pseudopotentials obtained from optical and photoemission analyses for a specific material were "transferrable". The transfers were done in several ways. First, for the case of crystals which can exist in more than one structural form, such as zincblende and wurtzite (ZnSe), analyses of the optical properties of one structure gave potentials which were appropriate for interpreting the optical spectrum of the other structure. Next, it was found that the potentials for the constituent elements in a compound semiconductor could be extracted. For example, in the case of InSb, the In and Sb potentials were determined from an optical analyses of InSb. This was surprising since InSb is

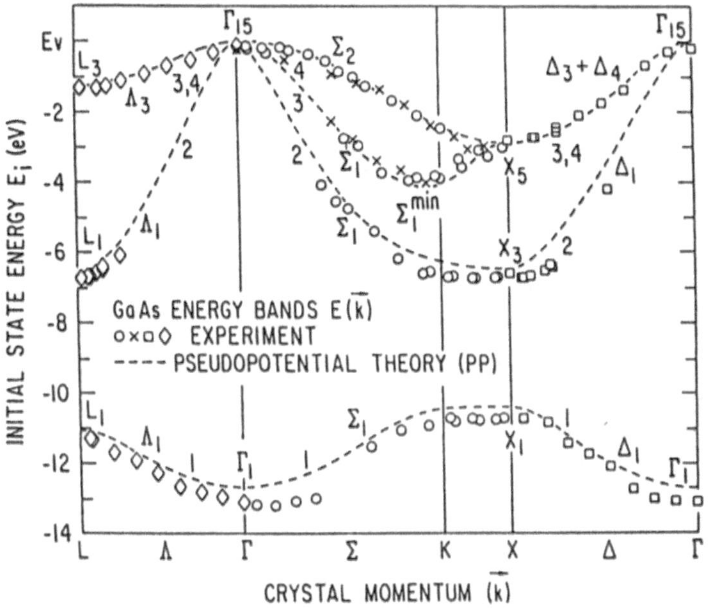

Fig. 5.1. Valence electronic energy bands for GaAs as determined from angle-resolved photoemission [5.9]

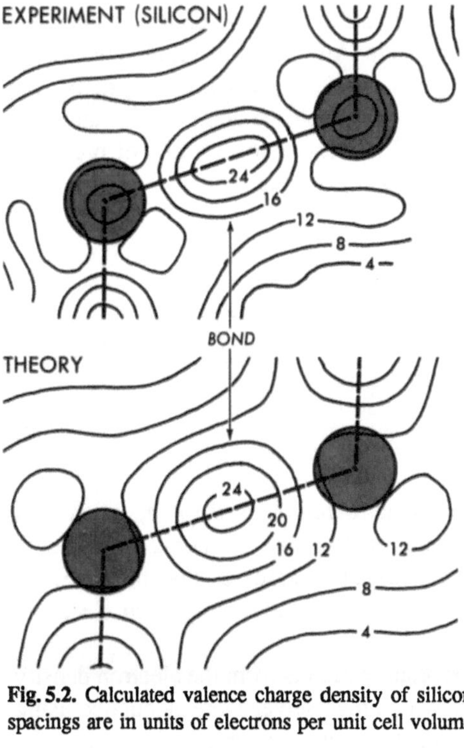

Fig. 5.2. Calculated valence charge density of silicon compared to X-ray data [5.10]. The contour spacings are in units of electrons per unit cell volume

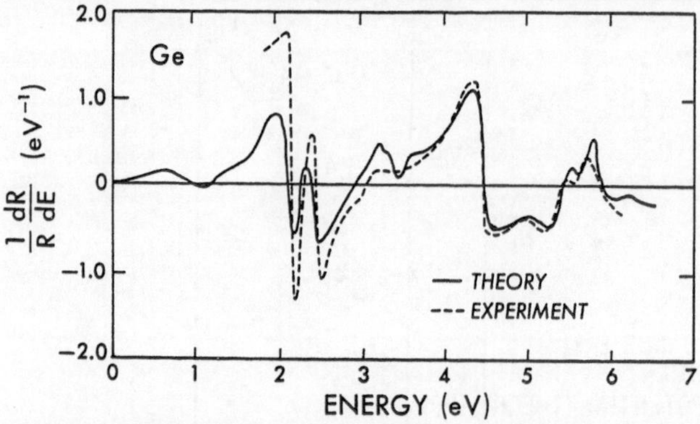

Fig. 5.3. Calculated derivative reflectivity spectrum for germanium. The experimental data is based on wavelength modulation spectra [5.11]

a semiconductor which bears little if any resemblence to metallic In and semimetallic Sb. However, the In and Sb pseudopotentials obtained in this way were sufficiently accurate to give useful information about In and Sb [5.12].

More important than the use of the atomic potentials extracted from semiconductor studies was the concept that this could be done. The fact that to lowest order the atomic pseudopotentials were transferrable and that a pseudopotential could be associated with an atom inspired proposals that tables of empirical pseudopotentials for all the atoms could be made and these could be used to explore electronic sturcture in a wide variety of crystals. Although this scheme works for similar materials and gives reasonably good results for dissimilar systems, the precision is not, in general, on the level of what could be achieved for the original systems for which the fits were made. The reasons for the loss of precision were well understood. In the EPM, the Schrödinger equation

$$-\frac{\hbar^2\nabla^2}{2m}\psi_{n,\boldsymbol{k}}(\boldsymbol{r}) + V(\boldsymbol{r})\psi_{n,\boldsymbol{k}}(\boldsymbol{r}) = E_n(\boldsymbol{k})\psi_{n,\boldsymbol{k}}(\boldsymbol{r}) \tag{5.1}$$

involves a potential $V(\boldsymbol{r})$ which was fit to give wavefunctions $\psi_{n,\boldsymbol{k}}(\boldsymbol{r})$ and energy levels $E_n(\boldsymbol{k})$ for band n and wavevector $\boldsymbol{k}$ which are appropriate to a specific electronic density $\varrho(\boldsymbol{r})$ where

$$\varrho(\boldsymbol{r}) = \sum_{n,\boldsymbol{k}} |\psi_{n,\boldsymbol{k}}(\boldsymbol{r})|^2 . \tag{5.2}$$

For metals, the charge density outside the core region is smooth, whereas it becomes peaked or almost vanishes in some regions between atoms in insulators and semiconductors. As described earlier, the EPM assumes that the potential $V(\boldsymbol{r})$ combines both the core potential and the potential arising from electron–electron effects.

An early attempt to separate out the screening effects from the electron density was done [5.7] using dielectric functions which themselves were calculated using the EPM. This study showed that by separating out a core contribution to the potential an even more transferrable pseudopotential emerged. The valence electrons which screened these potentials would have different distributions according to whether they were metals, semiconductors, or insulators. These concepts were already embodied in the earlier work on model potentials [5.12] particularly the contributions made by *Heine* and co-workers [5.13, 14].

In the 1970s, the EPM was in the enviable position of having explained optical and photoemission properties of solids with high precision. It spawned new concepts forming the foundation of our present view of semiconductor electronic structure, and it was extendable. However, the extensions were limited, and it was not a "first principles" theory. For more general applications to different atomic configurations and to make the scheme "ab initio", the electron–electron contributions to the potential need to be separated from the core contributions.

5.2 Surfaces and Interfaces

The problem of separating out the electronic contribution to the EPM pseudopotential is essential for a study of surfaces and interfaces. If the EPM were applied to a surface, the results for the electron density at the surface would resemble the end of an ideal solid with an abrupt cut. The electrons could not readjust to the presence of the surface because in the EPM model as it was originally constructed all atomic cores are identical and the electrons would experience the same potential at the surface as they do in the bulk. It is clear physically, however, that when a crystal is cut, the electrons near the interface do not remain in their original bulk configurations. The electronic charge density readjusts to account for the presence of the surface.

New schemes [5.15, 16] emerged in the 1970s to separate the potentials and examine the readjustment at the surface. The pseudopotential was explicitly separated into core and electron–electron contributions:

$$V(\boldsymbol{r}) = V_{\text{core}}(\boldsymbol{r}) + V_{\text{e-e}}(\boldsymbol{r}) \,. \tag{5.3}$$

Model potentials were used for V_{core}, and $V_{\text{e-e}}$ was taken to be a functional of the charge density $\varrho(\boldsymbol{r})$. There was still an empirical "flavor" to the calculations because some parameters were used in the model potentials designed to represent the effects of electronic and correlation contributions to $V_{\text{e-e}}$.

A major feature of the calculations for surfaces and interfaces was the use of self-consistency. Since $V_{\text{e-e}}$ depends on the charge density, which was computed from the wavefunction, which in turn depends on $V_{\text{e-e}}$, a self-consistency loop could be used. First, a configuration such as a surface or interface is chosen which fixes the structure and the positions of the cores. Then, the core potentials and a guess for $V_{\text{e-e}}(\boldsymbol{r})$ based on the $\varrho(\boldsymbol{r})$ generated with empirical potentials serve to start the calculation. Equation (5.1) is solved for $\psi_{n,\boldsymbol{k}}(\boldsymbol{r})$ to obtain $\varrho(\boldsymbol{r})$ using (5.2). Then, $V_{\text{e-e}}$ is computed and used for the next $\psi_{n,\boldsymbol{k}}(\boldsymbol{r})$. Usually six times through this loop gave the same input and output $\varrho(\boldsymbol{r})$ or "self-consistency".

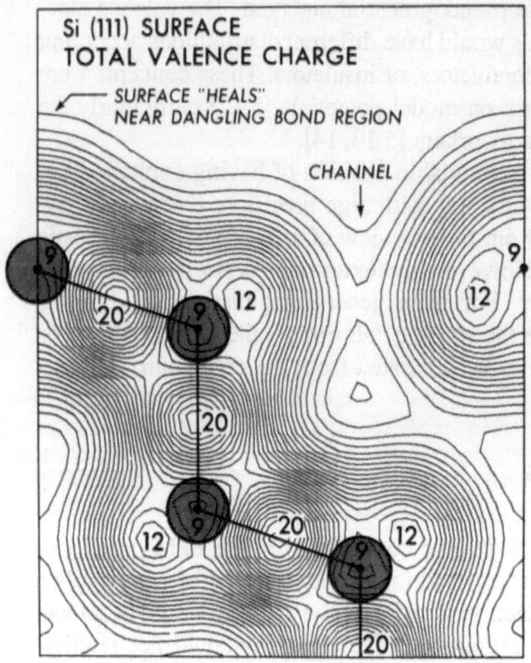

Fig. 5.4. Total valence charge distribution for an unrelaxed Si(111) surface. Charge is plotted as contours in a (110) plane intersecting the (111) surface at right angles. Plotting area starts in the vacuum and extends about 4 1/2 atomic layers into the crystal. Atomic positions and bond directions are indicated by dots and heavy lines, respectively. Contours are normalized to electrons per Si bulk unit-cell volume

Dozens of surfaces and interfaces were examined [5.17], and the calculations produced plots of the electronic charge density near the interfaces. A good example of the information which the theory could provide is given in Fig. 5.4. Here an ideal Si(111) surface is viewed from one side; that is, the plot is in a plane perpendicular to the (111) surface. The charge density $\varrho(\boldsymbol{r})$ is displayed on the plane, and the modification in the bonding near the surface is evident. After moving only a few layers into the crystal, one sees charge distributions and bonds which closely resemble the bulk as shown by a comparison with Fig. 5.2. However, near the surface the charge smooths out and "heals" the cut produced when the surface was formed. Several features of the ideal structure remain. For example, because of the crystal structure, there are long channels with low charge density which extend into the solid. It is likely that interstitial impurities travel along these channels when they enter the solid.

These calculations can also be used to explore the occurence of electronic "surface states" which were predicted in the 1930s. Because of the abrupt changes in the potential experienced by an electron near a surface or interface, localized electronic states can be bound in this region. These states are very sensitive to the surface conditions. Impurities, defects, and structural changes can greatly influence the properties of surface states. Because of this, reproducible data was scarce until the 1970s when high vacuum techniques were developed to a point where surfaces stayed clean long enough for detailed study. It was this development in the experimental studies which enabled meaningful comparisons between experiment and theory.

On the theoretical side, new tools were also necessary. Because of the nature of the calculational approach, it is convenient to treat a surface or interface with a "supercell" approach [5.18]. Most bulk crystalline calculations assumed translational invariance in which a unit cell is repeated. However, the presence of a surface breaks this translational invariance. Similar problems arise for impurity calculations and in general for local configurations. In this method, this difficulty is overcome by assuming that the surface is contained in a large unit cell or supercell of a crystal. This cell is then repeated infinitely which allows use of the standard techniques. For example, in the case of Si(111) shown in Fig. 5.4, the cell can be as thick as 12 Si layers but with half of it filled with Si and the other half with vacuum. The interface between Si and the vacuum has the properties of a Si surface. For a Schottky barrier [5.17], a cell is chosen with half semiconductor and half metal. For a heterojunction [5.17], the two halves are filled with two different semiconductors.

In addition to providing results for the charge densities near interfaces and surfaces, these calculations provided detailed densities of states, band gaps, and other information about the electronic properties of surfaces and interfaces. They did not give much information about the positions of the atoms at the surfaces. Surface structure was used as input. The fact that atoms rearrange at surfaces through relaxations and reconstructions had to be included as input. This situation changed in the 1980s after total energy calculations were developed to determine bulk and surface crystal structures.

5.3 Total Energies and Structural Properties

For EPM calculations and even for the computations for surfaces and interfaces where electron–electron and electron–core potentials were separated, the positions of the atoms were assumed to be given by experiment or by the specific model considered. Although some experimental probes such as low energy electron diffraction were available in the 1960s, the analyses of the data were complex and definitive surface structural determination was not generally available. The situation has improved with the advent of new techniques such as scanning tunneling microscopy in the 1980s. These have been valuable aids in determining surface structure. On the theoretical side, for many systems calculations were limited to comparing electronic properties for various structures with measured data such as those based on photoemission studies. The best agreement for the electronic structure was interpreted as confirmation for the surface structure assumed in the calculation.

For bulk solids, it was usually possible to obtain the atomic positions from X-ray analyses so these data were easy to input. However, the dependence of structure on high pressures has not been determined for many solids. A theoretical tool for exploring this question was needed. In addition, pseudopotential theory had become so refined that most disagreements between experiment and theory regarding the placements of energy levels were eventually decided in favor of theory. This meant that the underlying theory was sound and capable of being extended to more detailed and wider applications. An important goal became the determination of crystalline properties without resorting to any experimental input [5.13].

The first major success in this area was the study [5.19] of several crystal structures for Si and the determination of ground-state structural properties. The approach is based on comparing the total energy as a function of volume for Si in different structures. To do this, the pseudopotentials were refined [5.20–24], and an accurate and convenient scheme [5.25, 26] for calculating total energies was developed.

For the pseudopotential, the most fruitful approaches resembled the original scheme used by *Fermi* [5.27] when he invented the pseudopotential in 1934. Since most solid-state effects involve only the outermost parts of the electronic wavefunction, the core oscillations can be ignored for many applications. Hence, a nodeless approximation to the true wavefunction which is extrapolated to zero inside the core region and fit to be equal to the true wavefunction outside the core should yield good estimates of solid-state effects. The potential is also constrained to provide wavefunctions with the proper normalization. This scheme can be implemented in a number of ways and pseudopotentials are readily available or can be calculated relatively easily for most atoms once the atomic wavefunctions are known. All that is required is the atomic number.

The above procedure produces V_{core}. For $V_{\text{e-e}}$ a density functional approach is used [5.28, 29]. The simplest approach is to use a local density approximation (LDA) where the potential at point $\boldsymbol{r}$ depends on $\varrho(\boldsymbol{r})$ and nonlocal effects are not included. The LDA approach is appropriate for ground-state properties but not for excited states. This limitation will be discussed later. However, for structural properties the ab initio core potentials together with the LDA for $V_{\text{e-e}}$ and a scheme for calculating the total energy allows detailed studies of structural properties. Since the pseudopotentials require only the atomic numbers as input and the electron–electron interactions are computed from the charge density, the results for calculated solid-state properties depend only on information about the constituent atoms. However, the calculations are done for specific crystal structures and the nature of the structure must be included as input. The separation of the atoms is not needed, but the relative position of the atoms in the structure are required as described below.

A major objective of a total energy calculation is to compare energies for a subset of candidate structures. It is then assumed that for a given volume, Nature will choose the lowest energy structures among the subset. For example, if we use Si as the prototype solid, then it is known that Si at atmospheric pressure and ambient temperature is found in the dimond structure with a bond length of 2.35 Å. Hence, when a comparison is made between the total energies for Si calculated in various structures at ambient pressures and temperatures, the energy for the diamond structure should be lowest. In the following discussion, we will ignore temperature variations and compare the structures and structural properties as a function of pressure or volume.

Since it is assumed that within the pseudopotential model the components of the solid are positive cores and negative valence electrons, then for rigid cores the total energy is

$$E_{tot} = E_{kin,e} + E_{e-c} + E_{e-e} + E_{c-c} \ . \tag{5.4}$$

The kinetic energy of the valence electrons $E_{kin,e}$ can be evaluated by using the gradients of the calculated electron wavefunctions. Electron–core contributions, E_{e-c}, to the total energy are evaluated by using the pseudopotential V_{core}. The E_{e-e} term contains the Hartree energy, which assumes that each valence electron moves in an average potential generated by the other valence electrons. Energy E_{e-e} also contains the contributions of exchange and correlation which are evaluated within the LDA. Finally, the Coulomb interactions between the cores contribute to E_{e-e}, and this is evaluated as a Madelung energy using standard approaches [5.30].

Since all the contributions to (5.4) depend on volume, the total structural energy, E_{tot}, depends on the assumed volume. This means that a structure which has lowest energy at one volume may be higher in energy than other structures at compressed or expanded volumes. This is the origin of solid–solid structural phase transitions induced by pressure.

The results for several candidate structures of Si are shown in Fig. 5.5 as a function of volume normalized to the equilibrium volume (V_0) for the diamond structure. Among the structures considered, the diamond structure has lowest energy. However, for reduced volumes (volume structure $V/V_0 < 1$) other structures have lower energy. For example, the structural transition from diamond to white tin (β-Sn) has been studied experimentally.

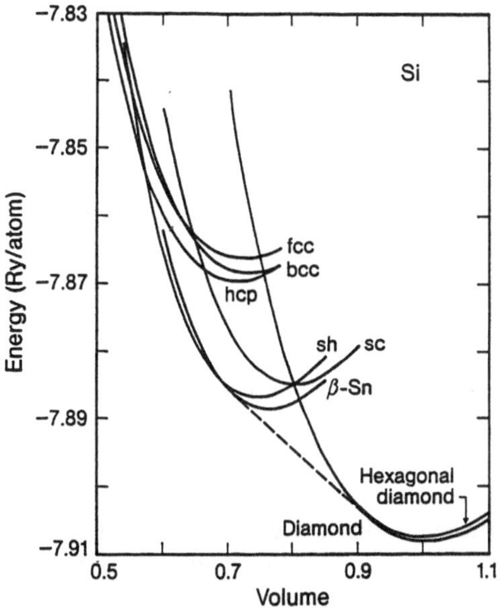

Fig. 5.5. Total energy curves of seven structural phases of silicon as a function of volume normalized to the volume of the diamond phase. The dashed line, which is the common tangent between the diamond and β-Sn phase, represents the path for this pressure-induced structural transition. Its slope gives an estimate of the transition pressure

At low pressures, Si remains in the diamond structure and then at a specific transition volume a mixture of the two phases appears. At a lower transition volume, the transition is complete and the system is in the β-Sn structural phase. The points of contact between the common tangent and the two $E_{tot}(V)$ curves gives the transition volumes. The slope of the common tangent provides an estimate of the pressure required to produce the transition. The calculated transition volumes and the transition pressure are in excellent agreement with experiment.

At higher pressures, other structures have lower energies than diamond or β-Sn. For example, the total energy calculations successfully predicted the existence of a hexagonal close-packed (hcp) phase in the range of 400 kbar. A face-centered cubic (fcc) phase [5.31] is predicted at even higher pressures and the properties of the lower pressure phases such as the β-Sn and simple hexagonal (sh) (or primitive hexagonal, ph) phases have been calculated in detail. In principle, other metastable structures which are not lowest in energy for a given volume can be formed. The BC-8 structure is an example of a structure of this kind. It is body-centered cubic with eight atoms in a unit cell. Calculations using the total energy scheme put the minimum energy for the BC-8 phase at volumes between the minima for diamond and β-Sn. Since the BC-8 minimum lies above the common tangent between diamond and β-Sn, it will not be formed by just reducing the volume. However, the calculations reveal that a transition path from β-Sn to BC-8 has a lower energy barrier to overcome than from diamond to BC-8. Hence, theory suggests that a possible scheme for forming BC-8 is to pressurize the β-Sn phase and then release the pressure. In practice, a higher degree of success in finding metastable phases is expected when both temperature and pressure are varied.

The above theoretical approach can be used to examine a large variety of crystal structures once suitable candidates are suggested. Not all structures are expected to form. For example, the possibility of the existence of graphitic Si has been raised often. However, the total energy calculations [5.32] indicate that for Si this structure has very high energy even though it is the lowest energy structure for C. If the Si lattice were expanded instead of compressed, then graphitic Si could be stable, but the required "negative" pressures would be very difficult, if not impossible, to achieve.

Once $E_{tot}(V)$ is available for a given structure, many ground-state properties can be computed. The volume at which $E_{tot}(V)$ has a minimum gives an estimate of the lattice constant, and the curvature of $E_{tot}(V)$ near the minimum gives the bulk modulus or its inverse, the compressibility, of the solid. Typical values for the calculated lattice constants are within 1% of experimental values. In cases where electron d-states are close to the valence electron states, the results are not as accurate since the pseudopotential approach assumes that the cores do not overlap and core–core interactions can be treated using only Madelung terms. A similar situation exists for calculations of the bulk modulus. Most calculations of the bulk modulus yield results which are within 6% of the measured values. Table 5.1 gives results for C, Si, and Ge in the diamond structure. These results are typical, and the agreement between theory and experiment is impressive when

Table 5.1. Static structural properties

	Lattice constant [Å]	Bulk modulus [Mbar]
Si		
calc.	5.45	0.98
expt.	5.43	0.99
% diff.	0.4%	−1%
Ge		
calc.	5.66	0.73
expt.	5.65	0.77
% diff.	0.1%	−5%
C		
calc.	3.60	4.41
expt.	3.57	4.43
% diff.	0.8%	−1%

one considers that only the atomic number and the structure have been used as input.

Another quantity of interest is the cohesive energy. Using Si as an example, a comparison of E_{tot} for Si in the diamond structure at the equilibrium volume can be compared with E_{tot} computed for the situation where the atoms are very far apart. If the distance between atoms in the latter case is sufficiently large, then they do not interact and this case can be taken as an isolated atom reference. The difference between E_{tot} calculated for the two cases gives a measure of the energy lowering which occurs when the solid is formed. This cohesive energy depends sensitively on the form of the V_{e-e}. Although the LDA gives a good measure of this value, improvements in the exchange and correlation potentials [5.33] give more precise estimates. In some cases, it is the inability of the LDA to estimate the energy of the isolated atoms which is the limitation. The improvements in V_{e-e} by going to approximations which are beyond the LDA appear to have an almost negligible effect on the calculated values of the lattice constants and bulk moduli.

Another important product of the total-energy pseudopotential approach is the determination of vibrational spectra. This calculation requires a knowledge of the atomic masses in addition to the atomic numbers of the constituent atoms. Once the structure is assumed, the atoms are displaced so as to represent the presence of a particular lattice vibrational mode in the crystal. This approach [5.34, 35] has been named the *frozen phonon* method. A calculation of E_{tot} for the distorted system is compared with the E_{tot} for the undistorted case and the increase in energy is related to the energy of the vibrational mode or to the phonon frequency at a given wavevector. Using the atomic masses, it is then possible, in principle, to calculate the phonon spectrum, that is, the frequency versus wavevector phonon dispersion curve. A limitation of this method is the fact that the wavevectors at which the frequencies of the phonons are calculated are discrete; however, if force constant models are used, it is possible in some cases to compute the dispersion curve for all wavevectors.

Electron–phonon coupling constants can be obtained in a similar way. If the electron–lattice potential is compared for the distorted and undistorted cases, then electron–phonon coupling constants can be evaluated [5.36, 37]. These constants are calculated for each branch of the vibrational spectrum separately and as a function of wavevector. An average over all wavevectors gives an estimate of the electron–phonon coupling constant λ, which is used for estimating superconducting transition temperatures. This will be discussed in Sect. 5.6.

Recently several improvements and additions have extended the applicability of the approach described above. One limitation on the above method is the calculation of excited states. This is not a limitation of the pseudopotential method, but of the LDA. As mentioned before, the LDA is appropriate for ground-state properties. The electronic energy levels computed using V_{e-e} based on the LDA are not expected to give accurate estimates for the electronic energy levels of the solid. Hence, excitation by optical or other probes which involve transitions from occupied to empty electronic levels are not described well in schemes using the LDA. Therefore, the energy separations which were used to establish the EPM and which are accurately reproduced by the EPM are not obtainable with precision when the LDA is used. Typical errors in energy band gaps are larger than 0.5 eV for group IV semiconductors and even larger for insulators. The LDA underestimates the band gap of Si by 50% and predicts a zero gap for Ge. Despite this shortcoming, the LDA works well in the regime for which it is intended: ground-state properties.

Recently, the excited electronic or quasiparticle states have been computed [5.38] successfully using a scheme which accounts for local field effects arising from the deviations of the charge density from a constant free-electron-like configuration and by the inclusion of many-body self-energy effects. This work is discussed by *Louie* in Chap. 3 of this volume so the details will not be repeated here. However, it is important to note that this addition to the total energy pseudopotential approach now allows the calculation of ground-state and excited-state properties using ab initio theory. In principle, the EPM is superceded; however, it should be noted that the full ab initio scheme is complex and requires relatively large amounts of computer time and memory. But it is a first-principles approach and ground-state and excited-state properties are computed on an equal footing.

Recent improvements [5.33] in the calculation of exchange and correlation energies can be used to obtain more accurate estimates of V_{e-e}. As mentioned earlier, a primary advantage of obtaining better estimates of V_{e-e} for systems like Si is the more accurate determination of the cohesive energy. These approaches involve the use of quantum Monte Carlo methods, and they are fairly complex. At present, much of the effort in this area has focused on establishing the validity of the approach. Hence, simple prototype systems such as Si are being studied. In the future, it is likely that this promising approach will have even more important applications for systems where correlation energies are a larger fraction of the electron–electron energies. Oxides, low density systems, magnetic materials, and transition metals are of particular interest.

A restriction of the total energy scheme is the requirement that a subset of candidate structures must be chosen for comparisons of E_{tot}. Since it is not possible to compute E_{tot} for all possible crystal structures, low energy structures may not be included in the subset chosen. One approach which has helped, but is also limited in its applicability, is to calculate the forces on atoms in a given structure when the atoms are moved in various directions. By judicious choices of directions to sample, new structures with zero forces on the atoms can be discovered. A more promising approach [5.39] appears to be "molecular dynamics" sampling of energy values. Considerable effort has been expended using these methods in the last few years. Although the approach is not at a point where new structures are routinely being discovered, some interesting insights have been obtained, particularly into amorphous and liquid systems.

In the discussion of surfaces, it was emphasized that the ideal geometry one obtains by cutting a bulk ball-and-stick model of a solid is rarely found for a real material. Relaxations and reconstructions occur often. The total-energy pseudopotential approach can be used for surface structure in a similar manner as in the bulk. Again, Si has been examined in detail, and various surface structures have been proposed. The GaAs(110) surface has also received a great deal of attention. In general, the results for the surfaces of Si, GaAs, Ge, and other materials are consistent with measurements and, in some cases, predictive.The force approach discussed earlier is particularly useful for surface structure. Finding "zero force" configurations has aided in choosing structures which are likely to be stable. Interfaces are explored in the same manner as for surfaces.

5.4 Compressibilities and Empirical Theories

Despite the great success of the total-energy pseudopotential approach, there remains an important role for empirical theories. Empirical approaches often yield analytic results which can be used for exploring trends in material properties. Although great precision is not usual and some of the underlying models are very approximate, there are advantages to these approaches as has been emphasized by *Pauling* [5.40], *Phillips* [5.4], *Harrison* [5.41] and others.

For the tetrahedral semiconductors discussed here as prototype solids, several empirical approaches have given information about trends in structural and electronic properties. A useful example is the approach introduced by *Van Vechten* and *Phillips* [5.4, 42]. In this scheme, the average energy separation between the valence and conduction bands is considered to be an average optical gap E_g. This gap has homopolar and heteropolar components E_h and C. For group IV semiconductors $C = 0$, but it increases in going from III-V to II-VI semiconductors. Hence, C is related to the ionicity of the semiconductor while E_h measures its covalency.

Using this approach, *Phillips* and *Van Vechten* explored the spectral properties of the semiconductors to obtain information about their structural characteristics.

From the measured electronic dielectric constant ε and the plasma energy E_p, the average gap E_g can be determined through the relation

$$\varepsilon = 1 + \left(\frac{E_p}{E_g}\right)^2 . \tag{5.5}$$

An empirical expression is then used to relate E_g, E_h, and C

$$E_g^2 = E_h^2 + C^2 . \tag{5.6}$$

By taking E_h to be constant for a row in the periodic table (such as Ge, GaAs, ZnSe), it is possible to determine C and produce an ionicity scale based on C where ionicity $f = C^2/E_g^2$. *Phillips* and *Van Vechten* showed that this scheme contained structural information by demonstrating that the occurrence of 4-fold and 6-fold coordinated structures depended on the value of ionicity as computed using their approach. For high ionicity, 6-fold structures such as the NaCl structure were common. A critical ionicity was found which did an excellent job of separating the 4-fold and 6-fold coordinated structures.

Some connections between the Phillips–Van Vechten scheme and pseudopotential calculations can be made, but, in general, attempts at developing a detailed correspondence between the two approaches have not been fruitful. An exception is the calculation [5.7] of the bond charge as a function of ionicity using the EPM. This calculation demonstrated that the strong covalent bond for the group IV materials becomes weaker in going to the III-V and II-VI semiconductors. Although this general behavior is expected, the advantage of the pseudopotential calculation is that the volume of the bond charge can be evaluated as a function of ionicity. Hence, a detailed description of the covalency change can be made. By extrapolating between the calculated values for the group IV, III-V, and II-VI bond charges as a function of ionicity, a critical ionicity is found at the point where the bond charge extrapolates to zero. This value is very close to the estimated critical ionicity for the 4-fold to 6-fold transition found by *Phillips* and *Van Vechten* and adds support to their view that the bond charge stabilizes the 4-fold coordinated structures.

Although the Phillips–Van Vechten scheme is empirical and semiquantiative like other empirical approaches, it does have the advantage of allowing an exploration of material trends. After the trends are understood, more quantitative schemes can result. if this insight is combined with the information gained from exploring a series of ab initio calculations and experimental trends, sometimes simple yet quantitative models can result. The scaling model of compressibility is an example of this kind as described below.

The bulk modulus or its inverse, the compressibility, is one of the most "obvious" properties of solids. Some solids are soft while others such as diamond are very hard. Defects, impurities, dislocations, and a variety of macroscopic properties influence hardness, but for pure defect-free materials hardness generally scales with bulk modulus. The bulk modulus in turn is a quantity which can be calculated from the microscopic bonding properties of the solid. As described earlier, the total energy pseudopotential approach gives accurate determinations of the bulk modulus B through the curvature near the minimum of an $E_{tot}(V)$ curve for a specific structure. Equations of state can be fit to the $E_{tot}(V)$ results, and the equilibrium B_0 and its pressure derivative B_0' can be evaluated. In what follows, a simpler approach is used based on an empirical model. This scheme gives accurate values of B_0 using the bond length and a crude approximation to the ionicity as input.

The approach [5.43] is based on the observation that for a free-electron-like system with concentration n and Fermi energy E_F, it can be shown that [5.30]

$$B_0 = \tfrac{2}{3} n E_F \,. \tag{5.7}$$

This expression can be viewed as a bonding energy divided by a bond volume. To determine a similar scaling expression for a covalent system, the bonding energy is assumed to be E_h, the homopolar or covalent energy. In the series Ge, GaAs, and ZnSe, $E_h = 4.3\,\mathrm{eV}$. It is interesting that there is negligible variation in the bond length for these semiconductors even though their ionicity is changing. This suggests that it is the covalent nature of the bond which dominates in the determination of the bond length. To estimate the bond volume, charge density plots reveal that bonds are roughly cylindrical and have volumes which are approximately $\pi(2a_B)^2 d$ where a_B is the radius and d is the bond length. The scaling discussed above then suggests that

$$B_0 = 45.6\, E_h d^{-1} \tag{5.8}$$

for B_0 in gigapascals (GPa), E_h in electron volts, and d in angstroms.

If E_h is taken from spectral data [5.4, 42], Eq. (5.8) gives reasonable agreement with experiment for B_0. However, it is known [5.4] that E_h scales as $d^{-2.5}$, hence a simpler scaling relation results which depends only on d:

$$B_0 = 1761\, d^{-3.5} \,. \tag{5.9}$$

Equation (5.8) makes no distinction between the group IV, III-V, and II-VI semiconductors, yet it is known that ionicity increases in this series and B_0 decreases. A parameter can be introduced into (5.8) which accounts for this change to lowest order. This parameter, λ, is chosen to be 0, 1, and 2 for group IV, III-V, and II-VI systems respectively. The final formula for B_0 becomes

$$B_0 = (1971 - 220\lambda) d^{-3.5} \,. \tag{5.10}$$

This expression gives excellent agreement with experimental values as shown in [5.44].

Although the general behavior of (5.8) has been derived [5.45] using more first-principles methods based on the total-energy pseudopotential scheme, the power law and details differ somewhat. The microscopic analysis yields analytical expressions which give good results for group IV and III-V semiconductors but are not in good agreement for the II-VI compounds. Overall, the empirical scaling approach of (5.8) is superior and gives results for specific materials which are comparable to those obtained with a full microscopic computation based on the total energy.

The results of the empirical approach using (5.8) are suggestive regarding the quest for materials with large B_0. In particular, as B_0 increases the bond length d and ionicity parameter λ are decreased. Diamond sets the present limit for the largest known B_0. It has a short bond length $d = 3.567\,\text{Å}$, $\lambda = 0$, and $B_0 = 443\,\text{GPa}$. Since C-N bonds are shorter than C-C bonds, it was suggested [5.43] that systems based on C-N and B-C-N should have very high values of B_0. For some materials, part of the tetrahedral symmetry which is implied in this analysis may be lost; however, for partially sp^3 bonding, B_0 should still be large. The empirical approach does not predict d or structure, but ionic radii can be used to make estimates of bond lengths.

A recent attempt [5.46] to exploit the suggestions of the empirical theory uses the known structure of Si_3N_4 to test the hypothetical compound C_3N_4. A total-energy pseudopotential calculation for this compound yields a B_0 with a range of values close to that of diamond. The scaling theory gives similar results if an "intermediate" value for λ is used because of the fact that the ionicity of the C-N bond is expected to be between that of IV-IV and III-V compounds. The calculation for C_3N_4 is suggestive as a model and as a potential real system. Modern diamond anvil techniques when used with lasers produce pressures and temperatures which are high enough to simulate conditions expected deep inside the earth. It may be possible to use these approaches to fabricate materials like C_3N_4 with short bond lengths and low ionicity.

Another empirical approach to study the structural properties of solids involves the use of atomic, ionic, or "core" radii. This approach evolved from older versions where the conceptual basis was the ball-and-stick models of crystal structures. For the most part, the ball-and-stick models have given way to more quantum concepts such as orbital radii which depend on the various angular momentum states for an electron in an atom or a solid. Traditionally, orbital radii are estimated empirically by dividing bond lengths among the component atoms. By investigating a large data base for materials with varying bonding properties, reasonably consistent radii are obtained for many elements [5.4, 30, 40].

More modern approaches begin with quantum models. A direct link between pseudopotentials and orbital radii has been made, and this area is an active one. Some of the schemes are discussed by *Phillips* in Chap. 11 of this volume, hence only a few illustrations are given here. One example is the use of a set of radii obtained from first-principles calculations by *Zunger* and *Cohen* [5.21, 45] to separate structures as was done by *Phillips* and *Van Vechten*, who used spectral data as was discussed earlier. These radii can also be used to determine bond lengths. The Zunger–Cohen s-orbital radii which are appropriate for estimating the size of zero angular momentum pseudo-atom orbitals can be used to determine bond lengths in diatomic AB compounds. Defining $r_s^+ = r_s(A) + r_s(B)$ where r_s is the Zunger–Cohen relativistic screened atomic s-orbital radius, the bond length d for 41 4-fold ZnS structure compounds is given by

$$d = S_1 r_s^+ + S_2 - c|\chi(A) - \chi(B)| \,, \tag{5.11}$$

where $S_1 = 2.606 \pm 0.038$, $S_2 = 0.907 \pm 0.059$ a.u., $c = 0.056 \pm 0.019$ a.u., and χ is the Pauling electronegativity [5.40]. The comparison [5.48] between theory and experiment is excellent for d. Similar expressions to that given by (5.11) have been derived [5.48] for 106 6-fold AB compounds with NaCl structure and 132 8-fold AB compounds with CsCl structure. In both cases, d is proportional to r_s^+, and in plots of d versus r_s^+, the NaCl and CsCl compounds are separated fairly well.

A more detailed study [5.49] of the determination of AB crystal structures using atomic properties demonstrates that by using s-orbital radii and the Pauling electronegativity χ, it is possible to distinguish between four types of AB compound crystal structures, ZnS, NaCl, CsCl, and MnP or NiAs. If expressions are used to relate χ to the s-orbital radii, a simple scheme can be obtained. The structural maps using r_s for the anion r_s^a and for the cation r_s^c are capable of making structural separations which from the total-energy point of view are very close in energy. An example of a structural map of this kind is given in Fig. 5.6.

Although there is no first-principles theory explaining the origin of the successes of structure separations using orbital radii, the results do seem consistent with "physical intuition". These schemes are not strictly empirical. They rely on theoretical concepts and experience with ab initio calculations. Even if direct proofs of their validity are not available, they will continue to be used and refined because of their usefulness for studies of trends and some detailed properties of molecules and solids.

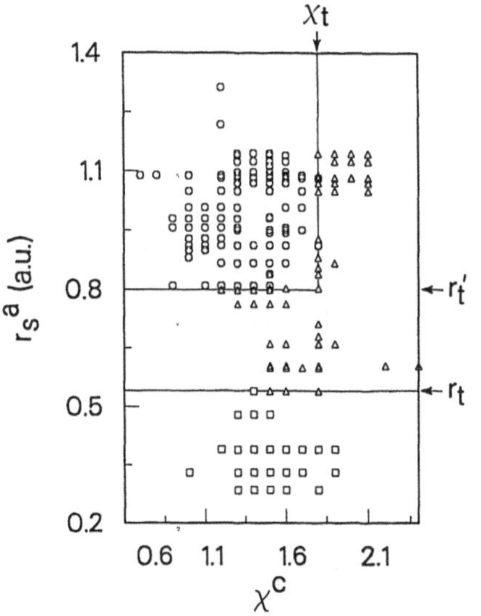

Fig. 5.6. Structure map for 255 AB suboctets: 38 NaCl (*squares*), 86 MnP or NiAs (*triangles*), and 131 CsCl (*circles*). The coordinate χ^c is the Pauling electronegativity, and r_s^a is the pseudopotential s-orbital radius of the anion. Structural separations occur at r_t, r_t', and χ_t

5.5 Metallic Clusters

Clusters of atoms ranging in number from a few to a few thousand atoms have been the subject of increasing interest. Chemists have long used these systems as special catalysts and for other useful purposes and electronic engineers and computer scientists view these systems as models for ultra-miniaturization. New techniques in materials science have allowed the fabrication of clusters on surfaces and, in some cases, as beams. The range in the sizes of the clusters is small so that systematic studies as a function of size can be done.

From a theoretical perspective, one can view clusters from an atom to molecule to cluster evolution, and theoretical chemists have often chosen this view. Condensed matter physiscists had added another approach based on the experience with bulk metals. The two approaches are compatible although language difficulties occur. We will describe the condensed matter physics perspective. We begin by discussing some aspects of the theory of bulk metals and then describe the use of these concepts for studying the properties of metallic clusters.

The total-energy pseudopotential method works as well for metals as it does for semiconductors and insulators. In addition to calculations of lattice constants, bulk moduli, and phonon spectra, applications to properties such as the Poisson ratio [5.50] and superconductivity [5.37] have been made. The use of more empirical approaches has also been successful. In fact, the concept that a metal can be treated as a gas of free electrons has produced one of the most important empirical models in solid-state physics. The quantum free-electron-gas model allowed Sommerfeld and others to solve many of the outstanding problems concerning solids. Earlier, classical free-electron models such as the Drude model yielded excellent results for the static conductivity and the optical properties of metals. The quantum model can be augmented by adding the effects of the lattice as a weak perturbation, resulting in the nearly free electron model, which gives good quantitative results. It is the nearly free electron model which evolved into the modern pseudopotential approach for studying metals.

If we step backwards and ignore the structural effects of the lattice, a simple yet very useful model emerges. This is the jellium model of metals, in which the electrons are treated as a quantum gas moving in a positive structureless background. The lattice of positive cores is considered to be smoothed out into a positive "jelly". This model is useful for exploring V_{e-e} since core effects are left out.

It is expected that, in general, metals will have relatively larger contributions from electron–electron interactions than semiconductors or insulators. Since their electron–core contributions are weaker, electron correlation effects are relatively larger. The parameter r_s characterizes the density of the electron gas. It is the radius measured in units of Bohr radii a_0 of the volume per electron. For an average density of ϱ_0 electrons per unit volume

$$\frac{4}{3}\pi(r_s a_0)^3 = \frac{1}{\varrho_0} . \tag{5.12}$$

For bulk jellium, the total electron energy E_e is

$$E_e = \frac{2.2099}{r_s^2} - \frac{0.9163}{r_s} - 0.094 + 0.0622 \ln r_s \;, \tag{5.13}$$

where the first term is the kinetic energy, the second is exchange, and the third and fourth terms arise from correlation effects. Higher-order correlation terms alter (5.13), and the assumption that the bulk volume is large eliminates the need to consider surface effects.

For clusters, the jellium model must include self-consistency to account for the "spill-out" at the edge of the jellium sphere. This complication is still considerably less difficult than doing a total-energy pseudopotential calculation. Since, in general, the structures of clusters are not known, the positions of the atoms need to be determined by the minimization of energies as described earlier for bulk crystals and surfaces. In the latter case, the computer limitations require small unit cells. Similar restrictions exist for clusters where the number of atoms in the cluster is analogous to the number of atoms in a unit cell for bulk and surface calculations.

If one views a cluster as a small piece of a crystal, then the jellium model appears to be a drastic over-simplification of cluster systems. If semiconductor clusters are considered, this would be a valid criticism since the covalent bonds and piling up of charge which characterizes semiconductors would not be adequately represented. The directional bonding properties which produce bulk semiconductor structures are expected to be important for clusters. At the opposite extreme, rare gas solids and clusters have small charge rearrangements, and their energies then depend on the geometrical arrangements and packing of the atoms. Hence, structure is again important. However, for simple metals, we will assume that, to lowest order, structure is not important and that it is the electronic contribution to the energy which is dominant. This contribution depends on the electronic density r_s and on the cluster size. The density of the cluster is fixed using the bulk solid value for r_s; the cluster sizes are fixed by the number of atoms in the cluster.

Although the jellium model for a cluster involves a sharp cutoff in the positive background [5.51], the electronic charge is allowed to spill out past the surface according to the constraints of self-consistency. However, even without self-consistency, the confinement of the electrons to the jellium cluster yields energy levels and quantization in a manner similar to that found for simple quantum models [5.51] such as harmonic oscillator potentials, square wells, and intermediate potentials between these as shown in Fig. 5.7. The above models also can be applied to study nuclei, and again, geometrical constraints are essential, as these can lead to shell structure.

For spherical symmetry, which would be appropriate for clusters with closed electron shell configurations, and a first-order approximation for partially occupied shells, the degeneracies coming from the spherical symmetry lead to shell structure. For angular momentum 0, 1, 2, ... ($s, p, d, \ldots$), the ordering and degeneracies (in parenthesis) are: $1s(2)$, $1p(6)$, $1d(10)$, $2s(2)$, $1f(14)$, $2p(6)$, $1g(18)$,

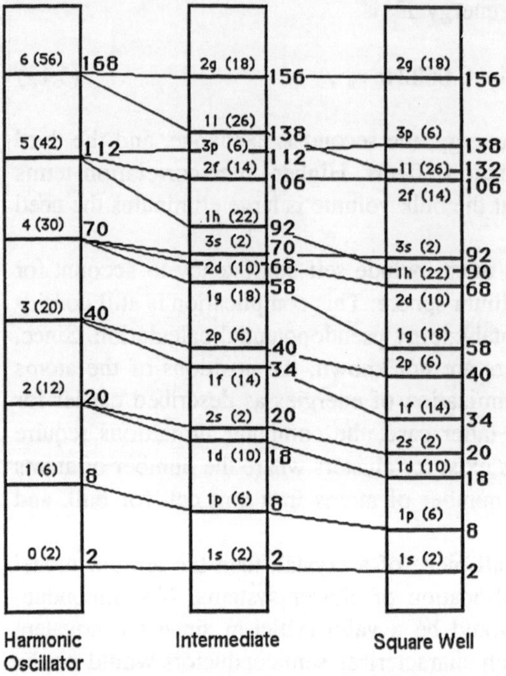

Fig. 5.7. Energy-level occupations for three-dimensional harmonic oscillator, intermediate, and square potential wells. (After [5.52])

$2d(10)$, $3s(2)$, $1h(22)$, etc. For monovalent atoms in an N-atom cluster, the elctrons fill each of the above energy levels. Special stabilities are found for the filled shells listed above. Hence, the total energies of clusters in this model are relatively low, and for $N < 100$, the calculations predict particularly stable clusters for $N = 2, 8, 18, 20, 40, 58$, and 92. The measured abundance spectrum [5.53] for Na shown in Fig. 5.8 is consistent with these results, and much of the fine structure is accounted for by including ellipsoidal [5.54] distortions using a model which is similar to the Nilsson model for nuclei.

The jellium model has been applied to clusters formed from other alkali metals, simple *sp* metals, and noble metals. Results [5.53] for Li, Na, K, Mg and Al are in good agreement with the experimental measurements of the abundance spectra. Sine the jellium model is solved using the LDA and the comparisons listed above involed calculations of the total energy for ground state configurations, good agreement is expected. However, for experiments which involve electronic excitations of the cluster, the LDA is not expected to give good results. There has been considerable theoretical work [5.55] lately which is aimed at producing schemes for calculating cluster polarizabilities, ionization potentials, optical absorption coefficients, and other properties related to experimental measurements involving electronic excitations. The problem of treating the excited states for clusters is somewhat more difficult than the same problem for bulk solids because of finite geometry effects.

At this point, despite the fact that experiments such as photoabsorption are not completely explained in terms of electronic models, the electronic shell structure

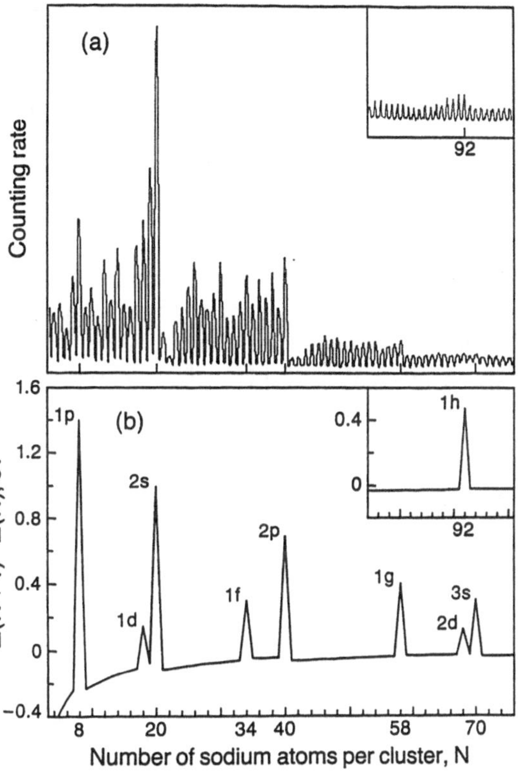

Fig. 5.8. Abundance spectrum for sodium clusters: (a) experimental and (b) theoretical

which is basic to the theory has been observed for a variety of metals and the concept appears to have a sound foundation.

5.6 Superconductivity

One of the most intriguing properties of matter is superconductivity. It took 46 years to explain the phenomenon microscopically. The discovery by *Kamerlingh Onnes* [5.56] of the phenomenon in 1911 began the mystery, and it was thought that the BCS theory [5.57] in 1957 ended it. However, the discovery [5.58] of high temperature superconducting oxides in 1986 may have added a new chapter to the saga and perhaps an entire new book. Here we will almost completely ignore the oxides and focus only on "BCS superconductivity" with pairing interactions caused by electron–phonon interactions.

Despite the sucess of BCS theory in explaining the properties of superconductors, it is difficult to use this theory to predict the existence of new superconductors or transition temperatures. The origin of this difficulty can be seen using the simplest theory of the transition temperature T_c based on the "BCS model". This model gives the expression

$$T_c = 1.14\, T_D e^{-1/NV} , \tag{5.14}$$

where T_D is the Debye temperature, N is the density of states of the Fermi energy, and V is the net attractive electron–electron potential. The parameter NV which is usually regarded as a single dimensionless term can be replaced by

$$NV = \lambda - \mu^* , \tag{5.15}$$

where λ is the attractive electron–phonon coupling constant causing the pairing and μ^* is an effective repulsive Coulomb constant which opposes the pairing. These two parameters are essential for estimating T_c; however, modern methods [5.59] for calculating T_c are more complex than (5.14) and are based on the Eliashberg theory [5.60]. The difficulty referred to above can be illustrated by the fact that for $T_D \sim 300$ K and $NV = 0.3$, then $T_c \sim 11$ K, but if NV is reduced to 0.03, then $T_c \sim 10^{-12}$ K. This estimate is made using (5.14), but a similar result is found even for the most complete models.

Therefore, despite the fact that a theory of T_c is available, estimates of T_c are extremely sensitive to λ and μ^*, and hence require a detailed knowledge of the normal state. It is this requirement which limits the determination of T_c. The Coulomb parameter μ^* varies less than λ and is often estimated from scaling relations which depend on the density of states at the Fermi energy. Consequently, a major focus in the quest for estimates of T_c is the determination of λ.

For simple metals, early model calculations gave estimates and trends in T_c, but the detailed electronic structure, vibrational spectra, and electron–phonon couplings were not known with sufficient accuracy to allow detailed calculations of T_c. In addition, the calculational techniques for obtaining E were not highly developed. For semiconductors, experimental determinations of band structure and electron–phonon couplings allowed some successful predictions [5.61] of superconductivity in degenerate semiconductors. In fact, one of the first superconducting oxides, $SrTiO_3$, [5.62] was predicted in this way.

As described earlier, the total energy approach using pseudopotentials is capable of predicting stable structures, vibrational and electronic properties, and electron–phonon couplings. Hence, it has all the ingredients necessary for predicting T_c. The Coulomb parameter μ^* is the only empirical input once the structure is chosen. Since μ^* is usually around 0.1 and scales with the density of states, this is not a critical limitation for this method.

The first applications were made for Si [5.36] and Al [5.37]. Since the latter system is a known superconductor, this calculation served as a test of the method. For Si, two structural phases were examined, sh and hcp. These were chosen because their charge distribution (Fig. 5.9) indicated that they are covalent metals, and covalency is an attractive feature for superconductivity [5.63]. Free-electron-like metals like the alkali and noble metals are not superconducting above 1 K, if at all. It is the tendency to form covalent bonds which increases the electron–phonon interaction, and the presence of local fields allows the attractive pairing interaction to be larger than the Coulomb interaction [5.63]. When the electron–

Fig. 5.9. Valence electron charge density contour plot in units of electrons/cell volume for simple hexagonal silicon. The plot is in the [10$\bar{1}$0] plane

phonon coupling constant λ becomes too large, lattice modes tend to be soft and often instabilities set in. This is the usual origin of the limit on T_c.

The recognition [5.64] that sh and hcp Si were covalent-like metals and hence could be superconductors was follwed by a detailed study [5.36] of their electronic structure, vibration spectra, solid–solid phase transitions, electron–phonon couplings, and superconductivity. It was predicted that sh Si would be superconducting in the range of 5–10 K and that T_c would be a function of pressure. It was expected that T_c would decrease with pressure in the sh phase and then increase when the pressure was close to the transition pressure to hcp. Finally, in the hcp phase, T_c was predicted to be $\sim$ 4 K. All of these predictions were verified.

Hence, the Si study is an example where the existence of the material and its electronic, vibrational, and superconducting properties were all predicted from first principles. Applications to Ge [5.65] and other phases of Si [5.31] were made. These predictions have not yet been tested. As diamond anvil techniques for achieving high pressures continue to improve, it is expected that more systems will be explored.

A novel system suggested by these calculations is a semiconductor superlattice. A good prototypical example is a GaAs-AlAs superlattice in which the variation of pressure and temperature can lead to a series of interesting systems. In general, many possibilities exist for a superlattice $ABABAB\ldots$ formed from two semiconductors A and B with different transition pressures P_t to the metallic state. Let $P_t^A < P_t^B$ and assume that in both cases the metallic phases are superconducting with $T_c^B < T_c^A$. Another variable of interest is the thickness of layers A and B denoted by t_A and t_B. We begin by keeping the temperature moderately high and varying the pressure.

For $T > T_c^A, T_c^B$, and $P < P_t^A, P_t^B$, both components are semiconductors, and the superlattice is a standard periodic array of heterojunctions. The properties depend on the components and their thicknesses t_A and t_B.

If $T > T_c^A, T_c^B$, but $P_t^A < P < P_t^B$, then material A has transformed from a semiconductor into a metal while material B remains semiconducting. For this system, we have a periodic array of Schottky barriers. The thicknesses t_A and t_B can be varied to give a wide range of physical effects. Because of modern molecular beam epitaxy techniques, this approach for making metal–superconductor interfaces and superlattices may be helpful.

For $T > T_c^A, T_c^B$ and $P > P_t^A, P_t^B$, both components become metallic, and again varying the thicknesses t_A and t_B provides an interesting system composed of a periodic array of metal–metal junctions.

Some interesting variations occur when the temperature varies. For $T_c^B < T > T_c^A$ and $P_t^A < P < P_t^B$, we have component A in a metallic state and superconducting while component B is semiconducting. If t_A and t_B are both much larger than a superconducting coherence length, then the system is composed of isolated superconducting and semiconducting components. As t_B gets shorter and becomes comparable with a superconducting coherence length, Josephson junctions can be formed. For very small t_A or t_B, proximity effects can be studied.

For $T < T_c^B < T_c^A$ and $P_t^A < P_t^B < P$, both components are superconducting and the superconducting superlattice can be made with reasonable precision unless interface changes occur because of the high pressure. Again, changes in t_A and t_B allow detailed studies of properties such as the proximity effect. Temperature variation should also be used as a tool for this case. For example, for small t_B and T slightly higher than T_c^B, there still might be a superconducting transition in component B because of the proximity effect.

Finally, if $T_c^C < T < T_c^A$ and $P_t^A < P_t^B < P$, then the normal metal–superconductor array which forms is very sensitive to t_A and t_B as discussed for the previous case.

For all the above cases, it is assumed that despite the pressure and temperature changes, the original superlattice will retain its periodic properties. For real materials, this may be difficult, and a host of other possible problems may make some aspects of this proposal difficult or impossible. However, it does serve as an example of a prototype system in which it may be possible to use a single superlattice formed from semiconductors whose properties are known or can be calculated to examine semiconductor–semiconductor, semiconductor–metal, metal–metal, normal metal–superconductor, semiconductor–superconductor, and superconductor–superconductor arrays. The variations are achieved by changing the pressure and temperature of the same heterostructure.

Although it is commonly believed that electron–phonon superconductivity can occur only at low temperatures, this does not appear to be a direct conclusion from theory. In particular, a major argument against an electron–phonon pairing mechanism for explaining superconductivity in materials like $YBa_2Cu_3O_7$ is that $T_c \sim 95\,K$ while the isotope effect is almost absent. Because of the properties of the Coulomb repulsion between electrons in a pair, the isotope effect can be reduced even when pairing arises from electron–lattice interactions. However, for three-dimensional systems, a vanishing isotope effect and $T_c \sim 95\,K$ leads to unphysical results. Hence, it is the extra restriction provided by the isotope measurement that is important here. Without this restriction, T_c can be very large when electrons are paired by phonons.

A particularly interesting case is solid hydrogen. It is expected that solid molecular hydrogen, which is an insulator, can be metallized at high pressure. Two types of transitions have been examined recently using the total-energy scheme described above, but with a Coulomb potential instead of a pseudopotential. The first case is the closing of the gap in the solid molecular phase, and the second is a higher pressure transition into a hexagonal metal. The gap closure is predicted to occur around 2.5 Mbar [5.66], and there is some experimental evidence [5.67] for darkening of the samples in this pressure range, which these authors interpret as arising from gap closure. For the higher pressure prediction, the structure is expected [5.68] to be a distorted sh structure with a threefold distortion. The pressure range for this transition is ~ 4 Mbar.

Just as in the case of Si, the electronic structure, phonons and electron–phonon couplings were computed. These were used to estimate the parameters λ and μ^*. A solution of the Eliashberg equations yields a $T_c \sim 150$ K. If the uncertainties in λ and μ^* are included, there is a range of about ± 80 K expected for the T_c calculation.

5.7 Conclusions

This review has covered a wide variety of subjects from clusters, ordinary semiconductors and metals, to a possible superhard insulator, and a possible very high temperature superconductor. All of this was done with a single theoretical approach as its basis. The details differ on how the approach is used depending on the problem, but for the most part, it is the total-energy-pseudopotential scheme which has made many of the applications described here possible. This theoretical approach is robust and is continually improving. The domain of materials studied and the range of properties appear to be continually increasing. There is every reason to predict a bright future for this area of material science.

Acknowledgements. This work was supported by National Science Foundation Grant No. DMR-8818404 and by the Director, Office of Energy Research, Office of Basic Energy Sciences, Materials Sciences Division of the U.S. Department of Energy under Contract No. DE-AC03-76SF00098.

References

5.1 M.L. Cohen: Nature **338**, 291 (1989)
5.2 M.L. Cohen: Science **234**, 549 (1986)
5.3 M.L. Cohen, J.R. Chelikowsky: *Electronic Structure and Optical Properties of Semiconductors*, 2nd ed., Springer Ser. Solid-State Phys., Vol. 75 (Springer, Berlin, Heidelberg 1989)
5.4 D. Brust, J.C. Phillips, F. Bassani: Phys. Rev. Lett. **9**, 94 (1962)

5.5 M.L. Cohen, T.K. Bergstresser: Phys. Rev. **141**, 789 (1966)
5.6 F. Bassani, G. Patori Parravicini: *Electronic States and Optical Transitions in Solids* (Pergamon, Oxford 1975)
5.7 J.P. Walter, M.L. Cohen: Phys. Rev. B **2**, 1821 (1970)
5.8 J.R. Chelikowsky, M.L. Cohen: Phys. Rev. B **14**, 556 (1976)
5.9 T.C. Chiang, J.A. Knapp, M. Aono, D.E. Eastman: Phys. Rev. B **21**, 3513 (1980)
5.10 L.W. Yang, P. Coppens: Solid State Commun. **15**, 1555 (1974)
5.11 L. Zucca, Y.R. Shen: Phys. Rev. B **1**, 2668 (1970)
5.12 M.L. Cohen, V. Heine: Solid State Phys. **24**, 37 (1970)
5.13 V. Heine, I. Abrankov: Philos. Mag. **9**, 451 (1964)
5.14 M. Appaillai, V. Heine: Tech. Rep. No. 5, Cavendish Laboratory, Cambridge (1972)
5.15 J.A. Appelbaum, D.R. Hamann: Rev. Mod. Phys. **48**, 3 (1976)
5.16 M. Schlüter, J.R. Chelikowsky, S.G. Louie, M.L. Cohen: Phys. Rev. B **12**, 4200 (1975)
5.17 M.L. Cohen: *Advances in Electronics and Electron Physics*, Vol.51, ed. by L. Marton, C. Marton (Academic, New York 1980) p.1
5.18 M.L. Cohen, M. Schlüter, J.R. Chelikowsky, S.G. Louie: Phys. Rev. B **12**, 5575 (1975)
5.19 M.T. Yin, M.L. Cohen: Phys. Rev. Lett. **45**, 1004 (1980)
5.20 T. Starkloff, J.D. Joannopoulos: Phys. Rev. B **16**, 5212 (1977)
5.21 A. Zunger, M.L. Cohen: Phys. Rev. B **20**, 4082 (1979)
5.22 D.R. Hamann, M. Schlüter, C. Chiang: Phys. Rev. Lett. **43**, 1494 (1979)
5.23 M.T. Yin, M.L. Cohen: Phys. Rev. B **25**, 7403 (1982)
5.24 S.G. Louie, S. Froyen, M.L. Cohen: Phys. Rev. B **26**, 1738 (1982)
5.25 J. Ihm, A. Zunger, M.L. Cohen: J. Phys. C **12**, 4409 (1979)
5.26 M.L. Cohen: Phys. Scr. T **1**, 5 (1982)
5.27 E. Fermi: Nuovo Cimento **11**, 157 (1934)
5.28 P. Hohenberg, W. Kohn: Phys. Rev. **136**, B863 (1964)
5.29 W. Kohn, L.J. Sham: Phys. Rev. **140**, A113 (1965)
5.30 C. Kittel: *Introduction to Solid State Physics*, 6th ed. (Wiley, New York 1986)
5.31 A.Y. Liu, K.J. Chang, M.L. Cohen: Phys. Rev. B **37**, 6344 (1988)
5.32 M.T. Yin, M.L. Cohen: Phys. Rev. B **29**, 6996 (1984)
5.33 S. Fahy, X.W. Wang, S.G. Louie: Phys. Rev. Lett. **61**, 1631 (1988)
5.34 M.T. Yin, M.L. Cohen: Phys. Rev. B **25**, 4317 (1982)
5.35 P.K. Lam, M.L. Cohen: Phys. Rev. B **25**, 6139 (1982)
5.36 K.J. Chang, M.M. Dacorogna, M.L. Cohen, J.M. Mignot, G. Chouteau, G. Martinez: Phys. Rev. Lett. **54**, 2375 (1985)
5.37 M.M. Dacorogna, M.L. Cohen, P.K. Lam: Phys. Rev. Lett. **55**, 837 (1985)
5.38 M.S. Hybertsen, S.G. Louie: Comments Cond. Mat. Phys. **13**, 223 (1985)
5.39 R. Car, M. Parrinello: Phys. Rev. Lett. **55**, 2471 (1985)
5.40 L. Pauling: *Nature of the Chemical Bond*, 3rd ed. (Cornell, New York 1960)
5.41 W.A. Harrison: *Electronic Structure and the Properties of Solids* (Freeman, San Francisco 1980)
5.42 J.A. Van Vechten, J.C. Phillips: Phys. Rev. B **2**, 2160 (1970)
5.43 M.L. Cohen: Phys. Rev. B **32**, 7988 (1985)
5.44 M.L. Cohen: Mater. Sci. Eng. **A105/106**, 11 (1988)
5.45 P.K. Lam, M.L. Cohen, G. Martinez: Phys. Rev. B **35**, 9190 (1987)
5.46 A.Y. Liu, M.L. Cohen: Science **245**, 841 (1989)
5.47 A. Zunger, M.L. Cohen: Phys. Rev. B **18**, 5449 (1978)
5.48 S.B. Zhang, M.L. Cohen, J.C. Phillips: Phys. Rev. B **38**, 12085 (1988)
5.49 S.B. Zhang, M.L. Cohen, J.C. Phillips: Phys. Rev. B **39**, 1077 (1989)
5.50 M.Y. Chou, P.K. Lam, M.L. Cohen: Phys. Rev. B **28**, 4179 (1983)
5.51 W.A. de Heer, W.D. Knight, M.Y. Chou, M.L. Cohen: Solid State Phys. **40**, 93 (1987)
5.52 M.G. Mayer, J.H.D. Jensen: *Elementary Theory of Nuclear Shell Structure* (Wiley, New York 1955)

5.53 W.D. Knight, K. Clemenger, W.A. de Heer, W.A. Saunder, M.Y. Chou, M.L. Cohen: Phys. Rev. Lett. **52**, 2141 (1984)
5.54 K. Clemenger: Phys. Rev. B **32**, 1359 (1985)
5.55 S. Saito, S.B. Zhang, S.G. Louie, M.L. Cohen: Phys. Rev. B **40**, 3643 (1989)
5.56 H. Kamerlingh Onnes: Akad. van Wetenschappen (Amsterdam) **14**, 113, 818 (1911)
5.57 J. Bardeen, L.N. Cooper, J.R. Schrieffer: Phys. Rev. **108**, 1175 (1957)
5.58 J.G. Bednorz, K.A. Müller: Z. Phys. B **64**, 189 (1986)
5.59 P.A. Allen, B. Mitrovic: Solid State Phys. **37**, 1 (1982)
5.60 G.M. Eliashberg: Sov. Phys. – JETP **11**, 696 (1960)
5.61 M.L. Cohen: Phys. Rev. **134**, A511 (1964)
5.62 J.F. Schooley, W.R. Hosler, M.L. Cohen: Phys. Rev. Lett. **12**, 474 (1964)
5.63 M.L. Cohen, P.W. Anderson: *Superconductivity in d- and f-Band Metals*, ed. by D.H. Douglass (AIP, New York 1972) p. 17
5.64 M.L. Cohen: In Proc. of the 17th Int. Conf. on the Physics of Semiconductors, ed. by D.J. Chadi, W.A. Harrison (Springer, New York 1985), p. 1571
5.65 J.L. Martins, M.L. Cohen: Phys. Rev. B **37**, 3304 (1988)
5.66 T.W. Barbee III, A. Garcia, M.L. Cohen, J.L. Martins: Phys. Rev. Lett. **62**, 1150 (1989)
5.67 H.K. Mao, R.J. Hemlsey: Science **244**, 1462 (1989)
5.68 T.W. Barbee III, A. Garcia, M.L. Cohen: Nature **340**, 369 (1989)

6. High-Temperature Superconductivity: The Experimental Situation

Allen M. Goldman

With 15 Figures

The various classes of oxide superconductors have critical temperatures which greatly exceed those of so-called conventional materials [6.1–5]. They may also have enormous critical magnetic fields as well, depending on the particular extrapolation employed. It is, however, as yet unknown whether large critical currents can be achieved for bulk material in a magnetic field. The prospects for a new generation of superconducting electronic devices and sensors are just beginning to be explored. The potential impact on technology of the new high temperature superconductors has been the driving force behind the enormous volume of work in the field over the last few years. The frenzied pace of research is consistent with its importance, given the words of the Soviet physicist *V.L. Ginzburg* [6.6]. He contends that the problem of high temperature (ultimately room temperature) superconductivity may be the *second* most important problem in physical science, behind that of controlled fusion, in terms of its potential impact on society.

Although oxide superconductors with low carrier concentrations and with transition temperatures above 10 K have been known for a number of years [6.7], the new materials are striking in that their transition temperatures are higher than any that might have been expected until well into the middle of the next century, given the evolutionary rather than revolutionary rate of progress prior to 1986. Figure 6.1, which contains a plot of transition temperature vs date, shows the remarkable change in the horizons of the field over a very short time [6.8]. In this chapter, an attempt will be made to survey the current status of research on high temperature superconductivity, concentrating on issues which relate to the nature of the superconducting state in the new materials and the underlying mechanism. Because of the very large number of papers published in this field, a review of this sort can have no pretensions of being comprehensive; it will necessarily be selective and subjective, and may exclude important late-breaking results. A fairly complete bibliography, current through the end of 1988, is available in the March 1989 issue of the *Journal of Superconductivity* [6.9].

The topics which will be covered are organized as follows: in Sect. 6.1, the structural and chemical nature of the new superconductors will be presented. Section 6.2 will be concerned with various macroscopic properties of the superconducting state in the new materials. Section 6.3 will be concerned with the experimental situation relating to the microscopic character of the superconducting state. Section 6.4 will discuss other elements of our understanding which are closely related to aspects of the theory, such as the role of magnetism and the insulating character of the undoped materials. This final section will also contain a

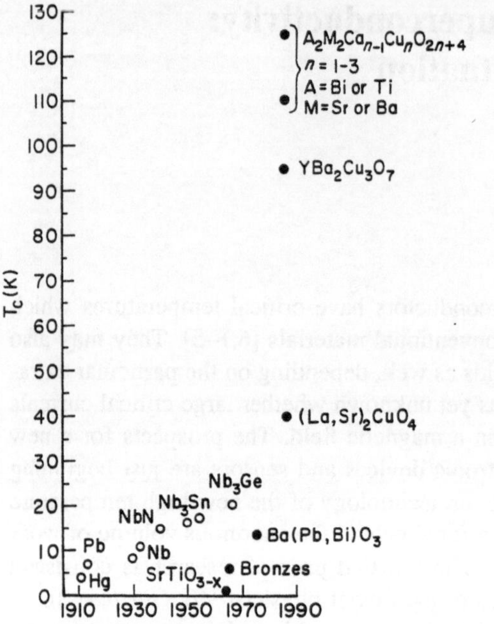

Fig. 6.1. Highest transition temperatures vs date. (From [6.8])

discussion of the issues critical to developing an understanding of the mechanism behind high temperature superconductivity. Throughout the chapter we will emphasize the central role that the fabrication of high quality and well-characterized materials plays in the elucidation of the mechanism. Due to the limitations of space, we will not attempt a discussion of the technological applications, even though they play a major role in driving research activities in this field.

6.1 Structural and Chemical Nature of the New Materials

Before discussing the structure and chemistry of the high temperature superconductors, it is important to note that *all* of the new materials exhibit the classic features of superconductivity such as zero electrical resistance and the Meissner–Ochsenfeld effect [6.10]. At this writing, there is no confirmed example of superconductivity at temperatures substantially exceeding 125 K [6.11]. The scientific literature and the press, however, have been filled with reports of anomalous behavior of the resistivity or some other property which has been attributed to superconductivity at extraordinarily high temperatures, in some instances in excess of room temperature. It is inappropriate in this review to discuss this subject at any length, but we emphasize that in considering the possible superconductivity of such systems it is essential that all of the standard criteria be satisfied. If they are not, and that if all that is observed is a *drop* in the resistivity, or even a slight signature of diamagnetism, then the authors are far from demonstrating superconductivity in these systems.

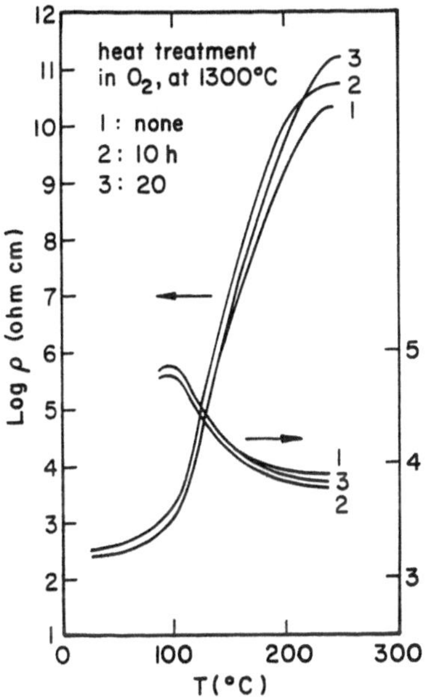

Fig. 6.2. The effect of heat treatments on the PTCR effect showing both the resistivity and the dielectric constant. (From [6.12])

What is frequently not appreciated is that a dramatic drop in resistivity may be attributable to a phenomenon other than superconductivity. Figure 6.2, in which the spectacular changes in electrical resistivity of materials which exhibit the positive temperature coefficient of electrical resistivity (PTCR) [6.12], illustrates an example of this. The phenomenon is found in certain semiconducting titanate ceramics in the vicinity of their ferroelectric Curie temperature. Its discovery emerged from investigations in the 1950s of controlled valency semiconductors aimed at the development of new thermistor compositions. Although the resistivity may change by seven or eight orders of magnitude, the mechanism is a grain-boundary rather than a bulk phenomenon, directly related to the ferroelectric phase transition in semiconducting barium titanate and in its solid solutions with strontium and lead titanates. Semiconductive properties are obtained by substituting trivalent donors (e.g., La) on the barium lattice sites, or pentavalent donors on the titanium sites. Possible dopants include Y, most of the rare earths through Er, and, in addition, Nb, Ta, W, Sb, and Bi. The rough similarity of these materials to those reported to be high temperature superconductors should be noted.

An important common feature of most of the presently known superconductors with transition temperatures in excess of 25 K is that they are copper oxides with perovskite crystal structures and contain CuO_2 planes. The main exceptions are the materials derived from $BaPb_{1-x}Bi_xO_3$, such as $Ba_{1-x}K_xBiO_3$, which has a transition temperature near 30 K, and exhibits a cubic, rather than a layered perovskite structure [6.5]. This compound does not contain copper, and has no known antiferromagnetic precursor compound, unlike all other high-T_c materials [6.13].

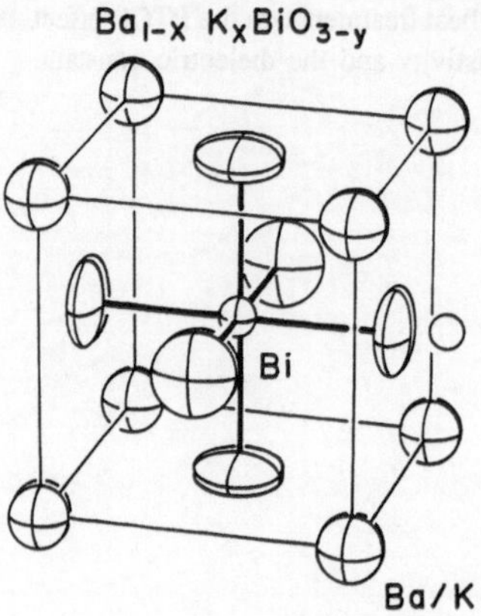

Fig. 6.3. Crystal structure of the cubic $Ba_{1-x}K_xBiO_{3-y}$ compound (from [6.5]). The figure itself is from [6.14]

The cubic $Ba_{1-x}K_xBiO_{3-y}$ material, the structure of which is shown in Fig 6.3, has a moderately low T_c of about 30 K, and exhibit a simple cubic perovskite structure with Ba or K at the corners, Bi in the center, and O in the faces. Superconductivity results when the nonmetallic parent compund $BaBiO_3$ is doped with K on the Ba site. Doping with Pb on the Bi site yields a superconductor material discovered some years ago, with a transition temperature of 13 K [6.7].

The compound which gave the first indication of superconductivity at temperatures of the order of 30 K, $La_{2-x}Ba_xCuO_4$ [6.1], is a member of the class of compounds with composition $La_{2-x}R_xCuO_4$ (with R = Ba, Sr, or Ca). Its structure corresponds to undoped orthorhomibic La_2CuO_4 or its high temperature tetragonal polymorph. This material can also be described as a layered compound where there is one plane of Cu atoms with four strongly bonded oxygen atoms in a square planar arrangement, with the addition of two oxygen neighbors above and below the Cu atom, such as in Fig. 6.4. The member of this family with R = Sr has the highest $T_c \approx 37$ K, which occurs for $x = 0.015$.

The so-called "123" compounds with composition $RBa_2Cu_3O_{7-x}$ (with R = Y, Nd, Sm, Eu, Gd, Dy, Ho, Er, Tm, Yb, and Lu) are by far the most widely studied high-T_c materials [6.13]. The reason for this is that $YBa_2Cu_3O_{7-x}$ was the first material studied with a transition temperature above liquid nitrogen temperature. Single-phase, single-crystal samples of these materials can be made readily. The structure of these compounds is shown in Fig 6.5. These materials have two copper oxide layers separated by a Y or R layer which contains no oxygen and have two BaO and one CuO layers which contain Cu-O chains. The Cu sites in the CuO_2 planes and the CuO chain layers are two distinct crystallographic sites. The crystal structure is orthorhombic as a consequence of the presence of ordered oxygen vacancies.

A formal oxidation state calculation requires an oxygen stoichiometry of 6.5. The fact that there are seven oxygens per formula unit implies that the carriers are associated with the excess oxygen. The superconductivity is very sensitive to this oxygen stoichiometry as the removal of oxygen atoms results in a decrease

Fig. 6.4

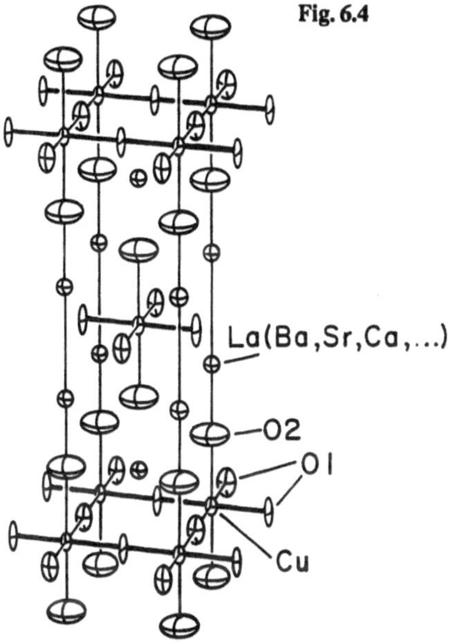

Fig. 6.5

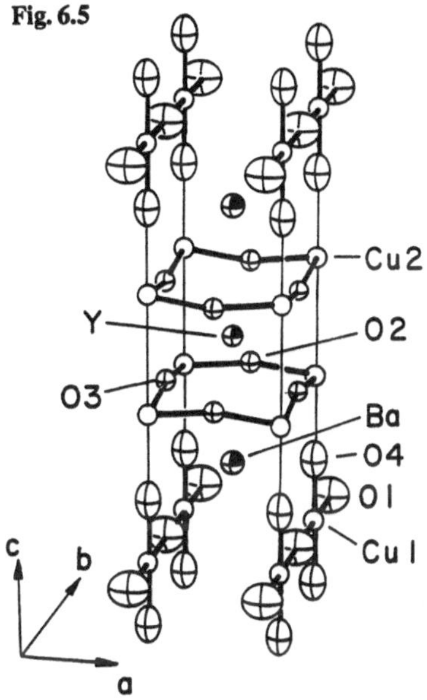

Fig. 6.4. Structure of the "single-layered" $La_{2-x}R_xCuO_4$ compound. (From [6.14])

Fig. 6.5. Structure of the "double-layered" $RBa_2Cu_3O_{7-x}$, "123" compound. (From [6.14])

in T_c with a transition to a tetragonal structure when the oxygen stoichiometry falls to 6.5, at which point T_c drops to zero.

The 100 K barrier was unambiguously broken with the discovery of the so-called multi-layer compounds, which involve various combinations of Bi, Ca, Sr, Cu, and O [6.3] and Tl, Ca, Ba, Cu, and O [6.4]. These two sets of elements form a hierarchy of structures which differ in the number of CuO_2 planes and the number of intercalated layers, as shown in Fig 6.6. A difficulty with experimental studies of these materials, and a problem for the determination of their structures, lies in the fact that many samples contain mixtures of the various phases. Although there appears to be a consensus as to the structures of these materials, the precise connection between structure features and electronic properties has not been identified in every instance.

The common feature of the layered perovskites is the fact that metal-site doping not changing the oxidation state of the Cu sites has a significant affect on superconductivity only if the substitution occurs on CuO_2 sites. This, together with oxygen stoichiometry experiments, suggests that qualitatively the superconductivity depends on the degree of charge transfer from the intercalated layers to the conducting CuO_2 planes, although the issue is far less clear for the "multi-layer" materials. Qualitatively, the transition temperature seems to increase with the number of closely spaced CuO_2 planes, although this may not be the critical factor.

Until quite recently, all copper oxide high-temperature superconductors were believed to be p-type. That is to say, all materials were doped so that the majority

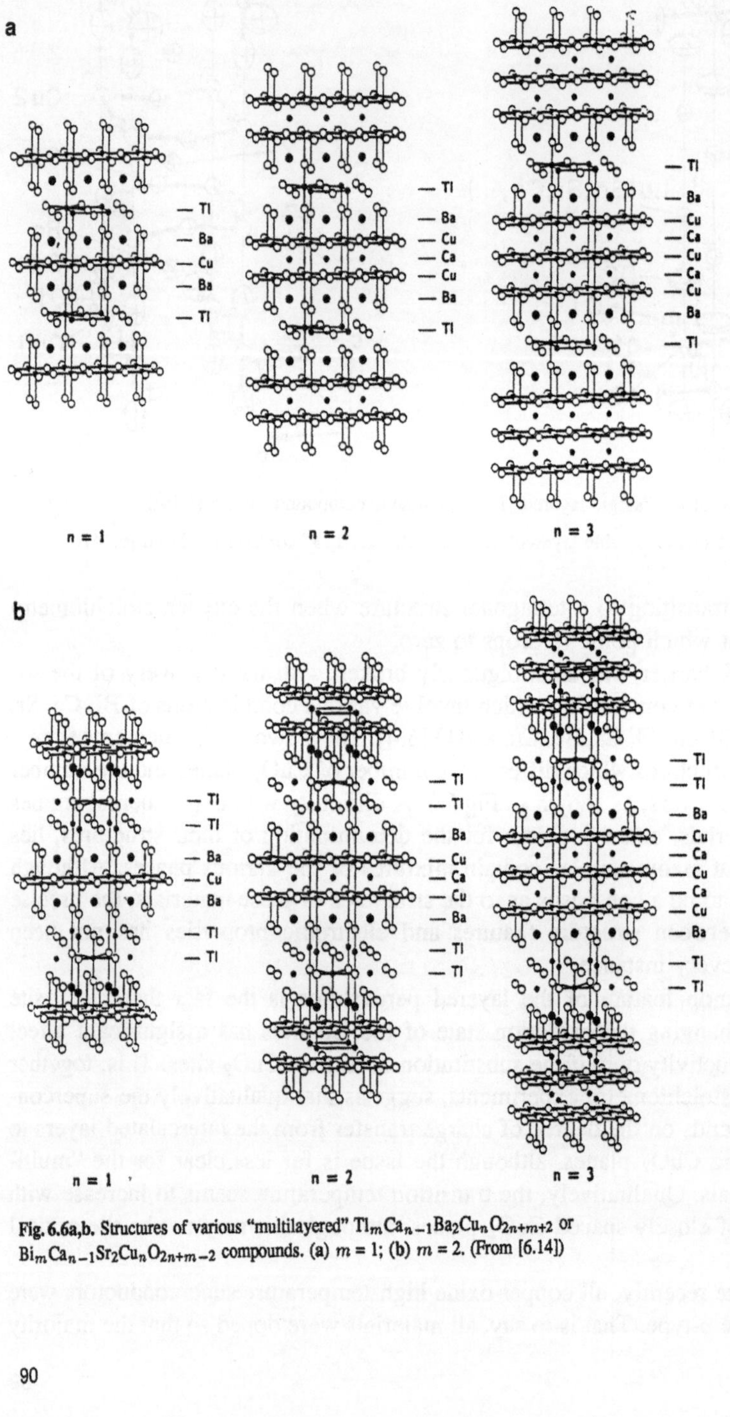

Fig. 6.6a,b. Structures of various "multilayered" $Tl_mCa_{n-1}Ba_2Cu_nO_{2n+m+2}$ or $Bi_mCa_{n-1}Sr_2Cu_nO_{2n+m-2}$ compounds. (a) $m = 1$; (b) $m = 2$. (From [6.14])

carriers were holes. Recently a class of superconductors comprising n-type materials was discovered [6.15]. The parent compound Nd_2CuO_4, cannot be doped with holes (normally provided by Sr^{2+}), but can be doped with electrons by adding Ce^{4+}. In particular it was discovered that $Nd_{2-x-z}Ce_xSr_zCuO_4$ has a superconducting transition temperature of about 30 K and $Nd_{2-x}Ce_xCuO_{4-y}$ of 20 K [6.16]. The experimental evidence for the electron character of the carriers in these materials comes from studies of their Hall and Seebeck coefficients, both negative, in contrast with the positive coefficients found for all of the other known high T_c cuprates [6.17]. Similar results have been found for Th-doped Nd_2CuO_4, [6.18].

6.2 The Superconducting State: Macroscopic Properties

One of the remarkable features of the superconductivity of the high-T_c materials is that they behave as ordinary superconductors in almost all of their properties. Electron pairs appear to be the carriers of supercurrent as determined in flux quantization [6.19] and vortex lattice density studies [6.20], and persistent currents are extremely long-lived. Because of the similarities with conventional superconductors, it is tempting to apply the Ginzburg–Landau model to these new materials. The corresponding parameters for $YBa_2Cu_3O_{7-x}$ have been determined from an analysis of a number of experiments [6.21]. The important point to note is that the coherence length of the Ginzburg–Landau model is very short. This implies that behavior other than mean-field behavior, in particular critical phenomena, may be observed in the high-T_c materials, in contrast with conventional superconducting materials. This extraordinarily short coherence length, which is found in other high-T_c compounds, also has profound consequences for some important macroscopic properties of these materials. In particular, flux pinning turns out to be rather different from what had been expected based on low-T_c materials. Finally, there is the matter of anisotropy. The high-T_c materials have manifestly anisotropic properties [6.22], including direction-dependent normal-state resistivities, critical currents and critical magnetic fields. Electric resistance studies in the vicinity of the superconducting transition indicate that it may be a Kosterlitz–Thouless transition [6.23–25], and transport studies above the superconducting transition offer ample evidence of two-dimensional character [6.26]. These issues will be addressed in some detail in this section.

The macroscopic properties of the new superconductors have been reviewed by Tinkham and Lobb [6.27], who have emphasized the effective granularity of these materials and the role that this property plays in determining the magnetic parameters. Here we will focus on the phenomena, and not emphasize any particular model. Superconductors in general can be divided into two categories type I and type II based on their magnetic properties. Type I superconductors exhibit a nearly perfect Meissner–Ochsenfeld effect, and can reversibly exclude flux up to the critical magnetic field. On the other hand, in type II superconductors, magnetic flux begins to penetrate the superconductor when the applied field exceeds a lower critical field, H_{c1}, and the superconducting properties persist up to a second critical field called the upper critical magnetic field, H_{c2}. Between the two fields, magnetic flux penetrates the superconductor in the form of a triangular lattice of quantized vortices. This state of field penetration is known as the "mixed" state, and the lattice of quantized vortices in called the Abrikosov flux lattice, after *A.A. Abrikosov* [6.28], who first described it using the Ginzburg–Landau model.

The high-T_c superconductors are type II in character. In bulk form they exhibit a variety of behaviors which suggests that they may be very different from the older type II materials under study before 1986. These differences appear in studies of the time-dependent response of the magnetization of the new superconductors in the presence of a magnetic field [6.29], in measurements of the temperature and magnetic field dependences of the resistivity [6.30], and in studies of the vortices either by direct decoration [6.20], or by investigation of the mechanical losses associated with forced oscillations in a magnetic field [6.31].

It is the inhibition of the motion of vortices by pinning that is responsible for zero electrical resistance in the mixed state. There is an electromagnetic Lorentz force exerted on the lattice when a current flows. As long as the lattice does not move the superconductor will exhibit zero resistance. Should the lattice become unpinned, then there will be a finite electrical resistance. Actually, in conventional type II materials, pinning forces are very large, and the motion of the flux is dominated by thermally activated hopping of the vortices out of the potential wells in which they are trapped. This process is known as flux creep, and is a rather different condition than flux flow, when the lattice moves continuously [6.10]. In contrast with the older superconductors, where the flux creep is small, in the new materials flux creep is very large, even at values of field and temperature at which the lattice would be stable for lower-T_c materials [6.32].

The possibility of unusual magnetic properties of high-temperature superconductors emerged rather early in the evolution of the field. *Müller* et al. [6.29] reported a so-called irreversibility line, which separates the region near T_c in the H-T plane in which the sample exhibits a reversible magnetization from a region in which it exhibits a history-dependent character. The irreversibility line was found to obey the relation

$$1-\frac{T}{T_c} \propto H^{2/3} . \tag{6.1}$$

Because this behavior was similar to phenomena observed in spin glasses, the line was called a *quasi–de Almeida–Thouless line* [6.33]. In addition to the irreversibility line, *Müller* et al. observed a logarithmic time dependence in the decay of the remnant magnetization of superconductors cooled below their transition temperature in a magnetic field, when the field was turned off. *Yeshurun* and *Malozemoff* [6.32] interpreted the same phenomena in the language of flux creep, pinning and flux flow in the superconducting medium. In this picture the reversibility line denotes the limiting conditions in which nonequilibrium supercurrents can persist over the duration of an experiment. This model of "giant flux creep" can explain apparent differences in values of $T_c(H)$ measured on the same samples as a consequence of the different measuring frequencies relative to flux creep rates. These facts lead to a situation where resistively measured values of the critical field exhibit discrepancies relative to the thermodynamically derived values. *Tinkham* has developed a model which appears to reconcile the two points of view [6.34].

From the point of view of the giant flux creep model, the energy barriers to flux creep in the new superconductors are about an order of magnitude smaller than in the old ones. This is due in part to the higher transition temperatures, but also to the smaller coherence lengths, which lower the pinning energies. This picture of flux creep in high-temperature semiconductors seems to be supported by measurements of the electrical resistance tail of superconductivity in which the resistance is found to decrease exponentially with temperature [6.30].

The result of this is that the various criteria employed to determine the upper critical magnetic field lead to substantially different curves and considerable ambiguity. This is illustrated in Fig. 6.7, where curves using different resistive and magnetic criteria are shown [6.27].

Another important aspect of the properties of type II superconductors is the flux line lattice itself, as indicated by magnetic decoration experiments [6.20]. This method involves coating the clean surface of a sample with fine energetic particles produced by a smoke technique. If the superconductor is in the mixed state then the particle will drift to the regions of sample with slightly larger fields,

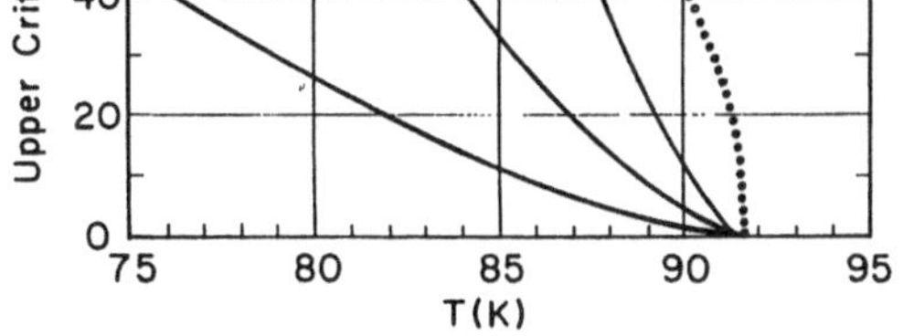

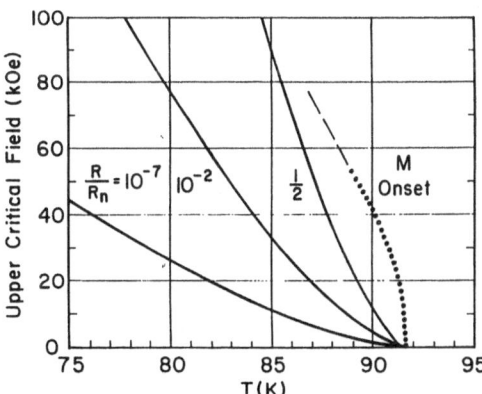

Fig. 6.7. Various values of the upper critical field of crystalline YBCO for the magnetic field parallel to the c-axis, with analysis and interpretation from [6.27]. The curve labelled M Onset represents the data of *Fang* et al. [6.35] for the field at which the linear rise in reversible magnetization begins, which probably marks the thermodynamic transition field. The data end at 89 K, and a linear extrapolation is shown as a guide to the eye. Curves labelled $R/R_n = 1/2$ and 10^{-2} respectively are taken from [6.36]. The curve labelled $R/R_n = 10^{-7}$ is meant to schematically represent the magnetic irreversibility line for the same crystal, for which experimental data are not available. For simplicity the transition width in zero magnetic field has been suppressed

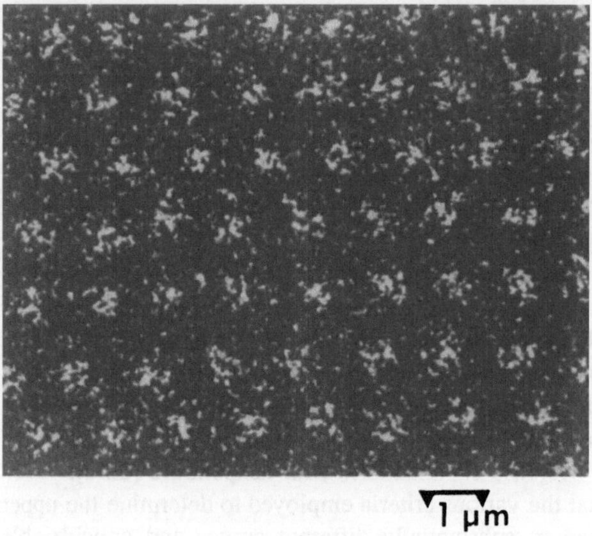

Fig. 6.8. Flux spots in $YBa_2Cu_3O_7$ sample decorated after cooling in a field of 13 G. (From [6.20])

and can delineate the pattern of vortices. This pattern can then be observed using a scanning electron microscope. The essential aspects of this technique were first used more than 20 years ago by *Essman* and *Traubel*, and by *Sarma* [6.37] to decorate the vortices of conventional type II superconductors. A result obtained by researchers at Bell Laboratories [6.20] is shown in Fig. 6.8. Although a vortex lattice was observed in $YBa_2Cu_3O_{7-x}$ at 4 K, such a structure was not observed at 77 K. This has lead to the conjecture that the vortex lattice melts at a temperature below the transition temperature.

This picture is supported by mechanical measurements of the shear modulus of the vortex lattice [6.31]. If a superconducting sample is attached to a mechanical oscillator, which is in turn placed in a field, the elastic modulus of the lattice will contribute to the force on the oscillator assembly if the latter is driven in a direction normal to the magnetic field. The rate of oscillation must, of course, be slow in comparison with the vortex lattice relaxation time. When the lattice melts, the lines will relax very quickly, and the lattice contribution, which causes a characteristic shift in the resonant frequency of the oscillator, will vanish. This disappearance of the shift is accompanied by a peak in the dissipation of the mechanical oscillator. This phenomenon has been observed in both $YBa_2Cu_3O_{7-x}$ (YBCO) and $Bi_2Sr_2CaCu_2O_y$ (BSCCO) samples. Unfortunately this experiments would have almost the same results if the lattice, rather than melting, were to become unpinned at a well-defined temperature, so it does not, of itself, serve as conclusive evidence of melting. The absence of a lattice in decoration experiments is also not conclusive evidence of melting since an unpinned lattice would not yield a stationary pattern. Whether melting actually occurs, and what its connection with the irreversibility line is, remain open questions at this point, along with other fundamental questions, such as whether the giant flux creep or the glass picture is correct, or if the two are manifestations of the same phenomenon [5.38].

It has become fashionable to conclude that these magnetic properties of the new high-T_c materials preclude their use for transport of technologically significant currents in useful magnetic fields. It should be noted that most of these phenomena have been studied in single-crystal samples, and are drastically modified by pinning in thin-film specimens. Indeed, scrutiny of the literature about conventional superconductivity of single-crystal materials would indicate very weak pinning and suggest that they too would not be useful. It was the development of an understanding of vortex pinning, and the development of techniques for the relatively controlled introduction of defects (which could serve as pinning centers) which led to the deveopment of a viable superconducting magnet technology. Presumably a similar path will be followed in the technological development of the new materials.

With the availability of single crystals it became possible to study the inherent anisotropy [5.22] of high-T_c compounds. Results on the temperature dependence of the electrical resistivity in YBCO are shown in Fig. 6.9 [6.39]. Similar results have been obtained for BSCCO single crystals [6.40]. It is seen that along the directions ***a*** and ***b*** the resistivity varies linearly with temperature, whereas along the ***c*** direction it varies as $1/T$. This result is an important input to any theoretical analysis. Critical currents and critical fields also show substantial anisotropies in all high-T_c materials in which they have been studied. The superconducting properties require an anisotropic generalization of the time-dependent Ginzburg–Landau model for phenomenological description [6.41]. This is a three-

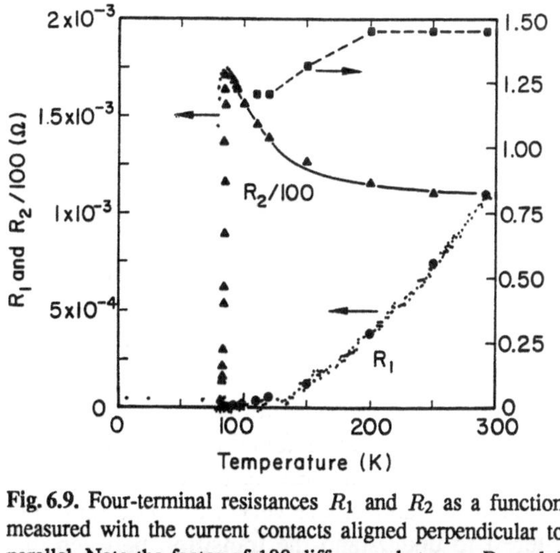

Fig. 6.9. Four-terminal resistances R_1 and R_2 as a function of temperature. The resistance R_1 is measured with the current contacts aligned perpendicular to the c-axis, and R_2 has them aligned parallel. Note the factor of 100 difference between R_1 and R_2 at room temperature. The small and large symbols represent data from different runs with currents of 500 μA and 20 mA respectively. Also plotted is the Hall number, which corresponds to 1-2–1.5 electrons per unit cell

dimensional theory with a large effective mass in the c direction, which gives the anisotropy in the coherence length. The zero-temperature coherence lengths in the c direction are the order of atomic dimensions, whereas the corresponding quantities in the ab-plane may be as high as 30 Å. These anisotropy effects in the normal state have been observed in most of the high-T_c materials that are not cubic.

The very short coherence lengths may imply that critical fluctuations [6.21, 42] play an important role in the character of the superconducting phase transition. In the case of conventional superconductors, there is evidence for Gaussian fluctuations in the order parameter, in the form of precursive behavior of the electrical resistance and the magnetic susceptibility [6.43]. There is very little direct experimental evidence of critical fluctuations, which should be ubiquitous in crystals with short coherence lengths such as the high-temperature superconductors. Indeed, estimates of the width of the critical region [6.21, 42] indicate that it may be quite wide and should be amenable to experimental study, although at this writing there is no conclusive evidence of critical fluctuations [6.44].

There are two aspects of fluctuations which have been studied in some detail on a number of different systems the temperature dependence of the precursive electrical conductivity above the transition [6.26], and the nature of the resistive transition, which appears to be topological phase transition or vortex unbinding transition of the Kosterlitz–Thouless type [6.23–25]. The resistance-reducing fluctuations appear to be "two-dimensional" above T_c over a rather extended range of temperatures in the materials with large anisotropy. It is as if the various CuO_2-containing planes were "uncoupled" fluctuating layers. There is no strong evidence of a crossover to three-dimensional behavior as the mean field transition temperature is approached from above.

This "two-dimensional" character of the transition seems to persist into the regime where the resistance falls to an unmeasurable value. Detailed studies of the temperature dependence of the resistance, the resistivity–magnetic field and the current–voltage characteristics of a number of thin-film and single-crystal systems are consistent with a vortex unbinding transition. In such a transition vortices are thermal excitations. Above the Kosterlitz–Thouless temperature the order parameter amplitude has a finite value, but there is a measurable resistance as there are many free vortices which flow in response to the Lorentz force. At the transition, vortices of opposite helicity are bound and the resistance gradually falls to zero as the number of free vortices decreases with decreasing temperature. The phenomenon is a consequence of the statistical mechanics of two-dimensional vortices, which interact through a potential logarithmic in their separation. The phenomenon has been studied in conventional thin-film superconductors. The surprise here is that the same effect is found even in essentially three-dimensional samples of the high-T_c materials, serving as further evidence of their intrinsic two-dimensional character and of the very weak coupling between the layers. It will be of interest to understand the transition between the two-dimensional behavior observed above and very near to the transition, and the three-dimensional behavior observed at lower temperatures. The study of the anisotropy of the macroscopic magnetic properties may hold the key as such properties depend on the c-axis effective mass whose temperature dependence near T_c should reflect the temperature dependence of the layer coupling in the c direction

6.3 Microscopic Superconducting Properties

6.3.1 Pairing

The great similarity of the macroscopic superconducting properties to those of conventional superconductors suggests that critical experiments to understand the mechanisms behind the high-T_c superconductivity will have to be microcopic rather than macrocopic in character. As mentioned earlier, studies of flux quantization [6.10] and measurements of the vortex line density [6.20] indicate that it is reasonably certain that the charge carriers in the superconducting state are electron pairs. The mechanism for the pairing is, however, not at all certain. Critical experiments or measurements have not yet been performed which lead us to distinguish between more or less conventional boson exchange mechanisms for superconductivity [6.45] and the more exotic schemes based on variants of the resonating valence bond model [6.46]. Somewhat more certain is the idea that the pairing is in the singlet state, although there certainly are theories [6.47] which suggest triplet pairing or pairing involving higher angular momentum states.

The very wide spread in the results of various experiments contributes in a significant manner to the current uncertain situation regarding the mechanism responsible for high-T_c superconductivity. This has resulted in an information explosion without an accompanying explosion of knowledge. The problem is that large numbers of measurements have been carried out on relatively poor materials with the results being greatly influenced by extrinsic factors such as impurities, structural defects, off-stoichiometry, polycrystallinity, and poor surface quality (in measurements such as tunneling, which are surface-sensitive). Because of the large number of nearly identical studies, all subject to the same systematic error, it is inappropriate to assume that the convergence of a particular set of numbers to a given value necessarily implies a scientific truth. This situation points out the great importance of control of material quality in achieving a fundamental understanding of high-temperature superconductivity. This same control should facilitate the development of materials suitable for applications.

Despite the above difficulties, there are a number of investigations which give rather clear results, and therefore a substantial body of information has been accumulated regarding many properties of the superconducting state of the high-T_c compounds. The most direct test for the nature of the superconducting charge carriers is a measurement of flux quantization and vortex line density. In a flux quantization experiment it is found that the flux associated with the persistent current that can be trapped in a cylinder of a conventinal superconducting material is quantized in multiples of $hc/2e$, where the number 2 in the denominator

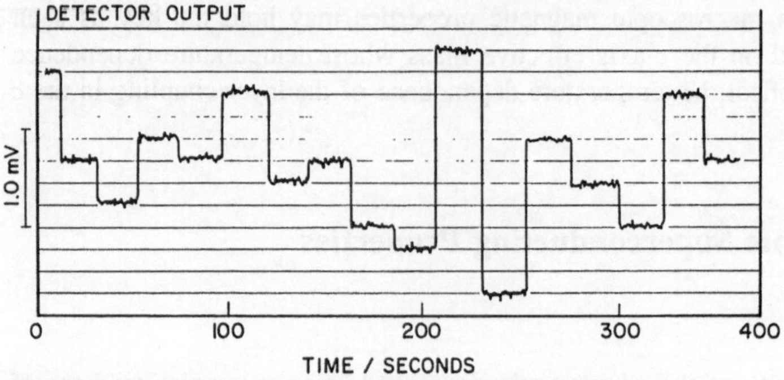

Fig. 6.10. Output of a SQUID magnetometer showing small integral numbers of flux quanta jumping in and out of a ring of $Y_{1.2}Ba_{0.8}CuO_4$. Step heights provide proof of the existence of pairs. (From [6.19])

follows from the fact that the electrons are paired, and h, c, and e are Planck's constant, the velocity of light and the electron charge, respectively. *Gough* and co-workers [6.19] have observed the step-like structure of the flux trapped in a YBCO ring as shown in Fig. 6.10. The persistent current states of the ring are not stable because the relatively large inductance of the loop used in the measurements leads to a small free energy difference between the levels. In effect the free energy barriers between the quantized levels are reduced, and this makes possible the noise or thermally activated transitions between the quantized states.

Decoration of the vortex lattice was also used to probe flux quantization [6.20]. From the vortex line density, and by knowing the value of the applied magnetic field, the magnitude of the flux quantum can be determined as being simply the total flux (i.e., the product of the field and the area) divided by the number of vortices in that area. The results are consistent with $hc/2e$.

A study of the ac Josephson effect [6.48] also supports the existence of pairs. This phenomenon is found in tunneling junction or weak link configurations which are biased on some voltage V. As a consequence of the bias, the transit of pairs between the two weakly coupled superconductors results in the emission of radiation at a frequency such that $h\nu = 2eV$, or 484 MHz for a 1 μV bias. This effect can be observed by irradiating a weak link, point contact or tunneling junction with a radiation to observe the beating of the ac Josephson effect with external radiation. This detection process results in constant voltage steps in the dc I–V characteristic of the link at rational multiples of a voltage given by $h\nu/2e$. The results of such a study are shown in Fig. 6.11.

A separate issue, still not completely clarified, is the orbital character of the pairing. For many years it was believed that the presence or absence of a dc Josephson effect in a tunneling junction consisting of two types of electrodes (one electrode in which the pairing was known to be singlet, s-wave, and a second electrode in which the pairing was of an unknown character) could be used to ascertain the character of the pairing in the second electrode. The dc

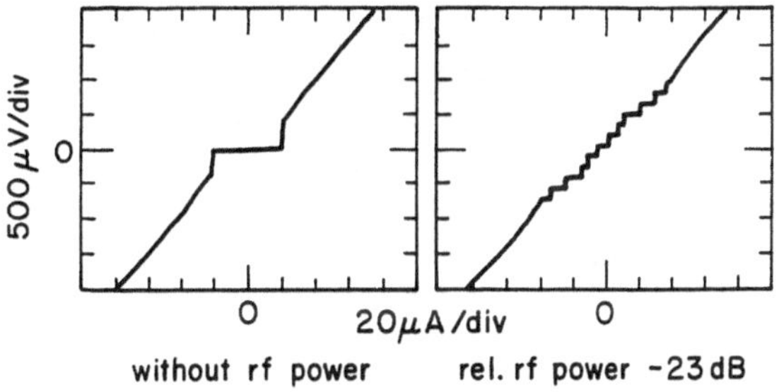

Fig. 6.11. Current–voltage characteristic of an $YBa_2Cu_3O_7$/PbSn point contact junction, with microwave radiation (70 GHz) (*left-hand photograph*), and without microwave radiation (*right-hand-photograph*); critical current $I_c = 20\ \mu A$; normal state resistance: $R_N = 20\ \Omega$. (From [6.48])

Josephson effect is a phenomenon in which a zero-voltage current flows between two superconductors separated by a thin insulating barrier. It was shown by *Akhtyamov* [6.49] in 1966 that such Josephson tunneling would not occur between triplet and singletstate superconductors, or between two superconductors whose pairing state were of different symmetry. This approach does not take into account the effects of spin–orbit coupling or the fact that for complex pairing the surface of the material may have a substantial admixture of singlet s-wave states even if the bulk involves more complex angular momentum configurations [6.45]. The surface of the material is the only region that is actually probed in a tunneling study of a short coherence length material.

It is believed that the absence of a dc Josephson effect between a conventional superconductor and a high-T_c material may indicate that the symmetries of the pairing states of the two materials are different. In evaluating such experiments a certain degree of care is necessary. There are many weak link or interface configurations which are not true tunneling junctions. In these, the voltage current may follow from the proximity effect, which may behave differently from the Josephson effect in the matter of the coupling between superconductors of different pairing symmetry. This issue arose in the interpretation of experiments on heavy fermion compounds, which are thought to have a pairing state which is unconventional, and remains unsolved [6.50]. In the previously mentioned study of the ac Josephson effect [6.48], a dc Josephson effect in a junction consisting of electrodes of YBCO and a Pb/Sn alloy was also reported. However, the geometry of a point contact junction and the actual nature of the proximity effect are, in general, not known, so that results for point contact junctions may not be as conclusive as those on insulating junctions, for which evidence of coupling would result in immediate conclusions relating to the pairing.

There are two classes of phenomena which seem to support conventional singlet pairing. The first is the relative insensitivity of the transition temperatures of high-temperature superconductors to impurities which scatter electrons. In conventional superconductors, magnetic impurities are strong pair-breakers, and nonmagnetic impurities have no pair-breaking effects. Thus, very small concentrations of magnetic impurities result in a rapid suppression of the transition temperature, whereas nonmagnetic impurities, unless they result in the localization of the electronic wavefunctions, have no effect. If the pairing were triplet, or involved p or d-states, nonmagnetic impurities would play a similar role. This phenomenon is not observed in high-T_c superconductors. In the YBCO family there is also an insensitivity to the presence of most magnetic rare earth ions on the Y site, but this appears to be the result of the lack of significant overlap of the wavefunctions of the charge carriers with the rare earth f electrons wavefunctions, which is essential for a scattering process to occur.

The second set of studies which support the idea of s-wave pairing involve measurements of the temperature dependence of the penetration depth [6.51,52]. There are specific predictions for singlet, s-wave materials, with higher angular momentum pairing, which may have line or point zeros of the order parameter in different directions in k-space. Muon spin resonance studies [6.51] which probe the spatial variations of the magnetic field inside superconductors, and investigations of the pentration depth by precision magnetic mesurements [6.52] are both consistent with s-wave pairing. The temperature dependence of the penetration depth as determined from dc magnetic measurements [6.49] is shown in Fig. 6.12. These result cast doubt on models of superconductivity which assume triplet pairing [6.47].

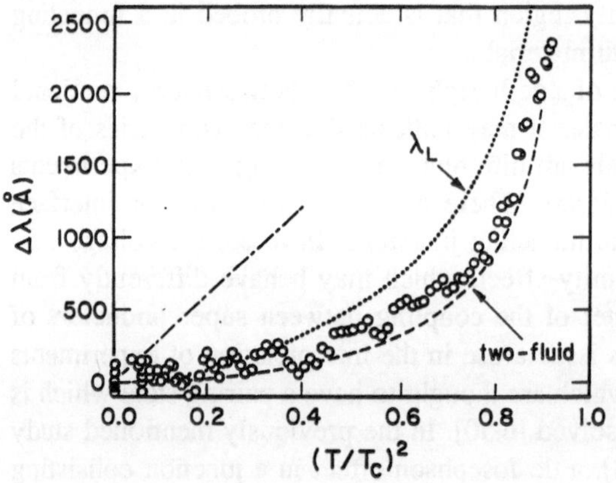

Fig. 6.12. The change in the penetration depth is shown as a function of reduced temperature. The dashed line is a two-fluid approximation with $\lambda(0)$ = 1400 Å and the dotted line represents the clean weak-coupling penetration depth λ_L for $\lambda(0)$ = 1400 Å. λ_L also fits the data if $\lambda(0) \approx 900$ Å. The dot-dashed line is the behavior seen in some polycrystalline samples and films. (From [6.52])

6.3.2 Pairing Mechanism

There are a number of experimental probes which lead to enhanced understanding of the mechanism for pairing in conventional superconductors. These include studies of the isotope effect; measurements of the specific heat; and measurements of electron tunneling and of infrared spectra. In principle, accurate studies of the parameter emerging from studies of this sort should allow distinctions to be made between conventional BCS theory, other Boson exchange models, or mechanisms based on variants of the resonating valence bond (RVB) model.

Some important results on normal state properties and the connection between magnetism, superconductivity and several photoemission results also have an important bearing on the mechanism. These will be discussed in a subsequent section.

Historically, studies of the isotope effect played an important role in identifying the electron–phonon interaction as a critical ingredient in the theory on conventional superconductivity. In the simplest models of the isotope effect in the BCS theory the transition temperature depends on the electronic mass through the relation

$$k_B T = 1.13\hbar\omega_c e^{-1/N(0)V} \tag{6.2}$$

where $N(0)$ is the single-spin density of states at the Fermi level, V is the average electron–phonon coupling constant (which is constant within the range $\hbar\omega_c$) and the frequency ω_c depends inversely on the mass M of the ions, so that the transition temperature depends on $M^{-\alpha}$ where $\alpha = 1/2$ for metallic elements which are conventional superconductors.

There are exceptions to this rule, such as the elements Zr, Mo, Ru, Os, and Re, where α ranges from 0 to 0.39. Because of these exceptions, a small isotope effect is not a "smoking gun" for a negligible role of phonons in the pairing. A detailed treatment of the exceptions involves consideration of the Coulomb repulsion in a manner more comprehensive than in the simple one-parameter BCS model. Indeed there might be an isotope effect in some non-BCS model if the interaction parameter depended in some manner on the ionic masses. Thus the presence of the isotope effect is really neither necessary nor sufficient evidence for phonon-mediated pairing. In the case of conventional superconductors, detailed studies of the electron–phonon coupling through quantitative tunneling investigations constitute the only truly conclusive evidence about the character of the pairing mechanism. This will be discussed below.

Most studies of the isotope effect have concentrated on the substitution of oxygen isotopes. This has usually been done by changing ^{16}O with ^{18}O diffusively at elevated temperatures, where the lattice oxygen in the chain sites is mobile. Although the effect of isotope substitution on phonon frequencies is claimed to be confirmed directly by Raman measurements [6.53], there is doubt as to the extent and the sites of exchange. The first experimental studies of YBCO at AT&T Bell Laboratories [6.53] and at Berkeley [6.54] reported $\alpha = 0$, with uncertainties of ± 0.02 and ± 0.027, respectively. Two later studies [6.55, 56] yielded small positive values of α, at the limit of the experimental resolution. Recent studies at Los Alamos [6.57], in which the starting materials were isotopically pure ^{18}O and ^{16}O, showed larger effects, probably due to differences in the processing with different isotopes. In the case of the LBCO system, an isotope effect with $\alpha \approx 0.2$ was reported, suggesting that phonons are of some importance in this system [6.58]. Recent studies of BiO superconductors by *Batlogg* and co-workers [6.58] yielded $\alpha \approx 0.2$–0.25, but these authors speculated that the effects are a consequence of "dressed" electronic excitations mediating superconducting pairing.

Experiments on the isotope effect are complicated because it is very difficult to be absolutely certain that processing variations do not affect the results. Although the current experimantal situations appears to be well defined, it is entirely possible that the last word has not yet been written about this very important measurement.

Measurements of the specific heats of superconductors historically have tied the superconducting transition to the Landau theory of continuous phase transitions. These measurements provided important information about the superconducting energy gap and serve as an important test of the microscopic theory. The problem with measurements of specific heats in high-temperature superconductors relative to conventional superconductors is that the lattice contribution is always important. It is difficult to extract unambiguously the electronic contribution from the data.

An important prediction of the BCS theory is that the specific heat discontinuity at the transition is $\Delta C = 1.43\gamma T_c$. Recent work on YBCO [6.59, 60] gave factors of 1.5 and 1.23 ± 0.08, respectively, values not far from that obtained by *Junod* and co-workers [6.61]. The difficulty is in the spread in the experimental values, which makes the use of such data for the testing of theoretical models somewhat premature.

Some of the details of the superconducting phase transition were mentioned earlier. It is certain that there is an excess specific heat associated with Gaussian fluctuations, but whether this result is evidence for a multicomponent order parameter is not yet clear [6.44].

A final issue concerning the specific heat is the survival of a linear term at low temperatures (i.e., far below the superconducting transition) at which an exponential decreasing value $\propto \exp(-\Delta/k_B T)$ is expected. Such a linear term could indicate the presence of so-called two-level systems or could be a consequence of impurities. It is not certain that the linear contribution is in fact present in BSCCO, or in other high-T_c materials. Its absence in the other materials would suggest impurities as the source of the effect in YBCO. Because such a term would provide important support for Anderson's RVB model [6.49], it is important that the issue of sample purity be resolved.

Single-particle electron tunneling in conventional superconductors has been used to determine the magnitude of the energy gap in the excitation spectrum, which is associated with a threshold for conductance in the I-V characteristic. In some instances, at a far more sophisticated level (using the results of the Eliashberg theory), tunneling has been used to ascertain the details of the electron–phonon coupling. A very detailed analysis of conductance–voltage characteristics of Pb tunneling junctions has been used to determine the electron–phonon spectral function [6.62]. Indeed, the latter data constitute quite direct evidence for the role of the electron–phonon interaction in the superconductivity of conventional superconductors. If the coupling between the superconducting electrons and the excitation responsible for the attractive interaction were weak, or if the mechanism did not involve boson exchange, then tunneling would perhaps not be quite as sensitive a probe of the mechanism for superconductivity. It would still be very useful, since determination of the gap, together with the transition temperature, can provide important information about the nature of superconductive coupling, and perhaps even its physical origin. The reason for this is that different values of the ratio $2\Delta/k_B T_c$, as well as being a key to the strong- or weak-coupling character of the interaction may be clues as to the nature of different mechanisms of boson-exchange superconductivity [6.63].

Because of this, it has been fairly clear, from the beginning of high-T_c superconductivity research that tunneling could potentially be an extremely useful quantitative probe of the superconducting state of high-T_c compounds. Unfortunately, because of the very small coherence lengths in these compounds (of the order of a few angstroms in the c direction, and 20–50 Å in the ab-plane) which determine the volume probed, tunneling is only a surface sensitive probe. Consequently the results are susceptible to structural and chemical disorder at the surface, or to changes in stoichiometry. Also, as a consequence of anisotropy, the tunneling characteristic is direction dependent, further complicating the task of determining reproducible results on polycrystalline samples, and in effect requiring that measurements be made on single crystals or oriented crystals. The experimental results reported to date have been extensive but not totally convincing. It may be inappropriate in any case to rely too heavily on consensus values of the superconducting energy gap obtained from tunneling investigations using the same methodology, as experiments which are essentially identical may be subject to the same systematic errors. In interpreting tunneling data it is important to keep an open mind, as there is no certainty that correct tunneling characteristics will look anything like those studied for ordinary superconductors. Attempts to discard data that do not fit the old criteria may be premature as it is likely that useful tunneling data will not be obtained until tunneling studies of different types are carefully integrated with other surface sensitive studies.

Tunneling studies of high-temperature superconductors have been carried out on bulk crystalline material, and on single-crystal, random polycrystalline, and ordered polycrystalline films. Junctions have been formed in several ways: by depositing a counter electrode on the material, relying on the formation of a surface insulating layer on the superconductor from oxygen depletion or some other phenomenon; by depositing a normal metal such as Ag or Au on the surface of the high-T_c material, and making a proximity junction; by the so-called break-junction technique; by using a tip in either a point contact tunneling geometry, or in through use of a scanning tunneling microscope (STM) configuration[1].

In almost every instance there are possible sources of systematic error. For example, it has been very difficult to prove that scanning tunneling microscopes are actually operating in a tunneling mode. In any measurements in which an electrode is in mechanical contact with the surface of a high-temperature superconductor, there is some danger that the tunneling measurement is actually invasive. The pressure exerted by a tip on the surface region can be high enough to change the transition temperature and other parameters. In addition, the sur-

[1] See [6.9] for an extensive list of tunneling investigations.

faces exposed by cleaving crystals in an ultra-high vacuum may not be ideal, reducing the credibility of results obtained using that procedure.

The major characteristics of the data obtained about single-particle tunneling to date is their great variety. Values of the ratio $2\Delta/kT_c$ ranging from less than 3.5 (BCS value) to more than 15 have been reported. The large spread may be associated with the so-called Coulomb blockade [6.64], which is a normal electron charging effect. This refers to electron tunneling between nearly microscopic conductors, which is strongly affected by the charging energy $E_c = e^2/2C$ of a single electron charge in a small capacitor C.

Both symmetric and antisymmetric I-V characteristics have been observed with the sign of the asymmetry varying. Some very recent theoretical work would imply that the asymmetry itself may reveal some features of the superconductivity mechanism involved [6.64]. However, this effect will have to be deconvoluted from less-important asymmetries, those which might be associated with the tunneling barrier, and which have been previously observed in junctions involving conventional superconductors. Multiple gap structures have been observed in some measurements. The latter may be a fundamental effect, or perhaps a consequence of the so-called Coulomb staircase effect [6.66], which can be observed during tunneling in a series combination of two nearly microscopic junctions. In this phenomenon the current increases abruptly each time the bias voltage increases by e/C.

A conductance, linear with voltage, has been a common feature of the normal state tunneling in many experiments. This is either an important signature of the features of the superconducting state relating to RVB theories [6.46] or an artifact of the tunneling geometry. At this point it is not possible to clarify this ambiguity, so the interpretation of the results must be considered very carefully until a more specific understanding of the experimental geometry emerges.

On the other hand it is useful to display some data which actually indicate that electron tunneling is ocurring between a superconductor and a counter electrode. This is shown in Fig. 6.13 where data is plotted from junctions between a Pb electrode and electrodes of YBCO [6.67]. Similar results have been obtained with a BSCCO [6.68] electrode. The important feature to note is that at low temperatures, i.e., below the Pb transition temperature, a Pb gap can be seen. Although this is no proof of the validity of the other aspects of the data, it is a promising result which suggests that in the future it may be possible to obtain quantitative results using tunneling.

Studies of the Andreev reflection and far infrared spectroscopy are two alternatives to tunneling for quantitative microscopic information. Andreev reflections [6.69] can be observed by tunneling, or during injection of electrons into the normal layer of a normal superconducting proximity sandwich. When an electron with an energy smaller than the energy gap reaches the interface it will not enter the superconductor as a quasiparticle excitation since there are no excitations with that energy. Instead, a Cooper pair propagates into the superconductor, and a hole propagates backwards from the incident electron. This hole travels back to the surface of the normal metal and gives rise to an excess current at the

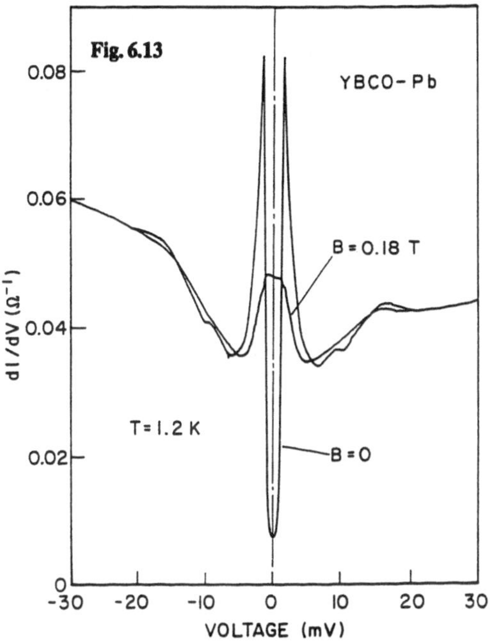

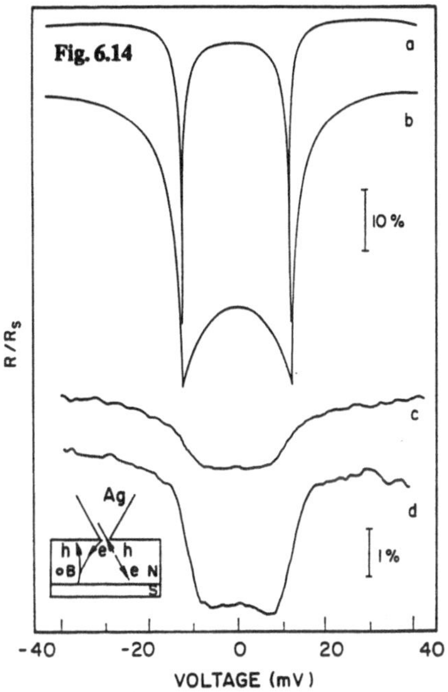

Fig. 6.13. Plot of dI/dV vs V of a YBCO-Pb junction with the Pb counter electrode in the normal and superconducting states. (From [6.67])

Fig. 6.14. Differential resistance of a point contact versus voltage. The upper curves (a and b) were calculated using the BTK model [6.70] and assuming a gap energy of 12.5 meV. The scattering potential is 1.4 and 0.4 for curves a and b, respectively. The lower curves show two measurements with point contact resistances of 25.30 Ω and 5.81 Ω, respectively, for curves c and d. The measurements were made on a 0.25–1 μm thick Ag–$YBa_2Cu_3O_{7-x}$ bilayer. Note the different scales for the calculations and the measurements. The inset shows the geometry of the single point-contact experiment with and without a magnetic field

injection point. The threshold energy for Andreev reflection is thus a measure of the energy gap of the underlying superconductor. Such experiments have been conducted using either point contact injection or tunneling. One such result is shown in Fig. 6.14. This method depends critically on the condition and sharpness of the normal–superconductor interface, and provides an alternative way of measuring the superconducting energy gap.

In principle, the measurement of far infrared reflectivity (FIR) should be a simpler probe of the microscopic features of the superconducting state than electron tunneling because the sampling volume is the order of the wavelength rather than the very short superconducting coherence length. Unfortunately, there have been numerous complications, and the current situation is inconclusive regarding the magnitude and even the existence of the energy gap. There are a number of reasons for these difficulties which have been reviewed by *Tinkham* and *Lobb* [6.27]. First, the spectrum is very complex, both in the superconducting and normal states. This results from the complex unit cell of the high-temperature superconductor, which leads to a complex spectrum of optically excited vibrational modes in the far infrared. These modes dominate c-axis reflectivity whereas in the ab-plane conductive response similar to that found in metals dominates. In polycrystalline composite samples spatial averaging involves all of these complications.

With the availability of single crystals and oriented film surfaces the data began to improve, but there were uncertainties relating to the interpretation of the normal state spectrum. Furthermore a reflectance feature at $500\,\mathrm{cm}^{-1}$ previously attributed to superconductivity, was associated by one group with a plasma edge [6.71]. In most experiments the measured reflectance fell short of 100% even at frequencies usually associated with energies below the gap, contrary to expectations for a zero resistance material. In materials in which the reflectance approaches 100%, a change in slope of the reflectivity curve is found. If this is interpreted as being due to the superconductive gap, the value of the gap would be very close to the value predicted by the BCS theory. An overriding problem which contributes to the overall uncertainty in the results in all of these experiments is that the absorption is computed from the difference of the measured reflectivities from 100%, and this involves calculating the difference of two large quantities.

Recently the study of Fermi-edge photoemission spectra has emerged as a potential tool for the study of details of the superconducting state [6.72]. Study of the near-edge line shape appears to yield the energy gap below the transition temperature [6.73].

6.4 Theoretical Considerations and Discussion

The quasi-two-dimensional character of the copper oxide planes appears to be very important for the electronic and magnetic properties of the superconducting oxides. Much of the theoretical effort to understand the mechanism for superconductivity has been focused on this feature. There is a widely held view that the mechanism in the high-T_c materials is different from that of previously known superconductors.

It is important to note that the approach to the theory of these materials involves two levels of reasoning, which must ultimately be combined. On the one hand, the microscopic electronic structure and valency set the stage for the actual detailed theory of the mechanism for superconductivity. This avenue of inquiry considers, for example, the role of oxygen in setting the character of the charge carriers, and the importance of the valency of the mulitvalent metallic constituent which is always present. These issues are of great importance because the existing data strongly imply that the normal-state properties of these materials are truly unique. On the other hand, the possibility that the spin liquid state of a Mott insulator is the correct context for discussing the theory of superconductivity is strongly suggested by magnetic studies, but is not yet fully accepted. Recent photoemission studies do not rule out such a state, although they appear to indicate the existence of a reasonably sharp Fermi edge [6.73].

In the past, the actual mechanism for superconductivity has been probed by measurements on the superconducting state. In the following we will be mostly concerned with the superconducting rather than with the normal state electronic structure. The latter, because of the complexity inherent in the high-T_c materials, is a complex subject in its own right.

The proposed mechanisms for high-temperature superconductivity are numerous. The problem with the theory is that so many models have been proposed that one might say that it is hard to understand how normal behavior near absolute zero could ever occur in any condensed system. For many models the existence of the observed antiferromagnetic parent compound [6.74] is crucial. Both La_2CuO_4 and $YBa_2Cu_3O_6$ are antiferromagnetic, however, the addition of oxygen to $YBa_2Cu_3O_6$ or Sr to La_2Cu_4 results in an extremely complex behavior. This evolution includes a two-dimensional magnetic phase which then evolves towards superconductivity. Hall effect studies indicate that the doping increases the carrier concentration, and that the majority carriers are holes for all the high-T_c systems discovered, except the most recent ones. (See Fig. 6.15 for a schematic of the phase diagram associated with the magnetic behavior.) Supporting this picture are the results of neutron scattering experiments in nonsuperconducting La_2CuO_4 which have revealed strongly correlated spin fluctuations of a two-dimensional nature [6.76]. These results are also supported by light scattering experiments [6.77]. In the superconducting state these experiments suggest the presence of magnetic excitations which may be important for the superconductivity itself [6.78].

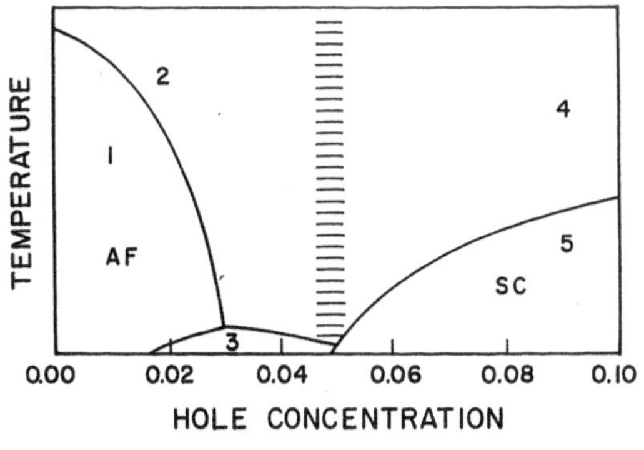

Fig. 6.15. Schematic phase diagram for $La_{2-x}Sr_xCuO_4$, $YBa_2Cu_3O_{6+y}$ and similar compounds as a function of temperature and number of holes per CuO_2 unit in a CuO_2 plane. Holes are added when x or y is increased from zero. In region 1 long range antiferromagnetic (AF) order occurs. The electrical conductivity in region 2 shows the 3D hopping character typical of localized electronic states. Region 3 appears to be a spin-glass-like state. Regions 3 and 5 are characterized by AF correlations on a length scale determined by the inter-hole distance. Regions 4 and 5 are characterized by extended electronic states. (From [6.75]; see also [6.74])

The ubiquitousness of these magnetic states has led to many theoretical models which use them as a starting point. The models proposed for high-temperature superconductivity include the view that superconductivity is of the conventional BCS type, and that the attractive interaction is produced by the exchange of a boson (other than a phonon) such as a plasmon, exciton, polaron or spin excitation (e.g., a spin wave). These models also assume that superconductivity cannot be explained within the conceptual framework of a Fermi liquid theory, and that a more revolutionary framework is needed [6.79]. This was provided by *Anderson* [6.45] who proposed the hypothesis of a resonating valence bond state for the CuO_2 planes. The basic feature of this state is that the electrons are described in terms of spin-singlet pairs, with excited states involving two chargeless quasiparticles, spin-1/2 Fermions (known as spinons), and positively charged spinless bosons (known as holons) which are paired below T_c.

The task of experiment is to impose constraints on the theoretical models of superconductivity, in the hope of ultimately ruling out all but one for a given materials class. It is not certain at this juncture that all high-temperature materials classes have the same mechanism. For example, cubic perovskites such as $B_{0.6}K_{0.4}BiO_3$ may very well be different. Unfortunately, despite the enormous number of publications to date, this standard program has not yet been carried out.

The reason for this is that many potentially critical experiments have produced ambiguous results, most likely as a consequence of uncontrolled features of the materials studied. The case of the superconducting energy gap has already been mentioned. In principle, the measurement of the ratio $2\Delta/k_B T_c$ could be very revealing as to the mechanism and even the character of the pairing. The results are unfortunately still quite inconclusive. The dependence of the gap on energy, in the spirit of the Eliashberg theory, is currently beyond the reach of any experimental technique. The linear behavior of the conductance with voltage in many tunneling junctions, which supports the RVB theory is unfortunately a sample-dependent effect. In the case of heat capacity, the very exciting possibility of a linear term in the temperature dependence of the specific heat at low temperatures – which was also believed to support the RVB picture – seems to be vanishing with improving sample quality, i.e., reduced impurity content. Sample inhomogeneities create difficulties with the unequivocal study of the critical region near the superconducting transition. Such studies may be very revealing of the nature of the pairing, among other issues.

There are a number of facts relating to the mechanism which are fairly well known. Pairing itself is not in doubt, as both vortex lattice imaging and flux quantization give strong evidence in favor of its existence. It is also likely that the pairing is s-wave in character, as determined from the temperature dependence of the penetration depth (Fig. 6.12). The coherence lengths are very short, however, of the order of 10 Å in some directions. This leads to speculations about real space pairs, which exist above T_c but would exhibit Bose condensation to give superconductivity below T_c [6.80]. This view is supported by the observation that the transition temperature of superconducting materials seems to be proportional to the carrier concentration, leading to the suggestion that the pairing interaction is nonretarded. The situation would then be rather different from the BCS case.

Whatever the actual theory is it is important to remember that it must have most of the attributes of BCS theory, an energy gap, a discontinuity in the heat capacity singlet pairs, trémendous anisotropy, and anisotropic superconducting vortices. Because of the short coherence length, critical fluctuations may be important, unlike what is found in conventional low-temperature superconductors. Whether critical currents in the bulk can ever be made large enough to be technologically relevant, will be determined only if a method of pinning vortices is realized.

The recurring problem with experimental studies of high-T_c materials has been sample quality. The early experiments, which were confined to polycrystalline ceramic samples, were relativley crude, and in many instances the results reflected the the structure of the material rather than an intrinsic superconducting ppp. Chemical impurities and second phases often complicated results which might have provided micoroscopic information. The single-crystal materials, although far superior, also suffer from inhomogeneities such as twin boundaries. One cannot yet take too seriously experimental consensus relating to essentially identical studies, since such studies are all susceptible to the same systematic error, and may all be incorrect for the same reasons.

Acknowledgements. The author would like to thank V. Kogan, M. Mecartney, J. Woods Halley, Jr., Oriol T. Valls, Stuart Wolf, Donald Gubser, David Larbalestier, and Guy Deutscher for helpful discussions. This work was supported in part by the Air Force Office of Scientific Research under Grant No. 87-0372, and by the Central Administration of the University of Minnesota.

References

6.1 J.G. Bednorz, K.A. Müller: Z. Phys. B **64**, 189 (1986)
6.2 M.K. Wu: Phys. Rev. Lett. **58**, 908 (1987)
6.3 C. Michel, M. Hervieu, M.M. Borel, A. Grandin, F. Deslandes, J. Provost, B. Raveau: Z. Phys. B **68**, 421 (1987);
H.G. von Schnering, L. Walz, M. Schwarz, W. Becker, M. Hartweg, T. Popp, Hettich, P. Muller, G. Kampf: Angew. Chem. Int. Ed. Engl. **27**, 574 (1988);
H. Maeda, Y. Tanak, M. Fukutome, T. Asano: Jpn. J. Appl. Phys. **27**, L209 (1988)
6.4 Z.Z. Sheng, A.M. Hermann: Nature (London) **332**, 138 (1988)
6.5 R.J. Cava, B. Batlogg, J.J. Krajedwski, R. Farrow, L.W. Rupp Jr., A.E. White, K. Short, W.F. Peck, T. Kometani: Nature (London) **332**, 814 (1988)
6.6 V.L. Ginzburg: Contemp. Phys. **9**, 335 (1968)
6.7 A.W. Sleight, J.L. Gilson, P.E. Bierstedt: Solid. State Commun. **17**, 27 (1975)
6.8 A.W. Sleight: Science **242**, 1519 (1988)
6.9 J. Talvacchio: Supercond. **2**, 1 (1989)
6.10 For a discussion of the standard properties of superconductors see: M. Tinkham: *Introduction to Superconductivity* (McGraw-Hill, New York 1975), reprinted by Robert E. Krieger, Malabar, FL 1980 and 1985
6.11 S.S. Parkin, V.Y. Lee, E.M. Engler, A.I. Nazzal, T.C. Huang, G. Gorman, R. Savoy, R. Beyers: Phys. Rev. Lett. **60**, 2539 (1988)

6.12 Bernard M. Kulwicki: "PTC Materials Technology, 1955–1980", in *Advances in Ceramics, Vol. 1, Grain Boundary Phenomena in Electronic Ceramics*, ed. by L.M. Levinson, D.C. Hill (American Ceramic Society, Columbus, OH 1981)
6.13 Compounds which are the lanthanide analogs of $YBa_2Cu_3O_{7-x}$ were found simultaneouly and independently at many laboratories worldwide. See: M.B. Maple, Y. Dalichaouch, J.M. Ferreira, R.R. Hake, B.W. Lee, J.J. Neumeier, M.S. Tricachvili, K.N. Yang, H. Zhou, R.P. Guertin, M.V. Kuric: Physica B **148**, 155 (1987), and references therein
6.14 I.K. Schuller, J.D. Jorgensen: Mater. Res. Soc. Bull. **14**, 27 (1989)
6.15 Y. Tokura, H. Takagi, S. Uchida: Nature (London) **337**, 345 (1989)
6.16 H. Sawa et al.: Nature (London) **337**, 347 (1989)
6.17 H. Takagi, S. Uchida, Y. Tokura: Phys. Rev. Lett. **62**, 1197 (1989)
6.18 J.T. Markert, J.T. Maple: Preprint, University of California, San Diego (1989)
6.19 C.E. Gough et al.: Nature (London) **326**, 694 (1987)
6.20 P.L. Gammel, D.J. Bishop, G.J. Dolan, J.R. Kwo, C.A. Murray, L.F. Schneemeyer, J.V. Wasczak: Phys. Rev. Lett. **59**, 2592 (1987)
6.21 A. Kapitulnik, M.R. Beasley, C. Castellani, C.D. Castro: Phys. Rev. **37**, 537 (1988)
6.22 The most striking studies of anisotropy come from torque magnetometer investigations, see: D.E. Farell, C.M. Williams, S.A. Wolf, N.P. Bausal, V.G. Kogan: Phys. Rev. Lett. **61**, 2805 (1988);
D.E. Farell, S. Bonham, J. Foster, Y.C. Chang, P.Z. Jiang, K.G. Vandervoort, D.J. Lan, V.G. Kogan: Phys. Rev. Lett. **63**, 782 (1989)
6.23 S. Martin, A.T. Fiory, R.M. Fleming, G.P. Espinosa, A.S. Cooper: Phys. Rev. Lett. **62**, 677 (1989)
6.24 N.-C. Yeh, C.C. Tsuei: Phys. Rev. B **39**, 9 (1989)
6.25 D.H. Kim, A.M. Goldman, R. Kampwirth: Phys. Rev. B **40**, 8834 (1989)
6.26 D.H. Kim, A.M. Goldman, J.H. Kang, K.E. Gray, R.T. Kampwirth: Phys. Rev. B **39**, 12275 (1989) and references therein
6.27 M. Tinkham, C.J. Lobb: In *Solid State Physics*, Vol. 42, ed. by H. Ehrenreich, D. Turnbull (Academic, New York 1989) p.91
6.28 A.A. Abrikosov: Sov. Phys.–JETP **5**, 1174 (1975)
6.29 K.A. Müller, M. Takashige, J.G. Bednorz: Phys. Rev. Lett. **58**, 1143 (1987)
6.30 T.T.M. Palstra, B. Batlogg, L.F. Schneemeyer, J.V. Waszczak: Phys. Rev. Lett. **61**, 1662 (1988)
6.31 P.L. Gammel, L.F. Schneemeyer, J.V. Waszczak, D.J. Bishop: Phys. Rev. Lett. **61**, 1666 (1988)
6.32 Y. Yeshurun, A.P. Malozemoff: Phys. Rev. Lett. **60**, 2202 (1988)
6.33 J.R.L. de Almeida, D.J. Thouless: J. Phys. A **11**, 983 (1978)
6.34 M. Tinkham: Phys. Rev. Lett. **61**, 1658 (1988)
6.35 M.M. Fang, V.G. Kogan, D.K. Finnemore, J.R. Clem, L.S. Chumbley, D.E. Farell: Phys. Rev. Lett. **37**, 2334 (1988)
6.36 Y. Iye, T. Tamegai, H. Takeya, H. Takei: In *Superconducting Materials*, ed. by S. Nakajima, H. Fukuyama, Jpn. J. Appl. Phys. Ser. 1 (JJAP, Tokyo 1988) p.46
6.37 U. Essmann, H. Trauble: Phys. Lett. **24A**, 526 (1967);
N.V. Sarma: Philos. Mag. **17**, 1233 (1968)
6.38 E.H. Brandt, P. Esquinazi, G. Weiss: Phys. Rev. Lett. (c) **32**, 2330 (1989);
R.N. Kleiman, P.L. Gammel, L.F. Scheemeyer, J.V. Waszczak, D.J. Bishop: Phys. Rev. Lett. (c) **62**, 2331 (1989)
6.39 S.W. Tozer, A.W. Kliensasser, T. Penney, D. Kaiser, F. Holtzberg: Phys. Rev. Lett. **59**, 1768 (1987)
6.40 S. Martin, A.T. Fiory, R.M. Fleming, L.F. Schneemeyer, J.V. Waszczak: Phys. Rev. Lett. **60**, 2194 (1988)

6.41 For a discussion of the Ginzburg–Landau Theory for short-coherence-length superconductors see: L.N. Bulajewskii, V.L. Ginzburg, A.A. Sobyanin: Zh. Ekspr. Teor. Fiz. **94**, 355 (1988) [English transl.: Sov.-Phys.–JETP **68**, 1499 (1988)]
6.42 C.J. Lobb: Phys. Rev. B **36**, 3930 (1987)
6.43 For a discussion of this issue see: D.H. Douglas: Phys. Rev. B **39**, 4748 (1989)
6.44 There are some specific heat results which probably are consistent with Gaussian fluctuations, although they may have been interpreted differently. See: S.E. Inderhees, M.D. Salamon, N. Goldenfeld, J.P. Rice, B.G. Payol, D.M. Ginsberg, J.Z. Liu, G.W. Crabtree: Phys. Rev. Lett. **60**, 1178 (1988)
6.45 A.J. Mills, D. Rainer, J. Sauls: In *Novel superconductivity*, ed. by S.A. Wolf, V.Z. Kresin (Plenum, New York 1987) p. 265
6.46 P.W. Anderson: Science **235**, 1196 (1987)
6.47 G. Chen, W.A. Goddard: Science **239**, 899 (1988)
6.48 J. Niemeyer, M.R. Dietrich, C. Politis: Z. Phys. B **67**, 155 (1987)
6.49 O.S. Akhtyamov: JETP Lett. **3**, 183 (1966)
6.50 For a review see: Z. Fisk, D.W. Hess, C.J. Pethick, D. Pines, J.L. Smith, J.D. Thompson, J.O. Willis: Science **239**, 33 (1988)
6.51 D.R. Harshman et al.: Phys. Rev. B **36**, 2386 (1987)
6.52 L. Krusin-Elbaum, R.L. Greene, F. Holtzberg, A. Malozemoff, Y. Yeshurun: Phys. Rev. Lett. **62**, 217 (1989)
6.53 B. Batlogg et al.: Phys. Rev. Lett. **58**, 2333 (1987)
6.54 L.C. Bourne et al.: Phys. Rev. Lett. **58**, 2337 (1987)
6.55 K.J. Leary, H.C. zur Laye, S.W. Keller, T.A. Faltens, W.K. Ham, J.N. Michaels, A.M. Stacy: Phys. Rev. Lett. **59**, 1236 (1987)
6.56 D.E. Morris, R.M. Kuroda, A.G. Markelz, J.H. Nickel, J.Y.T. Wei: Phys. Rev. B **37**, 5936 (1988)
6.57 K.C. Ott et al.: Phys. Rev. B **39**, 4285 (1989)
6.58 B. Batlogg, G. Kouvrouklis, W. Weber, R.J. Cava, A. Jayaraman, A.E. White, K.T. Short, L.W. Rupp, E.A. Reitman: Phys. Rev. Lett. **59**, 912 (1987);
T.A. Faltens et al.: Phys. Rev. Lett. **59**, 915 (1987)
6.59 M.V. Nevitt, G.W. Crabtree, T.E. Klippert: Phys. Lett. B **36**, 2398 (1988)
6.60 S.E. Inderhees, M.B. Salomon, T.A. Friedman, D.M. Ginsberg: Phys. Rev. B **36**, 2401 (1988)
6.61 A. Junod, A. Bezinge, J. Muller: Physica C **152**, 50 (1988)
6.62 W.L. Macmillan, J.M. Rowell: In *Superconductivity, Vols. I and II*, ed. by R.D. Parks (Marcel Dekker, New York 1968) p. 561
6.63 J.P. Carbotte, F. Marsiglio: In *Studies of High Temperature Superconductors*, ed. by A.V. Narlikar (Nova Science, Commack, NY 1989)
6.64 I. Giaever, H.R. Zeller: Phys. Rev. Lett. **20**, 1504 (1968);
E. Ben Jacob, Y. Gefen: Phys. Lett. **108A**, 289 (1985);
D.V. Averin, K.K. Likharev: J. Low Temp. Phys. **62**, 345 (1986);
This explanation was first suggested by M.D. Kirk et al.: Phys. Rev. B **35**, 8850 (1987)
6.65 F. Marsiglio, J.F. Hirsch: Preprint, University of California, San Diego (1989)
6.66 J.B. Barner, S.T. Ruggiero: Phys. Rev. Lett. **59**, 807 (1987);
L.S. Kuz'min, K.K. Likharev: JETP Lett. **45**, 495 (1987)
6.67 J. Geerk, X.X. Xi, G. Linker: Z. Phys. B **73**, 329 (1988)
6.68 M. Lee, D.B. Mitzi, A. Kapitulnik, M.R. Beasley: Phys. Rev. B **39**, 801 (1989);
M. Lee, M. Naito, A. Kapitulnik, M.R. Beasley: Solid. State Commun. **70**, 489 (1989)
6.69 H.F.C. Hoevers, P.J.M. van Bantum, L.E.C. van de Leemput, H. van Kampen, A.J.G. Schellingerhout, D. van de Marel: Physica C **152**, 105 (1988)
6.70 G.E. Blonder, G.E. Tinkham, T.M. Klapwijk: Phys. Rev. B **25**, 4515 (1982)
6.71 T. Timusk et al.: Physica C **153–155**, 1744 (1988)
6.72 Y. Chang et al.: Phys. Rev. B **39**, 4740 (1989)

6.73 Y. Chang et al: Phys. Rev. B **39**, 7313 (1989)
6.74 For a phase diagram see: J.B. Torrance, Y. Tokura, A.I. Nazzal, A. Bezinge, T.C. Huang, S.S.P. Parkin: Phys. Rev. Lett. **61**, 1127 (1988) and references therein
6.75 E.W. Fenton: In Proc. Int. Conf. on Transition Metals, Kiev, June 1988
6.76 G. Shirane, Y. Endoh, R.J. Birgeneau, M.A. Kastner, Y. Hidaka, M. Oda, M. Suzuki, T. Murakami: Phys. Rev. Lett. **59**, 1613 (1987)
6.77 K.B. Lyons, P.A. Fleury, J.P. Remeika, T.J. Negran: Phys. Rev. B **37**, 2353 (1988)
6.78 K.B. Lyons, P.A. Fleury, L.F. Schneemeyer, J.V. Waszczak: Phys. Rev. Lett. **60**, 732 (1988)
6.79 R.B. Laughlin: Science **245**, 525 (1988)
6.80 For a review of theories and this issue in particular see: V.J. Emery: MRS Bull. **14**, 67 (1989)

7. Surface Structure and Bonding of Tetrahedrally Coordinated Compound Semiconductors

Charles B. Duke

With 13 Figures

The purpose of this article is to provide an overview of the current state of knowledge of the surface structure and bonding of tetrahedrally coordinated compound semiconductors and to introduce the theoretical concepts and constructs which have emerged to give a quantitative, predictive description of experimental measurements on these materials. The article is introductory in character, building on comprehensive recently published reviews of both zincblende [7.1,2] and wurtzite [7.2,3] structure materials. Moreover, only the results of measurements and theoretical models are considered. Descriptions of the experimental techniques and model calculations may be found in the literature cited.

The first quantitative determination of a semiconductor surface atomic geometry was given in 1976 for the (110) surface of GaAs [7.4]. Since that time, such determinations have been reported for the (110) surfaces of all common zincblende structure compound semiconductors (i.e., AlP, GaP, GaSb, InP, InAs, InSb, ZnS, ZnSe, ZnTe and CdTe), the (2×2) structure on the cation (111) polar surfaces of GaAs, GaP, and InSb, and both the Ga and As (311) polar surfaces of GaAs [7.1,2]. In addition, due to the availability of theoretical predictions [7.5] of their geometries, the $(10\bar{1}0)$ and $(11\bar{2}0)$ cleavage surfaces of wurtzite structure materials have attracted interest. Following an early analysis of ZnO$(10\bar{1}0)$ [7.6], the geometries of both the $(10\bar{1}0)$ and $(11\bar{2}0)$ surfaces of CdSe have recently been determined by both Low Energy Electron Diffraction (LEED) [7.7,8] and Low Energy Positron Diffraction (LEPD) [7.8,9]. Structures have been predicted but not yet confirmed for the $(10\bar{1}0)$ and $(11\bar{2}0)$ surfaces of ZnSe and CdS, as well as the $(11\bar{2}0)$ surface of ZnO [7.5].

Additional perspective is afforded by experience with the surfaces of silicon which have been studied for 30 years [7.10,11]. During the past few years, the benchmark (7×7) structure of Si(111) was determined for the first time by *Takayanagi* et al. [7.12,13] using transmission electron diffraction. This proposed structure subsequently was refined by *Tong* et al. [7.14] using LEED and by *Robinson* et al. [7.15,16] using X-ray diffraction. Even quite complex surface structures are now susceptible to quantitative determination using modern surface science techniques. Such results suggest that since 1976 the determination of semiconductor surface structures by surface sensitive analytical techniques has become "routine" in that an arbitrary structure of interest can be obtained given adequate instrumentation and funding. Interest is now passing from the determination of these structures *per se* to the novel chemical bonding phenomena revealed by the analysis of classes of homologous structures for different materials. In the case of the large unit cell reconstructions on the (111) surfaces of Si and Ge, the resulting structures seem to originate from a delicate balance between dangling bond and strain energies [7.15]. Consequently, they fall into the generic category of epitaxically constrained surface bond rehybridization structures, a category which also contains the cleavage faces of tetrahedrally coordinated compound semiconductors [7.17].

We proceed by first reviewing some of the central concepts in semiconductor surface chemistry in the following section and then reviewing the situations for the cleavage surfaces of zincblende and wurtzite structure compound semiconductors, respectively, in the following two sections. The extension of the themes developed for clean surfaces to overlayer systems is indicated in the penultimate section. The chapter concludes with a synopsis.

7.1 Key Concepts in Semiconductor Surface Chemistry

When a semiconductor surface is formed, bonds are broken. This fact both puts the surface under stress and relaxes the bulk constraints on atomic movements to relieve this stress. A universal response to this situation by tetrahedrally coordinated semiconductors is to form new types of chemical bonds in the surface layer(s) (i.e., "surface bond rehybridization" [7.17, 18]) which, in turn, induce elastic distortions in the layers beneath. Analogous phenomena occur at metal surfaces [7.19], but the atomic motions in the surface layers in semiconductors are much larger and usually are interpreted in terms of chemical bonding concepts rather than as a response to inhomogeneous electrostatic fields.

An important concept for the interpretation of semiconductor surface structures is the notion of universality. If the structures of the same crystallographic surface of various materials are compared by scaling the atomic dimensions to the bulk lattice constants then, to a good approximation, each cleavage surface of tetrahedrally coordinated compound semiconductors is found to exhibit a common "universal" structure [7.5, 20]. The universality (i.e., lack of dependence on a particular material) of these structures leads, in turn, to the idea that they are caused by generic types of surface chemical bonding, in this case "epitaxically constrained surface bond rehybridization" [7.17]. The nature and origin of these types of bonding become subjects of prime interest.

In particular, the vacuum surfaces of tetrahedrally coordinated semiconductors exhibit large atomic rearrangements caused by the rehybridization of the surface chemical bonding to accomodate the extra charge which would have existed in localized directional "dangling" bonds at the surface of a truncated bulk solid [7.1]. The experimental indication of the surface (as opposed to material) specificity of this phenomenon is the fact that the (110) cleavage faces of zincblende-structure compound semiconductors exhibit the same atomic rearrangements for III-V (AlP, AlAs, AlSb, GaP, GaAs, GaSb, InP, InAs, InSb) as for II-VI (ZnS, ZnSe, ZnTe, CdTe) materials. Therefore the only distinction between the various surface structures is the experimentally measured [7.20] and theoretically predicted [7.21] linear scaling of the magnitude of the surface structural parameters with the bulk lattice constant a_0. In contrast to this situation for semiconductor surfaces, the small-molecule chemistries of III-V and II-VI compounds differ dramatically. For III-V molecules the anions prefer threefold distorted p^3 configurations (e.g., the As in AsH_3) and the cations planar sp^2 configurations (e.g., the Ga in GaH_3). Small molecules of the anion species alone form tetrahedra (e.g., P_4). The II-VI atomic species, on the other hand, form twofold-coordinated chain-like structures in either closed [e.g., S_6, S_8, monoclinic Se (Se_8)] or open (trigonal Se) geometries. Thus, if one envisages the surface chemical bonding of the cleavage faces of zincblende-structure compound semiconductors as intermediate between small molecule bonding and the bulk sp^3 tetrahedrally coordinated bonding, the occurrence of analogous surface geometries for III-V and II-VI materials is quite remarkable.

In order to describe the similarity between the III-V and II-VI surface structures, we require a theoretical construct which transcends simple inorganic coordination chemistry. For the (110) surfaces of zincblende-structure compounds, two quantum-chemical models have been reported in the literature: an extended tight-binding model, which has been applied to the III-V compounds [7.21] and to ZnTe [7.22], ZnSe [7.22, 23], ZnS [7.24], and CdTe [7.25]; as well as a first-principles pseudopotential model which has been applied to GaAs [7.26] and ZnSe [7.27]. Pseudopotential techniques are discussed in Chap. 5. These calculations reveal that the surface relaxation is driven by the lowering in energy of a band of surface states from near the middle of the band gap to the top of the valence band. These states change their character as the relaxation proceeds, going from dangling-bond character to surface-bond character in which added charge is deposited in the surface anion-cation bonds. Moreover, these surface states are observable by Angle Resolved Photoemission Spectroscopy (ARPES) so the predictions of the model calculations can be tested, as has been done for ZnSe [7.23, 28] and CdTe [7.25]. The major characteristics of these states are determined by the tetrahedrally coordinated bulk atomic geometry as manifested in the atomic connectivity at the surface. The details of the atomic potentials modify them only modestly; hence, the close relationship between the surface structure of III-V compounds (e.g., GaAs) and their isoelectronic II-VI counterparts (e.g., ZnSe). Surface chemical bonding is driven more by the substrate geometry (as reflected in the surface atomic connectivity) than by the small-molecule coordination chemistry of III-V versus II-VI species: i.e., the atomic geometry of the substrate "template" which the surface bonding must match epitaxically dominates the character of this bonding and gives rise to a new type of "surface" chemical bond which is unknown in either small molecules or bulk solids.

The conclusion to be drawn from this example is that a new type of "surface chemical bonding" is established by virtue of a successful validation of the predictions of the quantum-chemical models which must describe the experimental surface geometries (typically obtained via LEED) and the experimental surface state energetics (typically obtained via ARPES). More recently, surface phonon spectra have been reported [7.29] for GaAs(110) which also are in correspondence with the model predictions [7.30]. Therefore the experimental discovery [7.20] of the universality of the III-V and II-VI zincblende (110) surface structures ultimately has led to a new theoretical interpretation of the origin of these structures [7.5].

The final concept which it is useful to introduce in this section is that of metastability. In device applications semiconductor surface structures are not the equilibrium structures, but rather are those determined by the process steps used to fabricate the device. Thus, it is important to study surface structures as functions of processing conditions. In the case of the clean, cleavage surfaces, the equilibrium surface structure can be obtained either by vacuum cleavage or by ion bombardment and anneal cycles [7.11]. Polar surfaces exhibit several metastable structures with varying atomic composition. We return to this topic in Sect. 7.4 on adsorbate systems in which we find that, in general, the use of deposition and anneal cycles lead to various different atomic geometries, each of which is metastable at room temperature and below.

7.2 Zincblende (110) Surfaces

A picture of the relaxed surface atomic geometry of the (110) cleavage surfaces of a zincblende-structure compound is shown in Fig. 7.1 [7.3]. As noted earlier, the surface geometry is that expected for III-V compounds: the anion exhibits a distorted p^3 conformation and the cation a nearly planar sp^2 conformation. Second and deeper layer distortions occur, but are much smaller than those in the top layer and are less reliably extracted from the experimental measurements [7.1]. The central issue is why are the surface geometries of II-VI compounds so similar to those of III-V compounds?

To establish a quantitative measure of structural similarity, it is useful to define the concept of a bond-length-conserving top-layer rotation [7.1,4]. The

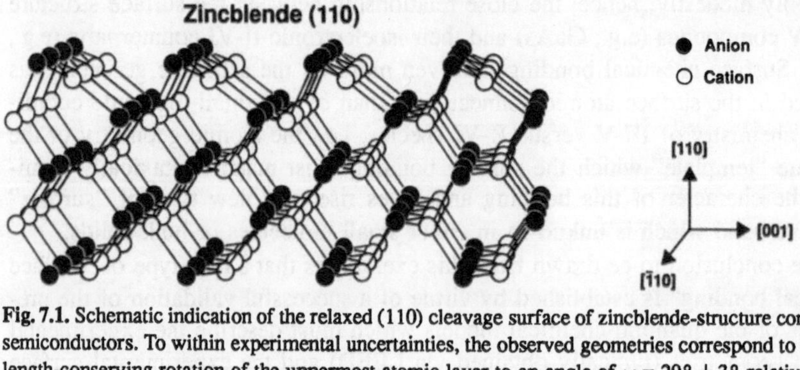

Fig. 7.1. Schematic indication of the relaxed (110) cleavage surface of zincblende-structure compound semiconductors. To within experimental uncertainties, the observed geometries correspond to a bond-length-conserving rotation of the uppermost atomic layer to an angle of $\omega = 29° \pm 3°$ relative to the unrelaxed surface plane. (Adapted from [7.3])

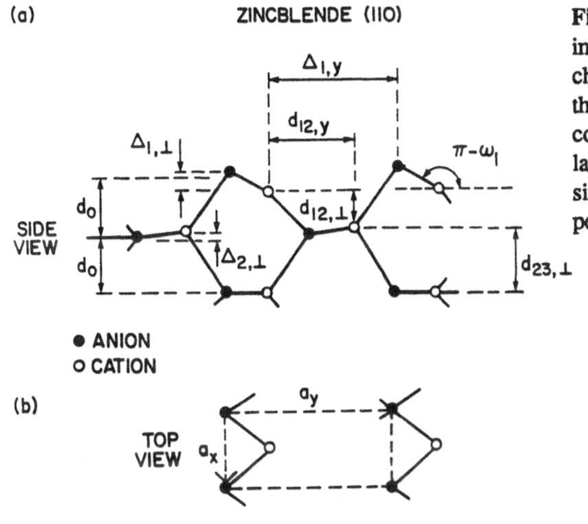

Fig. 7.2a,b. Schematic indication of the independent structural variables which characterize the atomic geometries of the (110) surfaces of zincblende-structure compound semiconductors. The top-layer tilt angle ω_1 is defined by $\omega_1 \equiv \sin^{-1}(4\Delta_{1,\perp}/a_y)$. (After [7.31], with permission)

independent surface structural variables are indicated in Fig. 7.2 [7.31]. The top layer can be relaxed in such a way that all bond lengths are held constant; the only independent structural variable is the tilt angle ω_1 between the plane of the chains of atoms in the relaxed top layer and the plane of the unrelaxed surface. The best fits to experimental LEED intensities are indicated in Table 7.1 for a variety of zincblende-structure compound semiconductors. All of the optimal structures occur in the range $26° \leq \omega_1 \leq 33°$. Moreover, only those of ZnS(110) and InSb(110) lie outside the range $27° \leq \omega_1 \leq 31°$. These structures can, of course, be refined further by permitting additional structural parameters to vary, but only small additional improvements in the quality of fit to the experimental intensites are achieved [7.1]. The results shown in Table 7.1 support the notion that to first order the atomic geometries of the (110) surfaces of all zincblende-

Table 7.1. Optimal bond-length-conserving top-layer rotated structures for the (110) surfaces of naturally occurring zincblende-structure compound semiconductors as determined from LEED intensity analysis. The structural parameters are defined in Fig. 7.2. The structures are obtained from [7.1].

Compound	Layer	Anion [Å]	Cation [Å]	$\Delta_{1,\perp}$ [Å]	$d_{12,\perp}$ [Å]	$d_{23,\perp}$ [Å]	$\Delta_{1,y}$ [Å]	$d_{12,\perp}$ [Å]	ω_1 [deg]
AlP	1	↑ 0.19	↓ 0.44	0.63	1.49	1.93	4.24	3.20	27.5
GaP	1	↑ 0.19	↓ 0.44	0.63	1.49	1.93	4.24	3.20	27.5
GaAs	1	↑ 0.20	↓ 0.49	0.69	1.51	2.00	4.42	3.34	29.0
GaSb	1	↑ 0.22	↓ 0.55	0.77	1.62	2.16	4.79	3.63	30.0
InP	1	↑ 0.21	↓ 0.52	0.73	1.55	2.08	4.60	3.48	30.0
InAs	1	↑ 0.22	↓ 0.56	0.78	1.57	2.13	4.74	3.60	31.0
InSb	1	↑ 0.23	↓ 0.65	0.88	1.64	2.29	5.12	3.89	33.0
ZnS	1	↑ 0.18	↓ 0.41	0.59	1.53	1.91	4.19	3.15	26.0
ZnSe	1	↑ 0.20	↓ 0.49	0.69	1.52	2.00	4.43	3.35	29.0
ZnTe	1	↑ 0.21	↓ 0.51	0.72	1.64	2.15	4.75	3.59	28.5
CdTe	1	↑ 0.23	↓ 0.58	0.81	1.71	2.29	5.08	3.84	30.0

structure compound semiconductors are the same and are characterized by $\omega_1 = 29° \pm 2°$, independent of the specific semiconductor material. Consideration of other experimental structure-analysis methodologies expands the uncertainties somewhat (e.g., ion scattering suggests that $\omega_1 = 29° \pm 3°$ [7.32]), but the basic conclusion remains unaltered.

As noted in Sect. 7.1, this "universal" structure of zincblende (110) surfaces has its origin in the behavior of filled anion derived surface states which are lowered in energy upon bond-length-conserving top-layer rotations and small additional relaxations. A quantitative expression of this concept is achieved using spectroscopically parametrized tight-binding models extended to encompass total energy calculations by inclusion of a phenomenological expression for electron–electron interactions determined by the measured elastic modulus, B, of the bulk crystals. Specifically, a sp^3s^* model developed for bulk III-V compounds [7.21–23] is used for these materials and for ZnSe [7.23], whereas sp^3 models developed from bulk spectroscopic data are utilized for ZnTe [7.22], ZnS [7.24], ZnO [7.34], CdTe [7.25], CdSe [7.35, 36] and CdS [7.35]. The total energy is written as the sum of an electronic part (band-structure energy E_{bs}) and an elastic part, i.e.,

$$E_{tot} = E_{bs} + \sum_{i<j} \left(U_1 \varepsilon_{ij} + U_2 \varepsilon_{ij}^2\right) , \tag{7.1}$$

in which ε_{ij} is the fractional change in the bond length away from its bulk crystalline value. E_{bs} is evaluated by summing the occupied states of the energy spectrum at "special points" [7.37]. The parameter U_1 is determined by the equilibrium conditions and U_2 is determined by the bulk elastic modulus B of the crystal, i.e.,

$$U_1 = -\left.\frac{\partial E_{bs}}{\partial \varepsilon}\right|_{\varepsilon=0} , \quad \text{and} \tag{7.2}$$

$$2U_2 = 9VB - \left.\frac{\partial^2 E_{bs}}{\partial \varepsilon^2}\right|_{\varepsilon=0} . \tag{7.3}$$

An eight-layer slab is used in most of the calculations to simulate a semi-infinite crystal. The surface atomic geometry is determined by minimizing the total energy in (7.1). The changes in the interatomic interactions due to the relaxation of the atoms in the top two layers are accounted for by use of the d^{-2} scaling law [7.38] for the orbital interactions. Details of the parametrizations and calculations may be found in [7.21–25, 34–36]. A comparison of the predicted structures with those obtained from Elastic Low-Energy Electron Diffraction (ELEED) intensity analysis is shown in Fig. 7.3 for the case of $\Delta_{1,\perp}$, see Fig. 7.2, which is the structural parameter most accurately extracted from the experimental intensities. Evidently, these models yield a quantitative prediction of the experimentally determined surface atomic geometries for the (110) surfaces of zincblende-structure compound semiconductors.

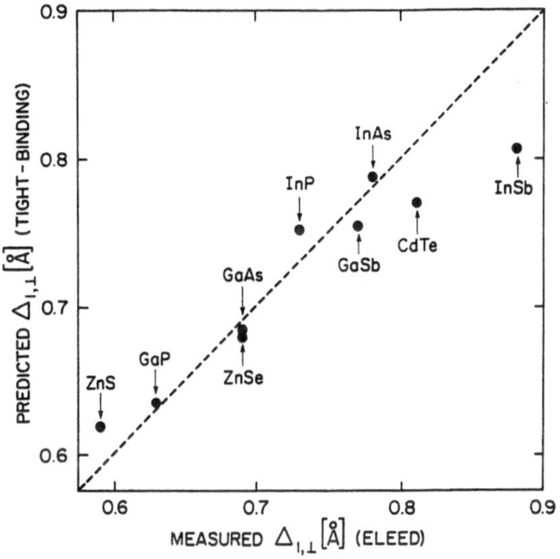

Fig. 7.3. Top-layer shear, $\Delta_{1,\perp}$, predicted by tight-binding total-energy minimization, plotted against the value determined by dynamical ELEED data analysis for the (110) surfaces of II-VI and III-V compounds crystallizing in the zincblende structure. (After [7.25], with permission)

The confirmation of the mechanism of the surface relaxation requires, however, the further prediction of the surface state eigenvalue spectra and their experimental confirmation, e.g., by ARPES. Using the tight-binding models, the surface electronic states and surface resonances are identified by examining the calculated symmetry-resolved charge densities and locating eigenfunctions which are localized within the top layers. The results are presented as surface-state energies along line segments in the two-dimensional surface Brillouin zone. For zincblende (110) the surface unit mesh and Brillouin zone are indicated in Fig. 7.4. The Brillouin zone is that area of momentum space for momenta parallel to the surface, $\boldsymbol{k}_{\|}$, which contains all of the independent values of $\boldsymbol{k}_{\|}$.

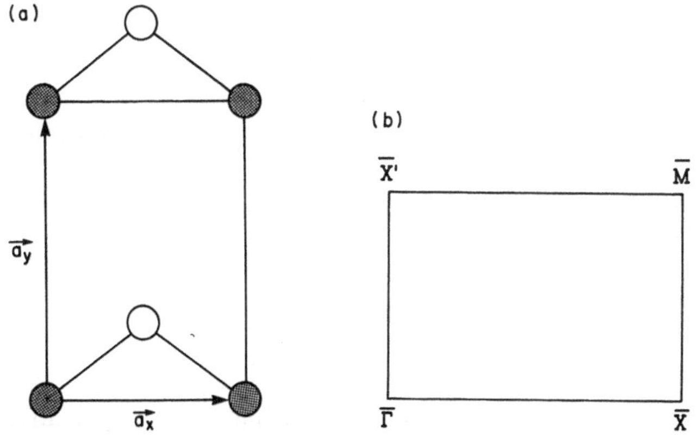

Fig. 7.4. Schematic indication of the surface unit mesh (**a**) and associated two-dimensional Brillouin zone (**b**) of the (110) surface of a zincblende-structure compound semiconductor with the surface atomic geometry specified as in Fig. 7.2

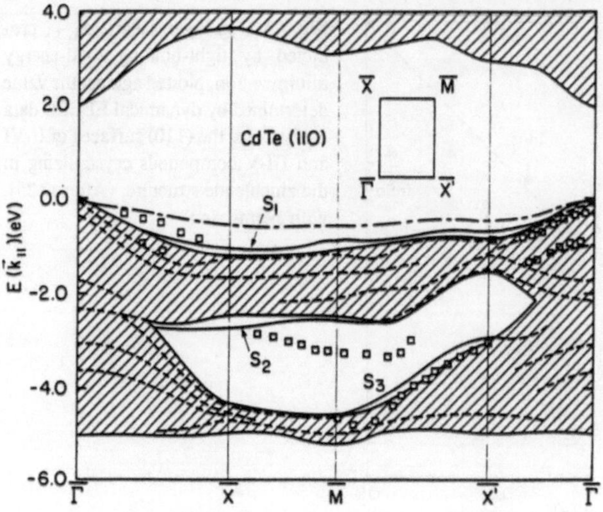

Fig. 7.5. Band structure of the (110) surface of CdTe. The surface Brillouin zone is indicated in the inset. The squares designate surface-state energies extracted from experimental angle-resolved photoemission measurements. The calculated surface-state and surface-resonance eigenvalues are shown by solid and dashed lines, respectively. The dot-dashed line designates the highest-energy occupied surface state S_1 of the ideal (110) face. (After [7.25], with permission)

Typically, energy eigenvalues versus $\boldsymbol{k}_{\parallel}$ are shown along the lines in momentum space which bound the Brillouin zone. Predicted eigenvalues for CdTe(110) are shown in this fashion in Fig. 7.5, in which they are compared with values extracted from ARPES measurements [7.25]. Shaded areas in the figure indicate the projection of the bulk energy bands on the surface Brillouin zone. Solid lines indicate eigenvalues associated with true (i.e., localized) surface states whereas dashed lines indicate those of surface resonances. The surface relaxation is driven by the lowering of the energies of the highest-energy occupied (anion derived) surface states ("S_1") from the dot-dashed line, corresponding to the unrelaxed surface, to the values indicated in the figure. The experimental surface-state energies extracted from ARPES results (designated by squares in the figure) are consistent with the predicted surface-state eigenvalues following this lowering, and hence with the proposed interpretation of the mechanism of surface relaxation. This was originally proposed and confirmed for GaAs(110) [7.1], and has been verified qualitatively for ZnSe(110) as well [7.23, 28, 39–41].

Another test of the validity of the model interpretation of zincblende (110) surface relaxation is afforded by the dynamics of surface atomic motions. In 1987, *Harten* and *Toennies* [7.29] reported the occurrence of a new "optical" branch of surface phonons for GaAs(110) along the $\bar{\Gamma}$–$\bar{X}$ line in the surface Brillouin zone with $\hbar\omega \cong 10\,\mathrm{meV}$ as shown in Fig. 7.6. Immediately thereafter the extended tight-binding model used to predict the surface atomic geometry and electronic structure of GaAs(110) [7.23] was shown also to predict both the energies and symmetries of these phonons [7.30], as noted via their label-

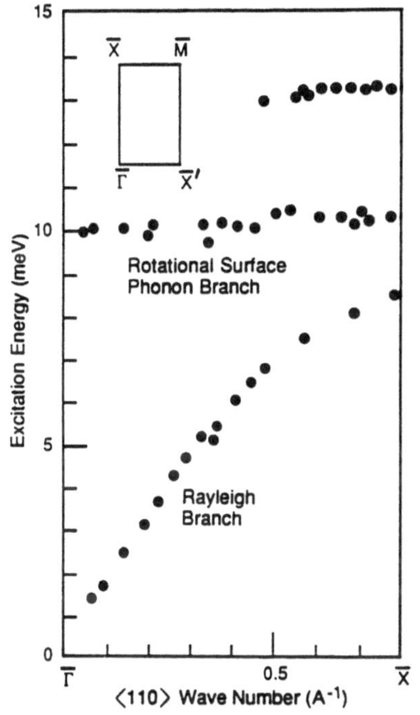

Fig. 7.6. Measured surface phonon dispersion curves for GaAs(110) along the $\bar{\Gamma}$–$\bar{X}$ line in the surface Brillouin zone. The surface temperature is 300 K. The surface Brillouin zone is shown in the inset. The predicted [7.30] rotational surface phonon energies are $\hbar\omega(\bar{\Gamma})$ = 10.7 meV and $\hbar\omega(\bar{X})$ = 12.6 meV. (Adapted from [7.29])

ing as "surface rotational phonons" in Fig. 7.6. Specifically, since the surface reconstruction is approximately a bond-length-conserving rotation, the total energy exhibits a valley along the rotation trajectory. The minimum of this valley determines the relaxed surface atomic geometry. Vibrations (rotations) along the trajectory around the energy minimum constitute the rotational surface phonons. The phonon frequencies can be determined from the total energy difference between the fully relaxed surface and the one slightly rotated from the equilibrium geometry via use of an expression obtained from the virial theorem, i.e.,

$$\tfrac{1}{2}M_a\omega^2u_a^2 + \tfrac{1}{2}M_c\omega^2u_c^2 = \Delta E_{tot} , \qquad (7.4)$$

where M_a, u_a and M_c, u_c are the mass and displacement of the top-layer anion and cation, respectively. ΔE_{tot} is the total energy change per surface unit cell; u_a and u_c are calculated from the rotated surface geometry. Hence, (7.4) determines the phonon frequency ω. The calculated range 10.7 meV $\leq \hbar\omega \leq$ 12.6 meV is in remarkably close correspondence with the results of *Harten* and *Toennies* shown in Fig. 7.6. A schematic diagram of the atomic motions associated with the rotational surface phonons is presented in Fig. 7.7.

In summary, a systematic study [7.1, 20] of the experimentally determined atomic geometries of the (110) surfaces of zincblende-structure compound semiconductors revealed the remarkable result that they all are approximately characterized as a single universal structure: a bond-length-conserving, top-layer rotation described by $\omega_1 = 29° \pm 3°$. Theoretical tight-binding models reproduced these experimental results [7.21, 23–25] and revealed that a surface-state-lowering

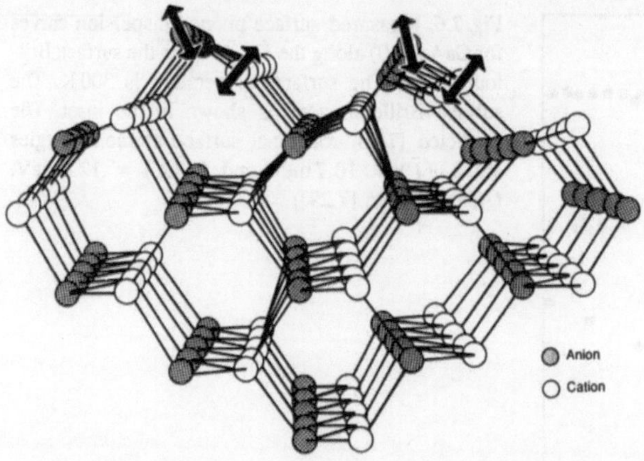

Fig. 7.7. Schematic diagram of the rotational surface phonons on the (110) surfaces of zincblende-structure compound semiconductors. The chains lie along the ⟨110⟩ direction in the surface and the normal mode displacements occur perpendicular to this direction. Motions of atoms below the top layer are much smaller than those of the top-layer atoms. (After [7.51], with permission)

phenomenon provided a unified interpretation of the commonality of the surface structures for III-V and II-VI semiconductors. The predictions of these models have been validated independently both by ARPES measurements of the energetics of the surface states in question [7.1, 25, 28, 39–41] and by the discovery via He atom diffraction measurements of a new branch of surface rotational phonons on GaAs(110) [7.29, 30]. Finally, for II-VI compound semiconductors, the resulting structures and electron energetics constitute a new type of chemical bonding driven by the constraint that the surface atomic layer must join epitaxically to the bulk.

7.3 Wurtzite Cleavage Surfaces

In contrast to zincblende-structure materials, for which (110) is the only cleavage surface, wurtzite structure compound semiconductors exhibit two cleavage faces: the $(10\bar{1}0)$ and $(11\bar{2}0)$ surfaces. Many common II-VI compounds exhibit the wurtzite structure, e.g., ZnO, ZnS, CdS and CdSe. Apart from a long historical interest in ZnO as a catalyst [7.42], a prototypical "ionic" semiconductor [7.34, 43–46], and a sensor material [7.47], attention to the surfaces of wurtzite structure compound semiconductors is recent in origin. In particular, they constitute an important test of the generality of the concepts developed in the preceding section for zincblende-structure materials, i.e., the universality of the surface atomic geometries and the surface-state-lowering mechanism for surface relaxations of II-VI compounds. Initial theoretical studies [7.5] predicted both phenomena for wurtzite cleavage faces, and recent experiments [7.3, 7–9, 36] are validating these predictions. We consider each of the two cleavage surfaces, in turn, from the perspective of ascertaining the extent to which the novel surface chemical bonding characteristic of zincblende-structure II-VI cleavage faces persists in the case of wurtzite structure II-VI compounds.

7.3.1 Wurtzite (10$\bar{1}$0)

The unrelaxed (10$\bar{1}$0) surfaces of wurtzite-structure compound semiconductors exhibit an anion-cation dimer structure with each surface species bonded to two atoms in the layer benath. Diagrams and descriptions of this surface may be found in early reviews [7.48, 49]. Upon cleavage, however, these surfaces relax due to the redistribution of dangling-bond charge via approximately bond-length-conserving top-layer rotations analogous to the zincblende (110) surface [7.3, 5]. A picture of the predicted [7.5] relaxed surface is given in Fig. 7.8. The independent surface structure parameters and associated reciprocal lattice (i.e., normal incidence LEED pattern) are indicated in Fig. 7.9. The qualitative features of the predicted relaxation have been confirmed by an early (1978) study of ZnO(10$\bar{1}$0) [7.6, 44] which embodies techniques and data that have been greatly improved in the ensuing decade. A study of CdSe(10$\bar{1}$0) was recently undertaken using both LEED [7.7, 8] and LEPD [7.8, 9]. The appropriate dimensionless tilt angle ω shown in Fig. 7.9, is predicted to be 18° for CdSe(10$\bar{1}$0) [7.35]. The LEED intensity analysis yields ω = 22° $\pm$ 4° [7.7, 8] and the LEPD analysis gives ω = 15° $\pm$ 5° [7.8, 9]. Both analyses give results compatible with the theoretical prediction.

An examination of the surface-state eigenvalue spectrum via ARPES also has been reported [7.36] with the results shown in Fig. 7.10 (using the conventions described in Sect. 7.2). The predictions are in satisfactory correspondence with the surface-state energies determined from the ARPES data, indicating that the surface-state lowering mechanism is appropriate for CdSe(10$\bar{1}$0) as well as for zincblende (110) surfaces. The rotational surface phonon energies have been

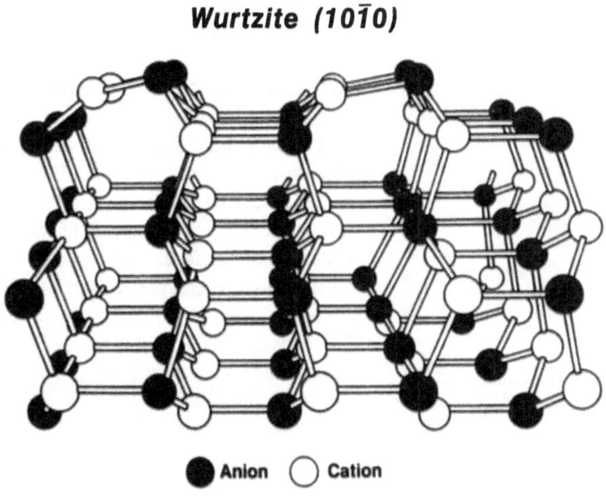

Fig. 7.8. Schematic indication of the relaxed (10$\bar{1}$0) cleavage surface of wurtzite-structure compound semiconductors. This structure has been confirmed experimentally for ZnO and CdSe. (After [7.3], with permission)

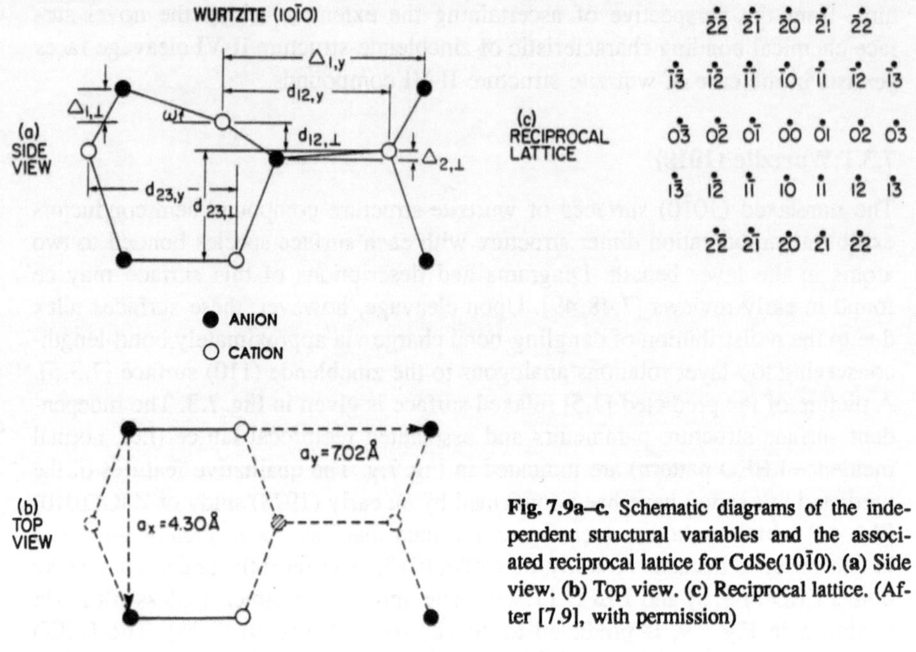

Fig. 7.9a–c. Schematic diagrams of the independent structural variables and the associated reciprocal lattice for CdSe(10$\bar{1}$0). (a) Side view. (b) Top view. (c) Reciprocal lattice. (After [7.9], with permission)

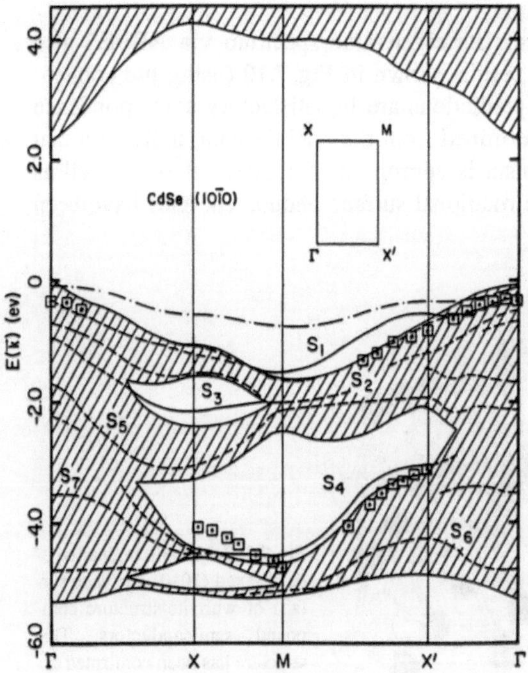

Fig. 7.10. Surface electronic band structure of CdSe(10$\bar{1}$0). The surface Brillouin zone is indicated in the inset. The surface-state eigenvalues are denoted by solid lines and the surface-resonance eigenvalues by the dashed lines. The dot-dashed line designates the energies of surface states S_1 characteristic of the unreconstructed surface. The squares are the surface-state energies extracted from angle-resolved photoelectron spectra. The states are labeled according to [7.35]. (After [7.36], with permission)

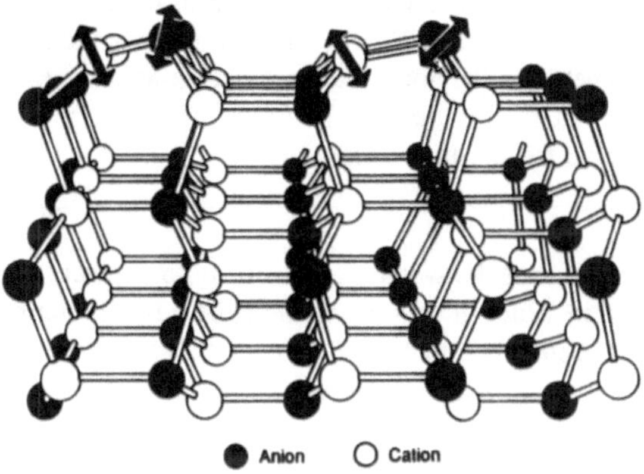

Fig. 7.11. Schematic indication of the rotational surface phonons on the (10$\bar{1}$0) surfaces of wurtzite-structure compound semiconductors. The normal mode displacements lie in the plane defined by the surface normal and the bonds between the surface anion-cation "dimers". (After [7.51], with permission)

calculated for the wurtzite (10$\bar{1}$0) surfaces [7.50], but have not yet been observed experimentally. A schematic diagram of the resulting motions is shown in Fig. 7.11 [7.51].

7.3.2 Wurtzite (11$\bar{2}$0)

The unrelaxed wurtzite (11$\bar{2}$0) surface is reminiscent of the zincblende (110) surface. It consists of anion-cation chains such that each surface species exhibits two surface bonds and one subsurface bond. In addition, however, wurtzite (11$\bar{2}$0) exhibits a new feature: four (rather than two) atoms per surface unit cell and a glide-plane symmetry which is observed via missing spots in the LEED pattern [7.44, 49]. Indeed, the observation of the missing spots led to early speculation that ZnO(11$\bar{2}$0) might be unrelaxed [7.44], although later calculations [7.34] revealed the occurrence of relaxations analogous to those of zincblende (110) and wurtzite (10$\bar{1}$0) which preserved the glide-plane symmetry.

The predicted relaxed surface structure for the wurtzite (11$\bar{2}$0) surface is illustrated in Fig. 7.12. The associated structural parameters and surface recipro-

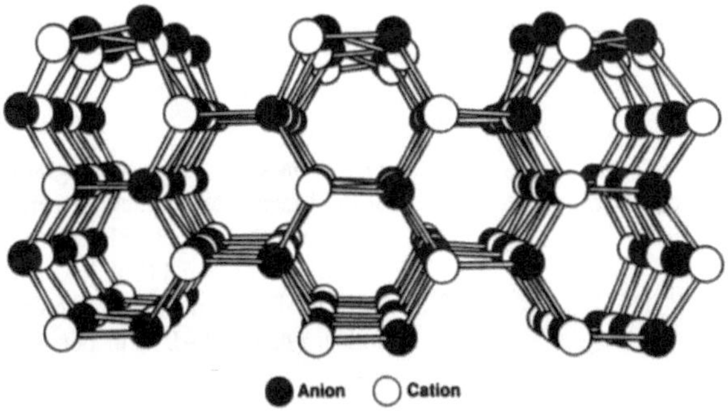

Fig. 7.12. Schematic indication of the relaxed (11$\bar{2}$0) cleavage surface of wurtzite structure compound semiconductors. This structure has been confirmed experimentally for CdSe(10$\bar{2}$0) [7.8, 9]. (After [7.3], with permission)

cal lattice are specified in Fig. 7.13. The bond-length-conserving relaxations are more complicated than for the $(10\bar{1}0)$ surface because the atoms in the top layer pucker, causing it to become nonplanar (Fig. 7.13). The dimensionless structural parameter in this case is the angle between the normal to each cation-anion-cation or anion-cation-anion triplet and that of the unrelaxed surface. If $\hat{\boldsymbol{n}}$ designates the former and $\hat{\boldsymbol{z}}$ the latter, we have for the polar angle ω

$$\hat{\boldsymbol{n}} \cdot \hat{\boldsymbol{z}} = \cos\omega \, . \tag{7.5}$$

This leads to a complicated dependence between $\Delta_{1,\perp}$ and ω to replace the simple expression $\Delta_{1,\perp} = d \sin\omega$, in which d designates the Cd-Se bond length, for the $(10\bar{1}0)$ surface. It is of the general form

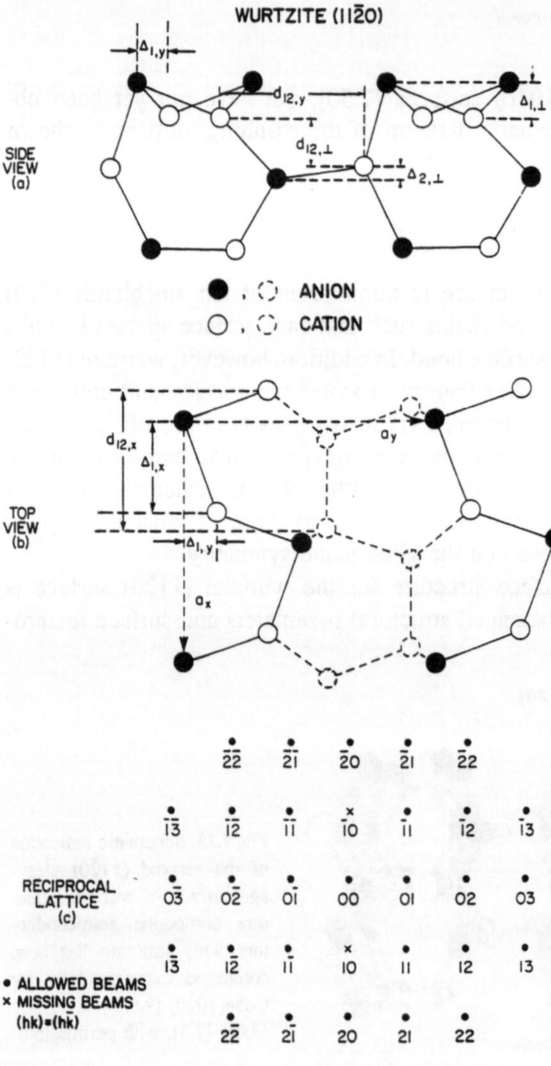

Fig. 7.13a–c. Schematic diagram of the independent structural variables and the associated reciprocal lattice for CdSe$(11\bar{2}0)$. (a) Side view. (b) Top view. (c) Reciprocal lattice. (After [7.9], with permission)

$$\Delta_{1,\perp} = f(\Delta_{1,x}, \Delta_{1,y}) \tan\omega \ , \qquad (7.6)$$

in which variations in $\Delta_{1,\perp}$, $\Delta_{1,x}$ and $\Delta_{1,y}$ are constrained to keep all the bond lengths constant. These constraints can be satisfied for $0 \leq \omega \leq 45°$. Predicted values occur in the vicinity of $\omega = 32°$ [7.9, 35]. The LEED analysis for CdSe(11$\bar{2}$0) yields $\omega = 33° \pm 4°$ [7.8] and the LEPD analysis gives $\omega = 27° \pm 7°$ [7.8, 9]. Therefore both LEED and LEPD give structures for CdSe(11$\bar{2}$0), which are in quantitative agreement with the concept that it exhibits a bond-length conserving top-layer rotational relaxation leading to a distorted p^3 conformation for the anion and sp^2 conformation for the cation. Since no structural data for wurtzite (11$\bar{2}$0) were available prior to the model predictions [7.35], the experimental results for CdSe(11$\bar{2}$0) constitute an *a priori* verification of the surface-state-lowering mechanism for the cleavage surfaces of tetrahedrally coordinated compound semiconductors. The predicted [7.5] near-invariance of ω with varying materials has not yet been tested experimentally.

Calculations of surface-state energies for the (11$\bar{2}$0) surfaces of ZnO [7.34], ZnS [7.24], ZnSe [7.23], CdS [7.35] and CdSe [7.35, 36] indicate that the surface states characteristic of the truncated bulk surface disappear entirely upon relaxation, becoming surface resonances. Although no ARPES measurements have been reported for (11$\bar{2}$0) compound semiconductor surfaces, angle-integrated photoemission energy distributions suggest the occurrence of strong surface resonances at the top of the valence band [7.36]. No surface phonon calculations have been performed for the (11$\bar{2}$0) surface. Surface rotational phonons analogous to those predicted for the (10$\bar{1}$0) surfaces are expected [7.50, 51].

7.4 Adsorption on Zincblende (110) Surfaces

Although there is an enormous literature on adsorbates on the zincblende (110) surfaces, relatively few of these systems exhibit epitaxical growth. Fewer still have been the subjects of quantitative surface structure analyses. In this section, we consider two of the most extensively characterized systems, i.e., GaAs(110)-p(1 × 1)-Sb and GaAs(110)-p(1 × 1)-Al, as examples of the limiting cases of self-passivating and reactive chemisorption, respectively. The latter example also illustrates the profound consequences of surface processing on surface structure via the generation of metastable systems. Our discussion follows that given earlier by *Duke* [7.18].

7.4.1 Epitaxically Constrained Adsorbate Bonding

In addition to novel surface chemical bonding on clean surfaces, geometrical constraints imposed by epitaxical growth also induce new types of adsorbate–substrate bonds [7.18, 52]. One of the most extensively studied semiconductor adsorbate systems is the p(1 × 1) Sb saturated monolayer structure on the (110) surfaces of GaAs [7.52–62] and InP [7.52, 63–65]. The Sb occupy surface sites in zig-zag chains roughly where the top-layer anions and cations would have been. The Sb species are larger than Ga and As, however, so that in order to accomodate the large Sb-Sb bond length (2.8 Å), the angle between neighboring Sb species closes to 91° from its value of approximately 109° in GaAs. In this structure, three of the Sb electrons participate in bonds and the remaining two generate an *sp* hybrid lone pair charge density which is analogous to the charge density for bonding to the substrate species but points out of the surface. Such bonding is similar to that of S_8 and Se_8. The valences of both surface Sb species are fully saturated at monolayer coverage so that the main driving force for the relaxation of the GaAs(110) substrate is eliminated by the Sb–substrate bonding. Indeed, both the Sb chains and the GaAs(110) substrate are nearly unrelaxed. The Sb bonded to the substrate As is lowered by 0.1 Å relative to that bonded to the substrate Ga and the uppermost As is relaxed downward by a corresponding 0.1 Å [7.54]. These are small distortions relative to the relaxations on a clean GaAs(110) surface, however, and reveal the profound influence of a saturated valence top-layer structure on the corresponding atomic geometry.

Unusual features of the Sb–substrate bonding are the p^2 local bonding conformation within the Sb chain (i.e., approximately 90° Sb-Sb-Sb bond angles) and distorted sp^3 local Sb conformation (approximately 105° bond angles) relative to the substrate. Since there are not enough independent p orbitals to support simultaneously both types of local valence, the detailed nature of the bonding is unlike that in either small molecules or the bulk tetrahedrally coordinated solids. A detailed study [7.52] revealed that the Sb–substrate bond is appropriately regarded as a hybrid between π orbitals of a p^2 bonded Sb chain and sp^3 dangling-bond orbitals of an undistorted GaAs substrate. The occurrence of this bond is driven by the combination of the large size of the Sb relative to the GaAs substrate, the saturated valence of the Sb monolayer, and the occurrence of epitaxial ordering of the monolayer relative to the substrate. Thus, the π-sp^3 hybrid bond is a unique type of surface bonding associated with the constraint of epitaxical monolayer growth on the compound semiconductor substrate. Direct confirmation of this type of bonding has been given during the past few years by observing the surface-state bands characteristic of the bonding charge density via angle resolved photoemission spectroscopy for $p(1 \times 1)$ Sb monolayers on GaAs [7.58–60, 62], GaP [7.60], and InP [7.63]. These states are the analogs for the Sb saturated monolayer systems of the back bonding rehybridized surface states which drive the surface rehybridization on the clean cleavage faces of tetrahedrally coordinated compound semiconductors [7.52].

7.4.2 Process-Dependent Bonding: Reactive Chemisorption

Al on GaAs(110) is by far the most extensively studied reactive chemisorption system for tetrahedrally coordinated compound semiconductors. Work performed prior to the end of 1983 has been reviewed by *Zunger* [7.66] from a theoretical perspective as well as by *Skeath* et al. [7.67] and by *Bonapace* et al. [7.68] from an experimental one. Herein we only recapitulate briefly the highlights of studies of the growth habits of Al on GaAs(110) and the consequences of the insight obtained from these studies for the design of injecting metall–semiconductor contacts.

At room temperature, the deposition of Al on GaAs(110) is thought to proceed via four steps with increasing coverage. At very low coverages, $\theta \ll 0.1\,\mathrm{ML}$, the aluminum exhibits mobile chemisorption, probably in twofold sites [7.69–72]. As the coverage increases, Al "clusters" form, which themselves are mobile [7.66, 67, 73–75]. Both of these processes compete with Al-Ga place exchange, which is hindered by a high kinetic activation barrier but which can occur at surface imperfections for low coverages. As the coverage is increased further (to $\theta \sim 1\,\mathrm{ML}$), the energy gained by the Al coalescing into clusters presumably provides enough energy to surmount this activation barrier locally and Al-Ga place exchange begins on a more extensive scale, resulting in free Ga appearing at the interface. At still higher coverages ($\theta > 1\,\mathrm{ML}$), the Al clusters coalesce to form islands which merge to form an Al overlayer (contaminated with Ga) because the "AlAs" interface buffer layer acts as a diffusion barrier which reduces the Al-Ga place exchange in the substrate [7.76]. At lower substrate temperatures (e.g., $T \sim 100\,\mathrm{K}$) this sequence of steps is altered by virtue of the lower surface diffusion of the adsorbed Al, leading to more homogeneous Al overlayers which distort the underlying atomic geometry of the substrate. Upon warming to room temperature at intermediate coverages ($1\,\mathrm{ML} \lesssim \theta \lesssim 2\,\mathrm{ML}$) the surface reverts to its room-temperature-deposition form of mobile islands surrounded by patches of relaxed bare GaAs(110). Annealing to 450°C produces Al-Ga place exchange and hence the formation of a few atomic layers of Al-rich $Al_xGa_{1-x}As$ [7.76–78]. The exact composition and structure of these layers depends on the initial Al coverage.

LEED studies of the morphology of Al adsorbed on GaAs(110) have been performed for both room-temperature [7.67, 76–78] and low-temperature [7.68] deposition. At room temperature the Al exhibits Volmer–Weber [7.79] growth, forming clusters at low coverages which with increasing coverage grow and coalesce to form (three-dimensional) islands. Even when the surface is nearly completely covered ($10\,\mathrm{ML} \lesssim \theta \lesssim 20\,\mathrm{ML}$) all ordered portions produce the relaxed GaAs(110) clean-surface LEED intensities, which could be characteristic either of free patches of GaAs(110) or of a disordered Al overlayer on a weakly perturbed GaAs(110) substrate. The coverage dependence of the intensities of substrate Ga and As Auger electron lines suggests the former rather than the latter, possibly accompanied by interdiffusion at some regions of the island–substrate interfaces. In contrast, low-temperature (100 K) deposition yields a much more homogeneous growth at low coverages, characterized by Al island growth on a more-or-less uniform interfacial Al overlayer [7.68]. At $\theta \lesssim 2\,\mathrm{ML}$, a (1×1) LEED intensity pattern is observed, but the intensities are characteristic of unrelaxed GaAs(110), indicating either that the interfacial overlayer has restored the GaAs(110) surface atomic geometry to its truncated bulk value, or that the interfacial layer is so disordered by the Al adsorption that only deeper layers of the substrate contribute to the LEED intensities. These truncated-bulk GaAs(110)

LEED intensities are observed for 2 ML $\lesssim \theta \lesssim$ 15 ML. For $\theta > 25$ ML only the LEED pattern of the epitaxical Al islands is observed.

Quantitative LEED intensity analyses have been given for the ordered GaAs(110)-p(1 × 1)-Al(θ) structures prepared by deposition of the nominal coverage of Al followed by vacuum thermal annealing [7.76–78]. Examination of the low-coverage ($\theta \lesssim 1$ ML) structures focused upon evaluating the numerous proposed models of chemisorbed Al geometries [7.80–86]. None of them proved compatible with the measured LEED intensities. The LEED intensity analysis revealed, instead, that for $\theta = 0.5$ ML Al replaced Ga in the second layer beneath the interface [7.76, 77]: a controversial conclusion which subsequently was confirmed both by energy-minimization calculations [7.69–71] and by core-level photoemission spectroscopy. The low-coverage analysis subsequently was extended to encompass a series of GaAs(110)-p(1 × 1)-Al(θ) structures for $0 \leq \theta \leq 8.5$ ML [7.76]. With increasing coverage, θ, Al first displaces Ga in the second layer, then in the third and finally in the first, forming a thin expitaxial layer of AlAs(110) on GaAs(110). A similar result occurs for Al on GaP(110) [7.87].

These results admit a simple interpretation which has provided the basis for a practical technology of metal source and drain contacts on thin film transistors [7.88]. Upon annealing Al deposited on GaAs(110), Al replaces the Ga first in the second atomic layer, next in the third, then in the first, revealing a complicated interplay between the energetically favored bulk Al-for-Ga replacement reaction and the kinetic limitations on Al and Ga diffusion. Moreover, the AlAs surface layer forms a cap on the GaAs substrate, hindering further diffusion in accordance with known activation energies [7.89] and measurements on Au/GaAs contacts [7.90]. Thus, we are led to the concept that the deposition and reaction of a few monolayers of reactive metal onto a semiconductor substrate could lead to a diffusion barrier at the interface and to profound changes in its electrical properties.

Studies of the electrical characteristics of Au contacts on InP(110) [7.91], GaAs(110) [7.91], CdS(10$\bar{1}$0) [7.92] and CdSe(10$\bar{1}$0) [7.92] confirm the expectations based on the insight gleaned from the examination of replacement reactions on GaAs(110). The presence of a few monolayers of reactive metal (e.g., Al or Ti) prior to deposition of the Au suppresses the diffusion of the semiconductor species through the Au overlayer and reduces the asymmetry of the I-V characteristic of the Schottky barrier, making it more "ohmic". Indeed, a few monolayers of Al on CdSe(10$\bar{1}$0) renders the resulting Au/Al/CdSe composite contact completely ohmic. The origin of the dramatic change in electrical characteristics is the fundamental difference between chemistry of the reactive metal surface compound (i.e., "Al_2Se_3") and that of the underlying substrate (i.e., CdSe). Thus, studies of semiconductor–metal interface structure led to a new technique for Schottky barrier height modulation and stabilization. This technique was used to fabricate the source and drain contacts on CdSe thin film transistor arrays for liquid-crystal displays in the early 1980s.

7.5 Synopsis

As techniques for surface structure determinations have matured, attention has focused increasingly on the consequences rather than the details of such determinations. The cleavage surface of tetrahedrally coordinated compound semiconductors have proven fertile ground for the discovery of new phenomena, including the universality of each cleavage surface's atomic geometry, the dominant role of surface states in creating surface reconstructions, the generation of new branches of surface phonons via surface reconstruction, the occurrence of new types of epitaxically constrained chemical bonding for II-VI compounds and overlayer systems, and the critical role of process conditions in producing the atomic geometries resulting from reactive chemisorption. These findings have emerged from the simultaneous confluence of four elements: quantitative surface structure analysis (mostly via LEED and ion scattering); a common theory of surface atomic geometry, surface atomic vibrations, and surface electronic structure; the ability to extract electronic surface state dispersion curves from angle-dependent photoemission analyses; and the measurement of surface phonon dispersion via He atom diffraction. Thus, it is the ability to make quantitative predictions of the results of a variety of surface-sensitive structural and spectroscopic measurements that permits the systematic accumulation of new insights concerning the nature and consequences of surface chemical bonding. This chapter has documented the value and viability of such an approach to the structure and chemistry of the cleavage surfaces of tetrahedrally coordinated compound semiconductors, thereby truly initiating "a new era of materials science".

Acknowledgements. This chapter was written while the author was at Pacific Northwest Laboratory, Richland, WA,which is operated by Battelle Memorial Institute for the U.S. Department of Energy under Contract DE-AC06-76-RLO 1830.

References

7.1 C.B. Duke: In *Surface Properties of Electronic Materials*, ed. by D.A. King, D.P. Woodruff (Elsevier, Amsterdam 1988) pp. 69–118
7.2 C.B. Duke: J. Vac. Sci. Technol. A **6**, 1957 (1988)
7.3 C.B. Duke, Y.R. Wang: J. Vac. Sci. Technol. B **6**, 1440 (1988)
7.4 A.R. Lubinsky, C.B. Duke, B.W. Lee, P. Mark: Phys. Rev. Lett. **36**, 1058 (1976)
7.5 C.B. Duke, Y.R. Wang: J. Vac. Sci. Technol. A **6**, 692 (1988)
7.6 C.B. Duke, R.J. Meyer, A. Paton, P. Mark: Phys. Rev. B **18**, 4225 (1978)
7.7 C.B. Duke, A. Paton, Y.R. Wang, K. Stiles, A. Kahn: Surf. Sci. **197**, 11 (1988)
7.8 T.N. Horsky, G.R. Brandes, K.F. Canter, C.B. Duke, S.F. Horng, A. Kahn, D.L. Lessor, A.P. Mills, Jr., A. Paton, K. Stevens, K. Stiles: Phys. Rev. Lett. **62**, 1876 (1989)
7.9 C.B. Duke, D.L. Lessor, T.N. Horsky, G. Brandes, K.F. Canter, P.H. Lippel, A.P. Mills, Jr., A. Paton, Y.R. Wang: J. Vac. Sci. Technol. A **7**, 2030 (1989)
7.10 R.E. Schlier, H.E. Farnsworth: J. Chem. Phys. **30**, 917 (1959)
7.11 A. Kahn: Surf Sci. Rep. **3**, 193 (1983)

7.12 K. Tagayanagi, Y. Tanishiro, S. Takahashi, M. Takahashi: Surf. Sci. **164**, 367 (1985)
7.13 K. Tagayanagi: In *Solvay Conference on Surface Science*, ed. by F.W. de Wette, Springer Ser. Surf. Sci., Vol. 14 (Springer, Berlin, Heidelberg 1988) pp. 55–58
7.14 S.Y. Tong, H. Huang, C.M. Wei, W.E. Packard, F.K. Men, G. Glander, M.B. Webb: J. Vac. Sci. Technol. A **6**, 615 (1988)
7.15 I.K. Robinson: J. Vac. Sci. Technol. A **6**, 1966 (1988)
7.16 I.K. Robinson, W.K. Waskiewicz, P.H. Fuoss, L.J. Norton: Phys. Rev. B **37**, 4325 (1988)
7.17 C.B. Duke: In *Solvay Conference on Surface Science*, ed. by F.W. de Wette, Springer Ser. Surf. Sci., Vol. 14 (Springer, Berlin, Heidelberg 1988) pp. 361–365
7.18 C.B. Duke: In *Atomic and Molecular Processing of Electronic and Ceramic Materials*, ed. by I.A. Aksay, G.L. McVay, T.G. Stroebe, J.F. Wagner (Materials Research Society, Pittsburgh, PA 1987) pp. 3–10
7.19 S.P. Chen, A.F. Voter, D.J. Srolovitz: Phys. Rev. Lett. **57**, 1308 (1986)
7.20 C.B. Duke: J. Vac. Sci. Technol. B **1**, 732 (1983)
7.21 C. Mailhiot, C.B. Duke, D.J. Chadi: Surf. Sci. **149**, 366 (1985)
7.22 D.J. Chadi: Phys. Rev. B **19**, 2074 (1979)
7.23 Y.R. Wang, C.B. Duke, C. Mailhiot: Surf. Sci. **188**, L708 (1987)
7.24 Y.R. Wang, C.B. Duke: Phys. Rev. B **36**, 2763 (1987)
7.25 Y.R. Wang, C.B. Duke, K.O. Magnusson, S.A. Flodström: Surf. Sci. **205**, L760 (1988)
7.26 A.C. Ferraz, G.P. Srivastava: Surf. Sci. **182**, 161 (1987)
7.27 A.C. Ferraz, G.P. Srivastava: J. Phys. C **19**, 5987 (1986)
7.28 R.P. Beres, R.E. Allen, J.D. Dow: Phys. Rev. B **26**, 769 (1982)
7.29 U. Harten, P.J. Toennies: Europhys. Lett. **4**, 833 (1987)
7.30 Y.R. Wang, C.B. Duke: Surf. Sci. **205**, L755 (1988)
7.31 C.B. Duke, C. Mailhiot, A. Paton, D.J. Chadi, A. Kahn: J. Vac. Sci. Technol. B **3**, 1087 (1985)
7.32 L. Smit, J.F. van der Veen: Surf. Sci. **166**, 183 (1986)
7.33 P. Vogel, H.P. Hjalmarson, J.D. Dow: J. Phys. Chem. Solids **44**, 365 (1983)
7.34 Y.R. Wang, C.B. Duke: Surf. Sci. **192**, 309 (1987)
7.35 Y.R. Wang, C.B. Duke: Phys. Rev. B **37**, 6417 (1988)
7.36 Y.R. Wang, C.B. Duke, K. Stevens, A. Kahn, K.O. Magnusson, S.A. Flodström: Surf. Sci. **206**, L817 (1988)
7.37 S.L. Cunningham: Phys. Rev. B **10**, 4988 (1974)
7.38 W.A. Harrison: *Electronic Structure and the Properties of Solids* (Freeman, San Francisco 1980)
7.39 C. Mailhiot, C.B. Duke, Y.C. Chang: Phys. Rev. B **30**, 1109 (1984)
7.40 A. Ebina, T. Unno, Y. Suda, H. Koinuma, T. Takahashi: J. Vac. Sci. Technol. **19**, 301 (1981)
7.41 T. Takahashi, A. Ebina: Appl. Surf. Sci. **11/12**, 268 (1982)
7.42 W. Göpel: Ber. Bunsenges. Phys. Chem. **82**, 744 (1978)
7.43 J.D. Devine, A. Willis, W.R. Bottoms, P. Mark: Surf. Sci. **29**, 144 (1972)
7.44 A.R. Lubinsky, C.B. Duke, S.C. Chang, B.W. Lee, P. Mark: J. Vac. Sci. Technol. **13**, 189 (1976)
7.45 I. Ivanov, J. Pollmann: Phys. Rev. B **24**, 7275 (1981)
7.46 W. Göpel, J. Pollmann, I. Ivanov, B. Reihl: Phys. Rev. B **26**, 3144 (1982)
7.47 W. Göpel: Prog. Surf. Sci. **20**, 9 (1985)
7.48 C.B. Duke: J. Vac. Sci. Technol. **14**, 870 (1977)
7.49 C.B. Duke: Crit. Rev. Solid State Mater. Sci. **8**, 69 (1978)
7.50 Y.R. Wang, C.B. Duke: Phys. Rev. B **39**, 5569 (1989)
7.51 C.B. Duke, Y.R. Wang: J. Vac. Sci. Technol. B **7**, 1027 (1989)
7.52 C. Mailhiot, C.B. Duke, D.J. Chadi: Phys. Rev. Lett. **53**, 2114 (1984); Phys. Rev. B **31**, 2213 (1985)
7.53 J. Carelli, A. Kahn: Surf. Sci. **116**, 380 (1982)
7.54 C.B. Duke, A. Paton, W.K. Ford, A. Kahn, J. Carelli: Phys. Rev. B **26**, 803 (1982)
7.55 C.M. Bertoni, C. Calendra, F. Manghi, E. Molinari: Phys. Rev. B **27**, 1251 (1983)

7.56 K. Li, A. Kahn: J. Vac. Sci. Technol. A **4**, 958 (1986)
7.57 W. Pletschen, N. Esser, H. Munder, D. Zahn, J. Geurts, W. Richter: Surf. Sci. **178**, 140 (1986)
7.58 M. Mattern-Klossen, R. Strümpler, H. Lüth: Phys. Rev. B **33**, 2259 (1986)
7.59 A. Tulke, M. Mattern-Klossen, H. Lüth: Solid State Commun. **59**, 303 (1986)
7.60 A. Tulke, H. Lüth: Surf. Sci. **178**, 131 (1986)
7.61 R. Strümpler, H. Lüth: Surf. Sci. **182**, 545 (1987)
7.62 P. Martensson, G.V. Hansson, M. Lähdeniemi, K.O. Magnusson, S. Wiklund, J.M. Nicholls: Phys. Rev. B **33**, 7399 (1986)
7.63 C. Maani, A.C. McKinley, R.H. Williams: J. Phys. C **18**, 4975 (1985)
7.64 C.B. Duke, C. Mailhiot, A. Paton, K. Li, C. Bonapace, A. Kahn: Surf. Sci. **163**, 391 (1985)
7.65 D. Zahn, N. Esser, W. Pletschen, J. Geurts, W. Richter: Surf. Sci. **168**, 823 (1986)
7.66 A. Zunger: Thin Solid Films **104**, 301 (1983)
7.67 P. Skeath, I. Lindau, C.Y. Su, W.E. Spicer: Phys. Rev. B **28**, 7051 (1983)
7.68 C.R. Bonapace, K. Li, A. Kahn: J. de Phys. **45 C5**, 409 (1984)
7.69 J. Ihm, J.D. Joannopoulos: Phys. Rev. Lett. **47**, 679 (1981)
7.70 J. Ihm, J.D. Joannopoulos: J. Vac. Sci. Technol. **21**, 340 (1982)
7.71 J. Ihm, J.D. Joannopoulos: Phys. Rev. B **26**, 4429 (1982)
7.72 R.R. Daniels, A.D. Katnani, T-X Zhao, G. Margaritondo, A. Zunger: Phys. Rev. Lett. **49**, 895 (1982)
7.73 A. Zunger: Phys. Rev. B **24**, 4372 (1981)
7.74 A. Zunger: J. Vac. Sci. Technol. **19**, 690 (1981)
7.75 P. Skeath, I. Lindau, C.Y. Su, P.W. Chye, W.E. Spicer: J. Vac. Sci. Technol. **17**, 511 (1980)
7.76 A. Kahn, J. Carelli, D. Kanani, C.B. Duke, A. Paton, L.J. Brillson: J. Vac. Sci. Technol. **19**, 331 (1981)
7.77 C.B. Duke, A. Paton, R.J. Meyer, L.J. Brillson, A. Kahn, D. Kanani, J. Carelli, J.L. Yeh, G. Margaritondo, A.D. Katani: Phys. Rev. Lett. **46**, 440 (1981)
7.78 A. Kahn, D. Kanani, J. Carelli, J.L. Yeh, C.B. Duke, R.J. Meyer, A. Paton, L.J. Brillson: J. Vac. Sci. Technol. **18**, 792 (1981)
7.79 R. Ludeke: J. Vac. Sci. Technol. B **2**, 400 (1984)
7.80 J.R. Chelikowsky, S.G. Louie, M.L. Cohen: Solid State Commun. **20**, 641 (1976)
7.81 L.J. Brillson, R.Z. Bachrach, R.S. Bauer, J. McMenamin: Phys. Rev. Lett. **42**, 397 (1979)
7.82 D.J. Chadi, R.Z. Bachrach: J. Vac. Sci. Technol. **16**, 1159 (1979)
7.83 E.J. Mele, J.D. Joannopoulos: Phys. Rev. Lett. **42**, 1094 (1979)
7.84 E.J. Mele, J.D. Joannopoulos: J. Vac. Sci. Technol. **16**, 1154 (1979)
7.85 J.J. Barton, C.A. Swarts, W.A. Goddard, III, T.C. McGill: J. Vac. Sci. Technol. **17**, 164 (1980)
7.86 C.A. Swarts, J.J. Barton, W.A. Goddard, III, T.C. McGill: J. Vac. Sci. Technol. **17**, 869 (1980)
7.87 A. Kahn, C.R. Bonapace, C.B. Duke, A. Paton: J. Vac. Sci. Technol. B **1**, 613 (1983)
7.88 C.B. Duke: Appl. Surf. Sci. **11/12**, 1 (1982)
7.89 R.M. Flemming, D.B. McWhan, A.C. Gossard, W. Wiegmann, R.A. Logan: J. Appl. Phys. **51**, 357 (1980)
7.90 L.J. Brillson, G. Margaritondo, N.G. Stoffel: Phys. Rev. Lett. **44**, 667 (1980)
7.91 L.J. Brillson, C.F. Brucker, A.D. Katnani, N.G. Stoffel, G. Margaritondo: Appl. Phys. Lett. **38**, 784 (1981)
7.92 C.F. Brucker, L.J. Brillson: Appl. Phys. Lett. **39**, 67 (1981)

8. Formation and Properties of Metal–Semiconductor Interfaces

John H. Weaver

With 45 Figures

The properties of interfaces formed by the deposition of metal atoms onto semiconductor surfaces have been studied extensively. The development of experimental techniques that allow investigations with high spatial and energy resolution has advanced the understanding of surface bonding, chemical reactions, intermixing, and Schottky barrier formation [8.1–26]. At the same time, our understanding of the processes and the mechanisms that give rise to the various atom profiles is still far from complete. Moreover, there is no consensus about the physical mechanisms responsible for the development of the Schottky barrier.

When describing metal–semiconductor junctions, it is critical to identify the existence of different bonding configurations parallel and perpendicular to the surface. One of the primary goals of this chapter is to examine atom distributions for metal–III-V semiconductor interfaces as a function of the substrate temperature, with and without annealing. Another is to identify chemical changes that are associated with the evolution of the interface, i.e., changes related to the number of adatoms present on a surface. A third is to examine Schottky barrier models in the context of surface chemical reaction, the processes of overlayer growth, temperature, and other system parameters.

In Sect. 8.1 we will review photoelectron spectroscopy and the information gained from it. Many of the conclusions of this chapter are drawn from analysis of synchrotron radiation and X-ray photoemission results, and Sect. 8.1 serves as a very brief summary of the technique as applied to interface evolution. In that section we will also describe experimental procedures, the specifics of the samples studied, and the procedures followed in data analysis.

In Sect. 8.2 we will examine the factors that control intermixing for a wide variety of metal/III-V systems, including very reactive, mildly reactive, and nonreactive but disruptive interfaces. We will show how intermixing and bonding configurations change as an interface evolves. We will use results obtained from different experimental techniques to demonstrate that concentration profiles are largely determined by the details of chemical reactions between the deposited metal atoms and the released anions. These reactions and the compounds that they produce are limited by the availability of sufficient numbers of metal and semiconductor atoms to form distinct reaction products.

In Sect. 8.3, we will investigate changes in interface formation associated with atom deposition onto semiconductor surfaces held at 20–60 K. By considering specific examples, we will demonstrate that the initial chemical processes are not significantly influenced by substrate temperature. At the same time, reduced temperature does significantly alter the redistribution of released semiconductor atoms in the evolving overlayer, particularly with regard to surface segregation and kinetic trapping. We will also show that interface morphology for systems where clustering is observed to occur spontaneously at 300 K, such as Ag/GaAs, can be kinetically constrained to grow in a more layer-by-layer fashion at low temperature.

In Sect. 8.4, we will examine band bending and the Schottky barrier height as a function of the number of adatoms deposited and their chemical activity. We will show that very different band bending behavior is observed at 20–60 K compared to 300 K and that this is independent of the overlayer reactivity. We will examine new results that demonstrate that surface band bending for III-V semiconductors induced by submonolayer amounts of metal is strongly dependent on the bulk dopant concentration (N), particularly at low temperature. In particular, these results show the importance of N and temperature (T) for controlling the position of E_F in the gap at the surface. These N- and T-dependent results cannot be understood in the context of existing Schottky barrier models. However, they can be understood when the quantum mechanical coupling of adatom-induced states with bulk states in the presence of a potential barrier is considered, and we will describe our dynamic coupling model.

In Sect. 8.5, we will examine changes in interface evolution induced by the deposition of metal ions instead of neutral atoms. We will show that the amount of disruption and the consequent redistribution of atoms for Ag/ZnSe(100), Ag/InP(110), and In/GaAs(110) is changed by depositing ions of relatively low kinetic energies, ≤ 400 eV. These results, and those from band bending studies, show that equivalent barrier heights are obtained despite significant morphology differences and defect densities.

In Sect. 8.7, we will discuss interfaces formed by a completely new method. This technique involves metal cluster deposition onto pristine semiconductor surfaces. These clusters are formed in a novel way that involves atom deposition onto an inert buffer layer of condensed Xe. Desorbing the Xe then lowers the clusters onto the semiconductor surface. In this way, complex processes associated with atom condensation onto semiconductor surfaces are completely avoided. These include single-atom bonding with a surface and the associated energy release, surface diffusion and adatom–adatom bonding (with energy release), and the formation of clusters or aggregates. Cluster deposition isolates these potentially disruptive, energy-releasing processes from the surface. When the preformed metal clusters are brought into contact with the surface, they are constrained by their bonding with other adatoms in a three-dimensional array. The resulting interface exhibits novel morphological and electrical properties, as will be discussed.

This chapter is written so that the reader can gain insight into the complex chemical and physical properties of interfaces and an appreciation of the processes of interface formation. It is not intended to be an extensive review of the literature since that would require a book in itself, particularly when dealing with Schottky barrier evolution. With apologies to the many who have contributed so much to this exciting field, we will emphasize a few key points and will draw heavily on the work that has been done at Minnesota. For more extensive literature reviews, the reader is referred to citations in the original papers.

8.1 Experimental Techniques and Analysis

8.1.1 Photoelectron Spectroscopy

The great power of photoelectron spectroscopy in studies of interfacial phenomena rests in its ability to reveal the electronic energy states of a system as they are modified by chemical processes. This can be done with a variety of different photoemission techniques, including those that emphasize valence band and core-level processes. Synchrotron radiation is the ideal radiation for such studies, as will be discussed [8.9, 27–32].

Photoemission can be described by a three-step model in which a photon is absorbed by an electron, the excited electron is transported to the surface, and the electron escapes from the solid into vacuum [8.33–36]. Within this simple model, it is assumed that the energy difference between the initial and final states of the electron is equal to the photon energy. In the photoabsorption process, the characteristic photon absorption depth, or skin depth, can be hundreds of angstroms or more. However, only those electrons excited within about 3λ of the surface are likely to escape without inelastic scattering, where λ is the photoelectron mean free path. Measurement of the number of primary electrons as a function of energy and emission direction in vacuum provides insight into the energy and momentum distribution of the electron in the solid. The emission spectrum closely resembles the density of initial states when the distribution of electron final states is independent of crystal momentum $\boldsymbol{k}$. This is the case in X-ray photoemission. In contrast, variations in the dipole matrix elements involving initial and final states must be considered when the photon energy is small ($\lesssim 100\,\mathrm{eV}$). In this case, the absorptive part of the dielectric function becomes

$$\varepsilon_2(\omega) = \frac{h^2 e^2}{3\pi^2} \sum_{i,f} \int_{\boldsymbol{k}} d^3\boldsymbol{k} |\langle \mathrm{f}|p|\mathrm{i}\rangle|^2 \delta(E_\mathrm{f} - E_\mathrm{i} - h\omega) f(E_\mathrm{i})[1 - f(E_\mathrm{f})] \,. \quad (8.1)$$

In (8.1) the photon operator has been approximated by the first term in the series expansion of the vector potential $\bar{A}$, the integral is over $\boldsymbol{k}$-space, and the sum is over all pairs of initial and final states $|\mathrm{i}\rangle$ and $|\mathrm{f}\rangle$. The δ-function assures conservation of energy, and only $\boldsymbol{k}$-conserving transitions are allowed. It is also required that the initial state be occupied and the final state be empty, consistent with the Fermi functions in (8.1). The derivation of (8.1) can be found in many intermediate discussions of the photoemission process [8.27–32].

In Fig. 8.1 we sketch the photoemission process. As implied by the inset, a photon of energy $h\nu$ is incident upon a sample, photoabsorption induces excitations from $|\mathrm{i}\rangle$ to $|\mathrm{f}\rangle$, and an electron is ejected. Excitations are sketched here

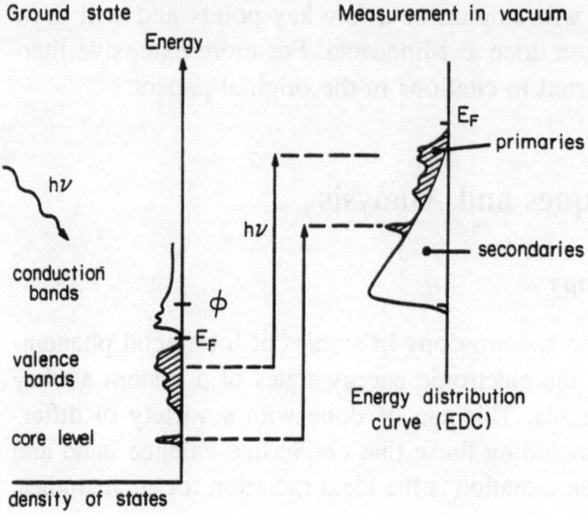

Fig. 8.1. Schematic of the photoemission process showing a photon of energy $h\nu$ incident upon a sample, electron excitation into higher energy states, and electrons escaping into vacuum, where they are energy-analyzed. The energy distribution curve at the right shows contributions from primary electrons and those which have scattered inelastically to produce a smooth background of secondaries

without dependence on the matrix elements of (8.1). The distribution of photoelectrons measured in vacuum is then depicted on the right of Fig. 8.1 where the energy extremes correspond to the work function and the Fermi level cutoff. Structure in the experimental energy distribution curves (EDCs) can be related to primary electron features.

Electrons excited within the solid have a certain probability of reaching the surface without being scattered. A substantial loss of energy occurs if the electron scatters by the creation of electron–hole pairs or by the generation of plasmons. Phonon scattering will influence the direction of the outgoing electron, but the associated change of energy is small, typically a few meV. If the inelastically scattered electrons that reach the surface have sufficient energy to overcome the work function barrier, they too can escape. These scattered secondary electrons contribute a background that is generally structureless, as sketched in Fig. 8.1. In studies of core levels of the sort to be discussed here, the background can be fit with an nth order polynomial or a Shirley s-type function [8.37]. This makes it possible to highlight the core level emission and investigate changes to to chemical processes or band bending.

One of the advantages of using photoemission to investigate surface and interface phenomena is that the inelastic scattering length of electrons in solids is short, typically ≤ 20 Å for electrons having kinetic energies in the range from 10 eV to several keV, giving probe depths of ≤ 60 Å. Compilations of inelastic scattering lengths for excited electrons in a wide variety of solids as a function of electron kinetic energy relative to the Fermi level show a so-called universal behavior in which λ exhibits a minimum for energies ~ 40 eV and increases rapidly for lower electron energies due to decreasing electron–solid interactions [8.38, 39]. The ability to vary the photon energy (and therefore the photoelectron kinetic energy and λ) is extremely important when information is sought concerning the spatial distribution of atoms of the same element but in distinguishable bonding configurations. At the same time, it should be noted that the effective probe depth can be changed by varying the photoelectron take-off angle, with maximum bulk sensitivity in normal emission and enhanced surface sensitivity in grazing emission.

In considering core level and valence band features, the starting place (and that depicted in Fig. 8.1) is the one-electron approximation. In this approximation, the binding energy of a core level referenced to the vacuum level E_b and the kinetic energy of the emitted electron E_k are related by $h\nu = E_k + E_b$, where $h\nu$ is the energy of incident photon. In many cases, this is quite adequate, even though it does not consider the fact that the remaining electrons can relax toward the hole created by the excitation of the photoelectron. In fact, then, photoemission measurements give the energy difference between the initial state of the N-electron system and the final state with $N - 1$ electrons [8.27–36, 40].

Complicating effects that should be considered in core level photoemission include multiplet splitting and shake-up and shake-off satellites. Multiplet splitting occurs when the system has unpaired electrons in the valence band. When the spin of the ejected electron is antiparallel to the unpaired valence band electron, the exchange interaction results in a lower energy for it than for photoelectrons having parallel spin. The resulting core level spectra show a doublet, and the peak separation will be a measure of the exchange interaction. Shake-up and shake-off satellite structures appear because the removal of a core electron from the closed shell perturbs the valence electrons. In valence band reorganization, an electron can be excited to a higher unfilled level and the energy associated with this transition leads to a shake-up satellite on the low kinetic energy side of the main core level peak. In a similar process, a valence electron can be excited to an unbound continuum state, leaving a hole in the valence band as well as one in the core level. This process is referred to as shake-off, and the outgoing core electron appears to have a lower kinetic energy than expected from the one-electron approximation.

The one-electron approximation also fails to account for the asymmetries observed in the core level spectra for metals due to the interaction of the positive core hole and the mobile conduction electrons. We shall return to this asymmetry when considering lineshape analysis procedures [8.41].

The core-level binding energies for an atom of a given element vary as the chemical environment of that atom changes, with nonequivalences arising from differences in oxidation state, molecular environment, or lattice site. The following simple analysis illustrates the origin of chemical shifts. They are critical to studies of interfaces because the goal is to determine changes in the environment. Bonding changes for a particular atom involve the spatial rearrangement of the valence charge, and there is a different potential that is created by the nuclear and electronic charges. The differences in binding energy for an atom in two different compounds A and B can be expressed as

$$\Delta E_i(A, B) = K(q_i^A - q_i^B) + (V_i^A - V_i^B) \tag{8.2}$$

in the charge potential model, where the first term describes the difference in electron–electron interactions between the core orbital and the valence charge and the second term describes the potential difference due to the surrounding atoms. If charge q_i is considered to produce screening, then the potential at the nucleus due to q_i is equal to q_i/r_v, where r_v is the average valence orbital radius. A change in the valence electron charge of Δq_i changes the potential by $\Delta q_i/r_v$. As a result, the binding energy of all core levels will change by that amount. The constant K takes into account the overlap between core and valence electrons. For cases where core–valence interaction is small, the core levels of a given atom exhibit a similar shift. This binding energy shift for equivalent compounds decreases as one descends a column of the periodic table because of the increasing value of r_v. The second term in (8.2) considers the influence of being bonded in molecules or solids. In ionic solids, the entire lattice contributes to this potential. In this case, the appropriate expression is of the form

$$V_i = \sum_{i \neq j} \frac{q_i q_j}{R_{ij}}, \tag{8.3}$$

which expresses the sum over potentials arising from ionic charges q_j centered at position R_{ij} relative to atom i.

This charge potential model has been successful for small molecules [8.42], but it fails for many compounds. A major simplification of the model is that energies that are involved in the rearrangement in charge configuration upon ionization of the atom are not included. No account is taken of the polarizing effect of the core hole on the surrounding electrons, both intra-atomic and extra-atomic. This involves a net flow of charge toward the hole created in the photoemission process in order to screen the sudden appearance of a positive charge. The energy of the screened hole state is lower than that predicted by the one-electron approximation. Hence, the measured binding energy is also reduced. The magnitude of the relaxation energy is expected to vary as the environment of the atom changes.

8.1.2 Experimental Procedures

Many of the conclusions of this chapter are based on experimental studies conducted with synchrotron radiation. These measurements have optimized surface sensitivity, using photon energies which typically gave photoelectron kinetic energies of ~ 40 eV and mean free paths of $\lambda \sim 4$ Å, as well as greater bulk sensitivity with $E_k \sim 10$ and $\lambda \sim 8$ Å. In the synchrotron radiation photoemission experiments, the core level and valence band spectra are collected with commercial electron energy analyzers. Maximum chemical information can be obtained by operating the analyzer and monochromator with high resolution, typically 0.2–0.3 eV. The experimental chambers are usually equipped with low energy electron diffraction (LEED) apparatus for structural analysis. A closed cycle helium refrigerator is often used for low temperature experiments. Typical operating pressures are 5×10^{-11} torr.

In the synchrotron radiation photoemission experiments discussed here, particular emphasis was placed on achieving high resolution results with high signal-to-noise ratios. Such results are needed if detailed lineshape analysis is to be successful. In turn, lineshape decomposition is necessary to establish the growth and attenuation of emission features for new interface components. It is particularly important in studies that focus on the movement of the surface Fermi level due to adatom-induced changes at the surface. In these studies, the shift in the kinetic energy of core-level electrons emitted from atoms bound in the bulk reveals changes in band bending. Reaction or disruption-induced changes alter the core level lineshape, as noted above. It is particularly important to separate electrostatic effects from chemical effects, and this requires fitting. In general, studies that rely on by-eye examination of core-level changes lack the precision needed for detailed investigations of Schottky barrier height ($\pm 50\,\mathrm{meV}$), particularly when the lineshapes are complex.

Additional measurements discussed here use angle-dependent X-ray photoelectron spectroscopy with $h\nu = 1486.6\,\mathrm{eV}$. These studies emphasized the spatial distribution of the Ga and As atoms in the evolving overlayer, particularly the concentration profile near the surface. In the sputtering experiments, the surface was alternately bombarded with 3.5 keV Ar^+ ions and the core-level spectra were measured. Sputter profiles were obtained by integrating the emission of the various core levels as a function of sputter time. The integrated areas were then normalized to data acquisition time.

High-resolution synchrotron radiation photoemission measurements (giving tunable $h\nu$ and variable surface sensitivity), XPS measurements (providing access to deeper core levels and variations in probed depth through changes in take-off angle), and sputter-profiling experiments (following the erosion of the interface), represent a powerful combination that provides insight into chemical bonding during interface formation. The results make it possible to examine the growth mode of the overlayer. In particular, layer-by-layer growth, clustering, atomic intermixing across the interface, and surface segregation give rise to distinctive results when the core level lineshapes are carefully examined and their intensities are determined as a function of the amount of material deposited.

The observations of the previous paragraphs notwithstanding, there are several caveats that should be carefully defined. First, photoemission results represent an average response of the probed region, both parallel and perpendicular to the surface. In systems where reaction and kinetically limited intermixing occur, there will be a concentration gradient perpendicular to the surface. Variations also occur parallel to the surface as reaction proceeds in an inhomogeneous fashion, but photoemission is not the optimal technique to investigate such processes on the nanoscopic scale. Second, the inherent surface sensitivity precludes studies of interfaces buried by many tens of angstroms of overlayer. Third, chemical shifts can be small and some may not be observable, despite changes in bonding configuration. Uncertainties in identifying chemical shifts can be reduced by careful lineshape analysis of high-resolution core-level results, combined with studies of both the anion and the cation for compound semiconductor surfaces. Fourth, sputter profiling is not free of pitfalls because of the destructive character of atom removal and complex interface morphologies. Preferential anion sputtering and cation enrichment is unavoidable for III-V semiconductors (or other binary systems), and atom distributions in the sputter profiles are smeared by the depth resolution associated with the mean free paths of the photoelectrons. Moreover, extended sputtering will alter the erosion rate by increasing the amount of surface roughness. While the intensity profiles should not be seen as being identical to concentration profiles, they provide clear qualitative information about species distributions normal to the surface. They are particularly useful for interfaces where systematics are sought.

8.1.3 Samples and Deposition Procedures

In most of the studies reviewed here, single crystals of GaAs(110) were cleaved at $\sim 5\times10^{-11}$ torr to produce atomically clean (110) surfaces. For the results that emphasize band bending changes, only those cleaves for which E_F was within 60 meV of the conduction band minimum (CBM) or valence band minimum (VBM) at 300 K were used. The n-type (p-type) samples were Si (Zn) doped, and the doping concentrations were $2\times10^{18}\,cm^{-3}$ (high doping, HD) and $1\times10^{17}\,cm^{-3}$ (low doping, LD). Results for p-GaAs with a doping concentration of $4\times10^{19}\,cm^{-3}$ (very high doping, VHD) were also obtained, as were results for n-type samples doped at $5\times10^{-16}\,cm^{-3}$ (very low doping, VLD). This wide range of N extends from the isolated donor or acceptor level regime to that characterized by a band of states derived from these levels (Mott insulator transition). The importance of knowing the magnitude of N will become apparent when we consider adatom-induced band bending changes at low temperature. For studies that emphasize interface chemistry, equivalent results have been obtained for n- and p-samples of different N, as expected.

Most of the results discussed here will be for GaAs(110), but some will be derived from studies of InP(110), InSb(110), and GaAs(100). Specific examples will be used to show the generality of the conclusions reached and to broaden the base of the understanding of interfacial phenomena. Experimental procedures for these other III-V's (and for II-VI and group IV systems as well) are equivalent to those described for GaAs(110) with the exception that sputter-annealing procedures were used to obtain clean GaAs(100) or ZnSe(100) surfaces.

Most of the overlayers were grown in situ by the condensation of metal atoms onto the clean surface. These metals were evaporated from resistively heated tungsten boats or baskets placed $\sim$ 30 cm from the sample's surface. Extensive degassing of the sources made it possible to form these interfaces at pressures consistent with the need to maintain ultraclean conditions (pressure during evaporation below 4×10^{-10} torr with operating pressure of $\sim 5\times10^{-11}$ torr). Typical deposition rates were $\sim$ 1 Å per minute, as measured with calibrated Inficon quartz oscillators placed near the samples (the rate was lower for the ultralow depositions). In all cases, the samples were exposed to the flux for timed periods after the evaporation rate had been stabilized. Angstroms and monolayer (ML) units will be used to describe the amount of metal deposition but uniform overlayer formation is not necessarily implied. For Co/GaAs(100) studies, 1.4 Å = 1 ML of Co in a structure commensurate with the GaAs(100) substrate since the Co atom density is twice that of the template. For reference, the surface atom density is 6.26×10^{14} atoms/cm^2 for GaAs(100) and 8.86×10^{14} atoms/cm^2 for GaAs(110). θ will be used to denote the amount of material deposited. To convert from angstroms to monolayers of GaAs(110), use 0.64 ML/Å for Ti, 1.03 for Co, and 0.66 for Ag.

For experiments conducted below room temperature, the samples were cleaved at 300 K and examined visually and with photoemission before cooling for overlayer deposition. This was important because E_F moves in the gap as a function of temperature and a partially pinned surface at 300 K can appear unpinned at low temperature. Two different cooling configurations were used. One involed attaching the sample to a braid anchored to the second stage of a closed-cycle helium refrigerator. The lowest temperature that could be achieved in this way was ~ 60 K. The second involed Ga-soldering of the sample holder into a specially designed Cu tank that was directly attached to the refrigerator cold head. In this case, it was possible to routinely reach 20 K. Temperatures between 20 and 350 K could be selected by using a heater to counter the cooling capacity of the refrigerator. Easy sample interchange was possible with either of the two low temperature systems. In both, the temperature was monitored with a Si diode or a Au-Fe/chromel thermocouple attached to the cold finger. Calibration tests established that the sample temperature was within 5 K of the diode or thermocouple temperature.

For the cluster deposition experiments, 200 L of Xe ($1 \mathrm{L} = 10^{-6}$ torr · s) were condensed onto the GaAs(110) surface to produce buffer layers ~ 30 Å thick prior to the deposition of metal. Each metal cluster deposition experiment began with a freshly cleaved surface. Photoemission measurements showed that Xe adsorption and desorption from GaAs(110) produced no changes in band bending or substrate modification. After cluster formation at 60 K, the sample was detached from the Cu cold finger and allowed to warm to 300 K. Substrate core-level ECDs were obtained following Xe desorption for sample temperatures $100 \leq T \leq 300$ K. With the exception of Ti/GaAs(110), the interfaces formed in this manner were stable.

8.1.4 Core-Level Lineshape Analysis

Detailed analyses are required for quantitative assessments of the atomic environments that form (and evolve) and to precisely distinguish electrostatic shifts from chemical shifts. Spectral lineshapes change as a function of the number of atoms deposited for a given overlayer, and they vary in distinctive ways from one overlayer to another. In lineshape analysis of evolving core level spectral for metal–semiconductor interfaces, it is critical to have an efficient and flexible deconvolution scheme.

The starting point for lineshape analysis of core levels for evolving metal–semiconductor interfaces is the treatment of the clean surface results. As is well known, differences in bond length, coordination number, and nearest neighbor positions generally result in surface shifted components, and high-resolution core-level spectra can distinguish between surface and bulk components. In fitting procedures, it is first necessary to determine the Gaussian and Lorentzian contributions for these bulk and surface components. Self-consistent intrinsic line widths and values for experimental resolution can be determined by calculating the experimental resolution and by iteratively adjusting the intensities, Gaussian parameters, and Lorentzian parameters for each core level at each core level photon energy [8.43].

In treating lineshapes for an evolving interface, the adjustable parameters include those that determine the background function; an intensity, energy position, and Gaussian width for each component; the Lorentzian width; the branching ratio; the spin-orbit splitting; and the asymmetry parameter. In our fitting, the spin-orbit splitting is held constant and is determined from the clean surface spectra. The branching ratio is $h\nu$-dependent near threshold for core level excitation, and changes in branching ratio can occur during interface evolution that are related to the density of final states into which the core levels are excited. Operationally, however, the branching ratio is usually held fixed.

The natural line width or core hole lifetime can be written as a Lorentzian of the form [8.44,45]

$$L(E) = \frac{1}{1+4(E - E_{\mathrm{L}}/\Gamma_{\mathrm{L}})^2}\,, \tag{8.4}$$

where E_{L} is the centroid and Γ_{L} is the full width at half maximum (FWHM). The Lorentzian width is independent of excitation energy. The instrument response function and phonon broadening of the core level can be represented by a Gaussian as

$$G(E) = \exp\left[-4\ln 2\left(\frac{E - E_{\mathrm{G}}}{\Gamma_{\mathrm{G}}}\right)^2\right]\,, \tag{8.5}$$

where E_{G} is the Gaussian centroid and Γ_{G} is the FWHM. Although the instrument response function does not depend on the sample or the temperature of measurement, phonon broadening depends on both sample and temperature. The Gaussian width can change during interface development due to variation in the phonon spectrum as released semiconductor atoms appear in new chemical environments. The relationship between phonon width of a core level and the chemical environment of the atom has been discussed by *Wertheim* and *Dicenzo* [8.46].

To fit the core level spectra requires that the Gaussian and Lorentzian functions be convolved, thereby producing a Voigt function. The Lorentzian lineshape has been shown to describe the core-level emission of insulators and semiconductors [8.44]. For metals, an analytic form was developed by *Doniach* and *Sunjic*

(DS) [8.41], where a variable was introduced to account for any asymmetry, α. For clean semiconductor surfaces, this asymmetry is zero. In general, however, the full form is needed to represent the emission from anions or cations that form metallic compounds or are dissolved or trapped in the metallic overlayer.

8.2 Interface Formation at 300 K

The typical boundary regions formed between transition metal overlayers and semiconductor substrates are characterized by disruption of the semiconductor that is accompanied by atom interdiffusion and compound formation. The spatial extent of this boundary region and its heterogeneity depend on the details of the system [8.1–11]. Unfortunately, there is as yet no clear way to predict the morphology, species profile, and stability of this region.

8.2.1 Co/GaAs

In the the following paragraphs, we focus on the formation of Co overlayers on cleaved GaAs(110) at 300 K and on sputter-annealed GaAs(110)-c(8x2) substrates. The goal is to examine chemical interactions at the interface and assess the spatial distribution of Ga and As atoms which are released from the substrate following Co deposition. Since earlier studies showed epitaxial bcc Co growth on GaAs [8.47–52] we were also interested in characterizing the structure of the Co overlayer at very low coverage. The details of those angle resolved Auger electron diffraction studies can be found elsewhere [8.53] and we cite here only some conclusions. The Co/GaAs system provides a nearly optimal system in which to combine chemical and structural mapping of the boundary region. We consider this system first because Co deposition induces limited substrate disruption, but in doing so, promotes Ga and As out-diffusion. At the same time, a template suitable for epitaxy is left intact and weakly ordered bcc Co forms. After considering this system and its relatively simple interface evolution, we will turn to more reactive interfaces.

In Fig. 8.2 we show representative Ga and As $3d$ core level EDCs for Co/GaAs(110)] [8.53]. These synchrotron radiation photoemission spectra have been background-subtracted and normalized to highlight changes in lineshape induced by the atom-by-atom deposition of Co at 300 K. As discussed above, the photon energies of 60 and 85 eV for Ga and As $3d$ states gave maximum surface sensitivity (95% of the signal originates from within 3 λ or $\sim$ 12 Å of the surface). Band-bending-induced (electrostatic) shifts of the Ga and As $3d$ features have been removed by aligning the substrate core level components (band bending will be discussed in Sect 8.4). The bottom-most EDCs of Fig. 8.2 are for the clean surface where emission from surface atoms appears at 0.28 eV higher binding energy for Ga and 0.38 eV lower binding energy for As.

Room temperature intermixing of Ga and As atoms with the Co overlayers is evident from from the evolving $3d$ core level emission of Fig. 8.2 taken from

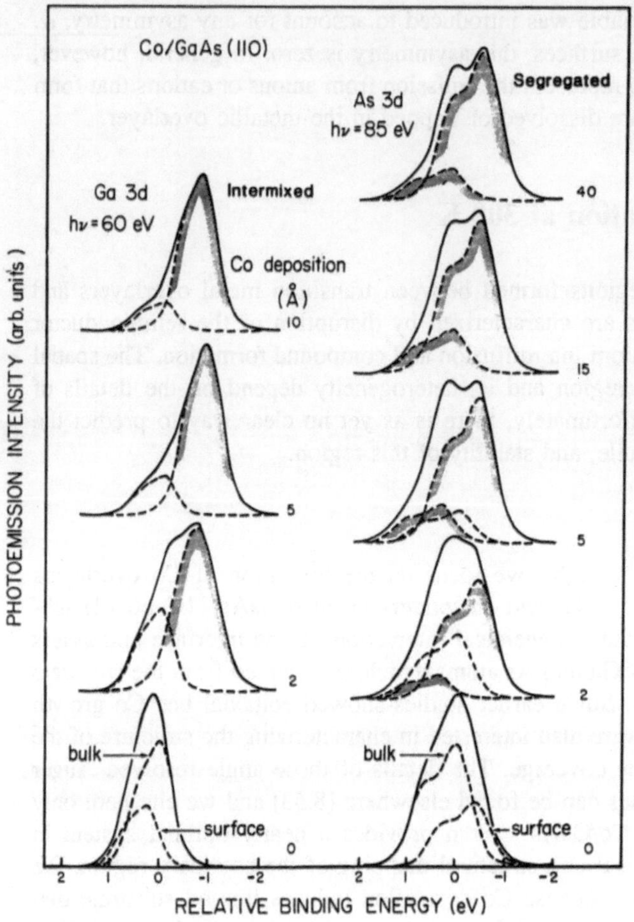

Fig. 8.2. Ga and As 3d core level spectra for 300 K atom deposition of Co onto GaAs(110) showing the surface and bulk components for the clean surface and the growth of disruption-induced features as the overlayer thickens. (From [8.53])

[8.53]. For Ga, a shoulder appears at sub-angstrom Co depositions on the low binding energy side of the main line opposite the surface-shifted component. Simultaneously, emission from the surface component diminishes rapidly. The new Co-induced feature appears at lower binding energy for increasing coverages and stabilizes at −0.9 eV. Significantly, it is dominant for coverages as low as 2 Å, indicating that substrate disruption produces substantial amounts of Ga intermixing in the overlayer, and the undisrupted substrate retreats from the surface.

The smooth variation of the Ga 3d core level energy indicates that a Co-Ga solution forms, rather than a specific bonding configuration, and lineshape analysis indicates sharpening for the intermixed Ga component as the amount of deposited metal increases. This behavior is rather typical because examination of the Ga $3d$ core level behavior for many systems rarely exhibits a distinct, invariant binding energy during overlayer growth. Likewise, the Ga core-level shifts measured at high coverage vary directly with the Pauling electronegativity difference between Ga and the overlayer metal [8.54, 55].

In Fig. 8.2 we also show representative As $3d$ core level EDCs and their decompositions. The new, chemically shifted component at −0.45 eV overlaps the surface component at −0.38 eV, and it is difficult to unambiguously separate the diminishing surface-atom contribution from the increasing reacted-As contribution at submonolayer coverage. Indeed, little or no lineshape change is evident for $\theta_{Co} \leq 0.5$ Å. Insight into the rate of loss of the As surface contribution can be gained by comparing with the submonolayer Ga behavior, where the surface shift was +0.28 eV. Strong Co-induced As emission appears on the low binding energy side of the substrate component by 1 and 2 Å Co coverage, and this provides clear evidence for intermixed As. At higher coverage this shifted component dominates the EDCs. At the same time, a second Co-induced component is evident at +0.2 eV and it can be associated with As atoms segregated *on* the Co surface while the component shifted to lower binding energy corresponds to As in solution *within* the Co matrix and coordinated by Co, i.e., intermixed.

In Fig. 8.3, we show attenuation curves for the Ga and As $3d$ emission intensities as a function of the amount of Co deposited, defined as $\ln[I(\theta)/I(0)]$, where θ is the coverage. The intensities for the different components were obtained from core-level decompositions. The rate of attenuation of the total signal is much faster for Ga than for As, and the As emission persists to high coverage. The component-specific profiles show that emission from substrate Ga and As atoms is attenuated with, 1/e values of $\sim$ 1 Å during the earliest stages of formation. This is because substrate disruption occurs and the undisturbed substrate retreats from the surface, i.e., the distance between the surface and the undisrupted GaAs substrate increases. During this process, Ga and As atoms previously bound in Ga-As bonds form new bonding configurations that are readily distinguishable. By $\theta_{Co} \cong 3$ Å, the decay rate for the substrate components slows to a 1/e value of $\sim$ 4 Å, consistent with photoelectron propagation through a growing overlayer that forms in an approximately layer-by-layer fashion. Figure 8.3 also shows that emission from Ga atoms released from substrate bonds diminishes exponentially with the same 1/e length as the unperturbed substrate for $\theta_{Co} \geq 5$ Å. From Fig. 8.2, we conclude that the environment of the released Ga atoms becomes increasingly dilute as Ga is dissolved in the Co-rich overlayer. The results of Fig. 8.3 show that Ga does not diffuse far from the original interface and that the solution is covered by further Co deposition. The concentration of Ga in solution changes with coverage so as to remain reasonably homogeneous within the probed region.

The photoemission results demonstrate that the behavior of the As atoms is very different from Ga once released from GaAs substrate bonds. In particular, emission from the intermixed and surface-segregated components, which have distinct binding energies, persists with little attenuation as the Co overlayer grows. Since there is no evidence from the substrate attenuation curves of

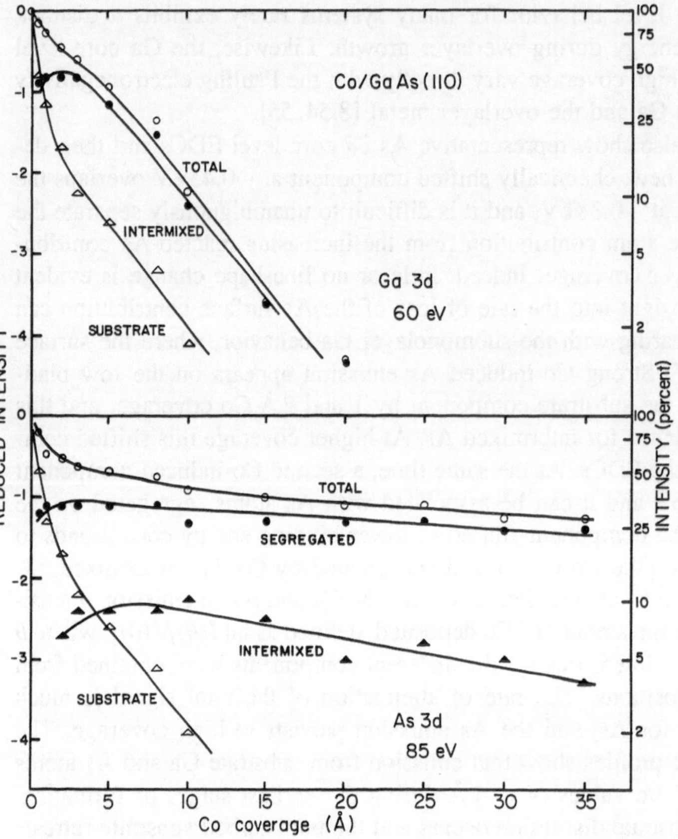

Fig. 8.3. Total and component-specific core-level emission intensities from Fig. 8.2 normalized to those of the clean surface for Co/GaAs. Disruption and retreat of the substrate is completed during the early stages of deposition, and substrate emission then decays exponentially as the overlayer thickens. The persistence of emission from As atoms reflects their segregation to the surface and near-surface region of the Co film. (From [8.53])

continued substrate disruption after it is covered up, we must conclude that As atoms released during the early stages of deposition are only partially trapped at the interface. That is to say, we find no evidence for extended Co-As compound formation. Instead, the released As atoms are displaced or expelled from the growing Co region to the surface, where they are only partially coordinated with Co (chemical shift 0.2 eV). While they could, in principle, be incorporated within the Co matrix as a solution or supersaturated solution, the solubility of As in Co is low. The fact that two different As components appear at high coverage and attenuate at approximately the same, slow rate indicates that this expulsion process continues.

The spatial distribution of Ga and As normal to the interface can be quantitatively determined with angle-dependent XPS in which the polar angle is varied from grazing to normal emission. These results (shown in [8.53]) demonstrate

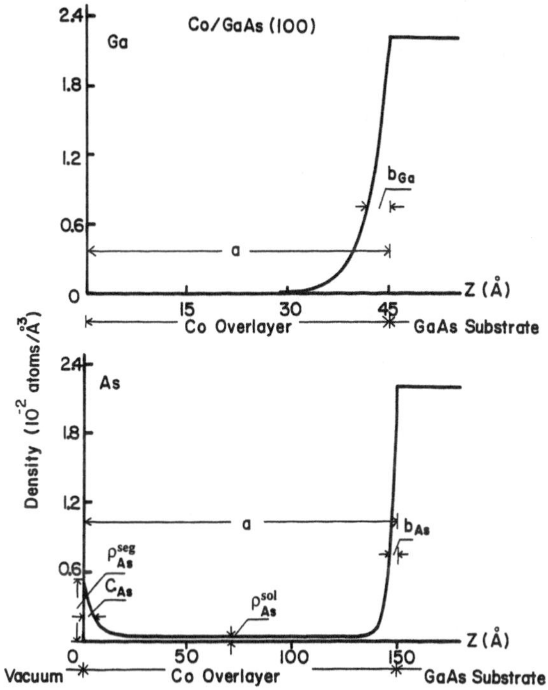

Fig. 8.4. Model profiles that describe the semiconductor atom concentrations in Co overlayers of nominal thickness 45 Å for Ga (*top*) and 150 Å for as (*bottom*). The concentration of Ga falls off exponentially with distance into the Co layer from the buried GaAs surface. For As, a bimodal distribution is needed to account for surface segregation, and there is also evidence for As in solution. (From [8.53])

that the Ga intensity detected at any angle decreases rapidly with coverage. For As, the intensity decreases only slightly with increasing coverage, again indicating the persistence of outdiffused As in the surface region. Moreover, the normalized intensity from As increases sharply at grazing emission for a given coverage. Polar-angle vs intensity results for different thicknesses allow a test of any particular form of $\varrho(z)$, the density distribution of atoms away from the surface.

In Fig. 8.4 we reproduce the density distributions that lead to the best fit of the experimental results for Ga and As, as derived in detail in [8.53]. For Ga, optimal agreement is found when the characteristic decay length of the Ga concentration into the Co overlayer (away from the buried interface) is $b_{Ga} = 3$ Å and the form of the concentration profile is exponential. These results also indicate that the total quantity of Ga released from the substrate is equivalent to 1.1 ML of the GaAs(100) surface. The small magnitude of b_{Ga} supports the observation that epitaxial Co growth can occur while making the case that a step function in the concentration profile at any interface is not likely.

To fit the As $3d$ polar profile results, the characteristic decay distance from the free surface, c_{As}, was found to be the same for all coverages and the quality of fit was very sensitive to the choice of c_{As}. The best fit was obtained by using $c_{As} = 5$ Å, again with an exponential form, as shown in Fig. 8.4. We also note that the amount of segregated As, ϱ_{As}^{seg}, decreased with increasing thickness. This is again consistent with the gradual dissolving of As in the thickening Co layer, a process which is not energetically favored because of the low As solubility. Detailed analysis presented in [8.53] indicates that the optimal value of ϱ_{As}^{sol} corresponds to about 0.26 at.% of the overlayer. This is a reasonable solubility for a polycrystalline overlayer. Analysis also indicates that ϱ_{As}^{seg} = 0.0084 at. Å^3 or 9.5 at.% at 80 Å; 0.0075 at./Å^3 or 8.4 at.% at 100 Å; and 0.0052 at./Å^3 or 5.8 at.% at 150 Å. Finally, we can estimate that the value of b_{As} for As is approximately the same as for Ga. The fact that c_{As} remains constant indicates that the decay profile of As at the vacuum surface does not change, even though the surface concentration of As diminishes with increasing film thickness. In turn, this indicates that the amount of As that can be tolerated in the overlayer diminishes sharply (exponentially) with distance from the surface. This can be understood qualitatively by recognizing that the layer-by-layer local density of states of a solid converges to that of the bulk over a distance of several layers.

In Fig. 8.5 we show in a very schematic way the evolution of an interface when compound formation does not occur. The deposition of adatoms onto the surface represents an energy-releasing process where the largest component of the energy comes from the formation of new chemical bonds. As a result, substrate atoms can be released from surface semiconductor bonds and appear in a new disordered surface array. The stability of this array, which is dictated by local bonds, changes as the number of metal atoms changes relative to the number of semiconductor atoms. For Co/GaAs, the evidence indicates that the progressive addition of Co leads to nucleation of bcc Co as the most favorable surface configuration rather than formation of Co-As or Co-Ga compounds or alloys. As this occurs, however, the Ga and As atoms trapped in the growing Co region increasingly resemble impurity or solute atoms and the energy cost of dissolving

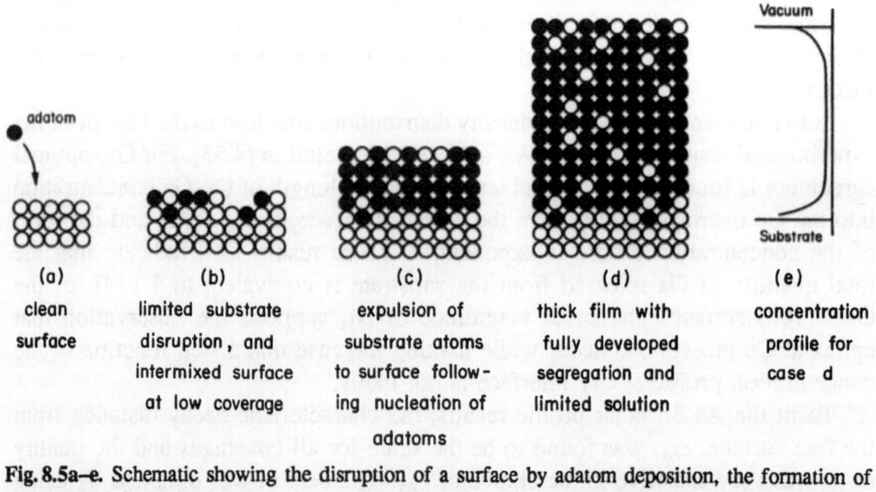

Fig. 8.5a–e. Schematic showing the disruption of a surface by adatom deposition, the formation of a disordered, intermixed array at low coverage, but then the nucleation of metal and the expulsion or solution of the semiconductor atoms. For Co(GaAs, the Co layer forms a weakly ordered bcc structure that is epitaxial on GaAs. Arsenic is expelled to the surface of this layer, as depicted in the concentration sketched in (e) and modeled in Fig. 8.4

the Ga and As atoms changes as the nuclei grow. The sketch of Fig. 8.5 shows that the energetically favorable process of metal nucleation leads to continued segregation of the solute. This continued segregation is experimentally observed for As, and the As concentration profile is bimodal once the Co layer is thick enough to have a bulk-like part and a surface part. In contrast, the Ga solubility is much higher and a bimodal distribution is not established. It is then clear that the formation of the interface is evolutionary because atom distributions change as the number of metal atoms available to nucleate increases, as the nuclei evolve from nanoscopic to microscopic crystals, and as energies and energy balances change.

Support for this model of semiconductor atom distributions across the vacuum/Co/GaAs boundary can be found in the results obtained when the interface is eroded by Ar bombardment. In Fig. 8.6, taken from [8.56], we show these sputter profiles for a 50 Å layer of bcc Co grown on GaAs(110). The distribution of Ga and As in the overlayer close to the buried interface is smooth, without structure that would reflect compound formation or the preferential trapping of one or the other semiconductor species. Significantly, the sputter profiles show atoms to be present in the surface and near-surface region, and they are readily removed by limited sputtering. The persistence of emission for Co atoms after the buried interface has been reached is likely to be an artifact related to knock-in effects. Likewise, the growth of Ga emission relative to As is due to preferential sputtering of the anion.

Studies of the epitaxial Fe/GaAs system show behavior that is similar to that for Co/GaAs [8.57]. The Fe/GaAs system possesses a higher degree of structural regularity (is better ordered) while also promoting substrate atom outdiffusion and surface segregation of As. These two transition metal systems are the exception rather than the rule as far as their interface properties are concerned. They are examples of simple systems in which strong reaction is not found and where the driving force for anion expulsion comes from the tendency to form Co or Fe

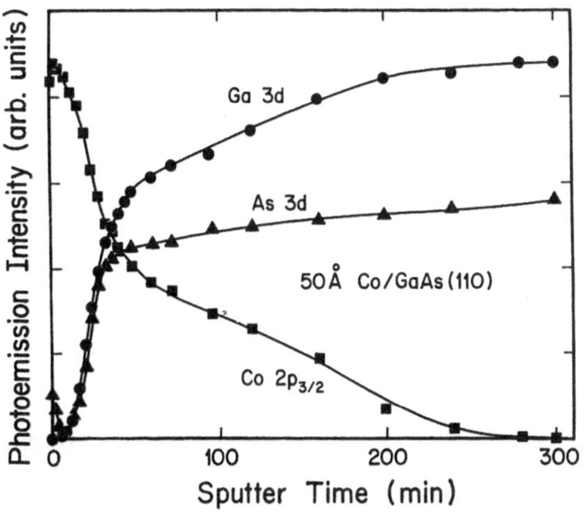

Fig. 8.6. Sputter profiles for 50 Å Co/-GaAs formed and studied at 300 K. Light sputtering removes the surface segregated As atoms. The concentration of both Ga and As increases without structure as the buried interface is reached. The persistence of Co after extensive sputtering is an artifact related to sputter-induced knock-in. (From [8.56])

nuclei. For most transition metals, the energetics favor the formation of metal–As-based phases at the buried interface. As we will see, however, the behavior observed for Ga is very similar to that observed for other GaAs interfaces. It is likely that analogous Ga profiles are established as a function of distance from the buried interface for a broad class of metal/GaAs interfaces.

8.2.2 Reactive Interfaces

In the following we focus on reactive metal–semiconductor interfaces, first considering Ti/GaAs(110) and then comparing with results for several other systems to draw general conclusions. In contrast to Co/GaAs, where the heats of formation for the arsenide or gallium alloy were relatively small, we find that there is a strong tendency for Ti/GaAs to form Ti-As bonds.

Many of the general features of reactive Ti/GaAs(110) interface development can be determined from the evolution of the Ga 3*d* and As 3*d* core-level emission, together with Ti and the valence-band EDCs [8.58–60]. Here, we emphasize the semiconductor atoms. In Figs. 8.7 and 8, we reproduce the Ga and As 3*d* core emission for $0 \leq \theta_{Ti} \leq 60$ Å (from [8.58]). The core-level positions were adjusted to compensate for band bending (to be discussed in Sect. 8.4). For Ga, results taken with $h\nu = 30$ eV are compared to those taken at $h\nu = 60$ eV to demonstrate the different surface sensitivities (approximately 10 Å at 30 eV but ~ 4 Å at 60 eV). Inspection of the Ga results for $\theta_{Ti} = 10$ Å shows very different lineshapes because the substrate component is much reduced in a relative sense in

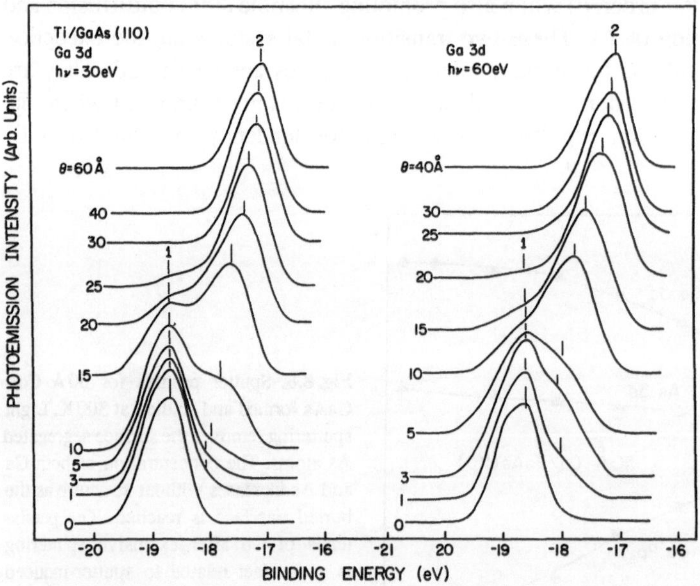

Fig. 8.7. Ga 3*d* emission acquired under bulk sensitive (*left*) and surface sensitive (*right*) conditions during overlayer formation for Ti/GaAs(110) at 300 K. Component 2 corresponds to Ga atoms intermixed in the overlayer. (From [8.58])

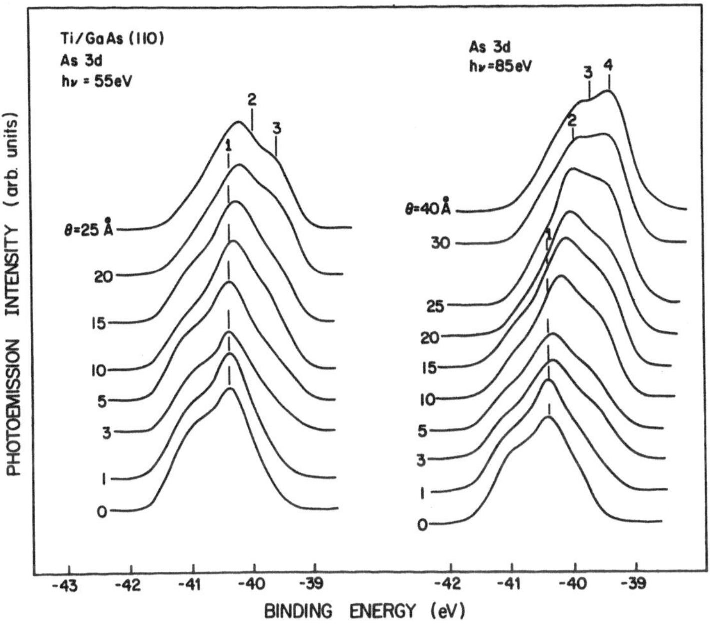

Fig. 8.8. As $3d$ emission for Ti/GaAs(110) as in Fig. 8.7. The persistence of As to high coverage is indicative of As segregation to the surface. (From [8.58])

the surface sensitive spectra. The Ti-induced emission on the low binding energy side of the main Ga $3d$ line is enhanced for $h\nu = 60\,\mathrm{eV}$ for equal coverages. It is clearly evident in spectra taken at coverages as low as $\theta_{\mathrm{Ti}} = 0.25\,\mathrm{Å}$. In this sub-angstrom range, the As $3d$ core emission broadens and a shoulder grows and replaces the surface component on the low binding energy side of the main line (Fig. 8.8). The rapid loss in Ga and As surface-shifted emission suggests that Ti atoms are dispersed on the surface, consistent with the randomness of the deposition process and chemical bonding that minimizes surface diffusion.

From Fig. 8.7 it is clear that the Ti-induced Ga component becomes more prominent as the amount of Ti deposited is increased. Moreover, the Ga $3d$ emission from the substrate is negligible by $\theta \cong 30\,\mathrm{Å}$ when probed with these photon energies. For $5 \leq \theta_{\mathrm{Ti}} \leq 60\,\mathrm{Å}$, the new Ga $3d$ component shifts steadily to lower binding energy but its full width at half maximum is not significantly broader than the bulk component of the clean surface. This shows that Ga atoms are in environments that change during the different stages of interface evolution and that the range of Ga environments at a given coverage is relatively small. As for Co/GaAs, this indicates progressive dilution of Ga in a matrix that is increasingly metal-rich.

In Fig 8.9 we show the total attenuation curves derived from the Ga and As $3d$ results for Ti/GaAs(110). As in Fig. 8.2, the substrate emission attenuates rapidly while the total Ga emission decays much more slowly for Ti than for Co overlayers. Comparison of the experimental attenuation lengths with predicted

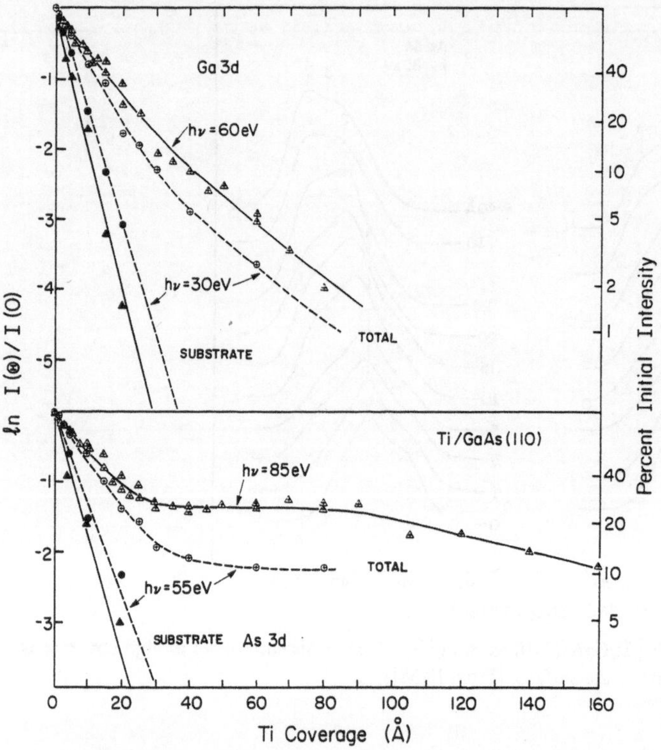

Fig. 8.9. Ga and As $3d$ attenuation curves for Ti/GaAs(100) derived from the results of Figs. 8.7 and 8.8 showing the rapid attenuation of the substrate emission but continued presence of Ga and As in the thickening film. The results for As demonstrate that an As solution in Ti is not as favorable as As expulsion to the surface. (From [8.58])

photoelectron mean free paths shows that the Ga $3d$ substrate component decays more rapidly than would be expected if the surface were simply being covered up by Ti metal. These results demonstrate that Ti deposition induces substrate disruption and the retreat of the undisrupted buried interface. For $\theta_{Ti} \geq 30$ Å, the rate of attenuation for the Ga emission is less rapid than at lower coverages as outdiffusion occurs during overlayer thickening. While the amount of Ga is not large, it is much greater after equivalent amounts of Ti deposition than observed for Co deposition.

The As lineshapes shown in Fig. 8.8 continuously change during interface formation and exhibit an intricate evolution behavior. Two distinct reacted components were needed to produce satisfactory fits for intermediate coverages and a third one was needed at higher coverages ($\theta \geq 15$ Å). The first two Ti-induced features correspond to Ti-As reaction products while the persistence of the third one to very high coverages is indicative of As segregation. This evolving interface is then very complex with the simultaneous existence of multiple bonding configurations and outdiffusion for As and a single, increasingly Ti-rich environment for Ga atoms expelled from the substrate.

The total attenuation results shown in Fig. 8.9 again indicate the rapid decrease of As $3d$ emission for $\theta_{\mathrm{Ti}} \leq 25$ Å but very little decrease for $30 \leq \theta_{\mathrm{Ti}} \leq 90$ Å. The broad, high plateau region demonstrates that As atoms move to the surface as Ti is added and that very little As is trapped. It is not until quite high coverages that the As content of the probed region diminishes significantly. The observation of anion segregation for a metal as reactive as Ti generated considerable discussion when first reported because chemical trapping [8.5–7] was expected to retain the anions near the buried interface. These results can be understood by recognizing that Ti-As compound formation can only occur when sufficient numbers of atoms of Ti and As are present for nucleation. When too few As atoms are present in a Ti-rich environment, they represent solute atoms. At a growing interface, their number easily exceeds the equilibrium solubility limitation and segregation occurs.

Sputter profiling experiments for Ti/GaAs make it possible to further investigate the atom distributions for Ga and As in the Ti overlayers. In the top panel of Fig. 8.10 we show the sputter depth profiles for the 100 Å Ti/GaAs(110) system (from [8.56]). These concentration curves show several very interesting features. First, the As content of the surface region is relatively high, consistent with the attenuation curve results of Fig. 8.9. However, gentle sputtering removes the As and the amount observed thereafter is small until the reacted region is reached. Second, the anion distribution exhibits a maximum after ~ 120 min of sputtering. Third, the cation profile exhibits a distinct valley-like structure that corresponds to the anion maximum. This valley reflects cation depletion near the buried interface and the maximum in the anion concentration shows that Ti-As formation leads to cation expulsion in front of the Ti-As reacted region. From Fig. 8.10 we estimate that the buried interface is reached after ~ 150 min of sputtering. Additional measurements [8.61] with different Ti overlayer thicknesses above 30 Å as well as for a reconstructed GaAs(110) surface showed no differences except for an increase in separation between the Ga peak and the vacuum surface. We conclude that the profile near the buried interface is established during overlayer formation but does not change as the thick film is grown. Analogous results were obtained for GaAs(100) and GaAs(110) surfaces. As for the Co/GaAs interface, we attribute the persistence of Ti emission after extensive sputtering to knock-in effects.

The results for Ti/GaAs shown in Fig. 8.10 reflect the atomic profile for a system grown at room temperature. The results of [8.62] show that heating enhances Ti-As reaction and expands the region over which compound formation occurs, resulting in the retreat of the buried interface. There is also enhanced Ga expulsion from the region where Ti-As forms, and there is promotion of Ga surface segregation. Indeed, the Ga concentraton close to the buried interface is only a few atomic percent after annealing at 365°C for 2.5 h, and the Ga density is very high in the near-vacuum region. This distribution has been confirmed by both in situ X-ray photoemission [8.62] and the X-ray diffraction measurements of *Wada* et al. [8.63].

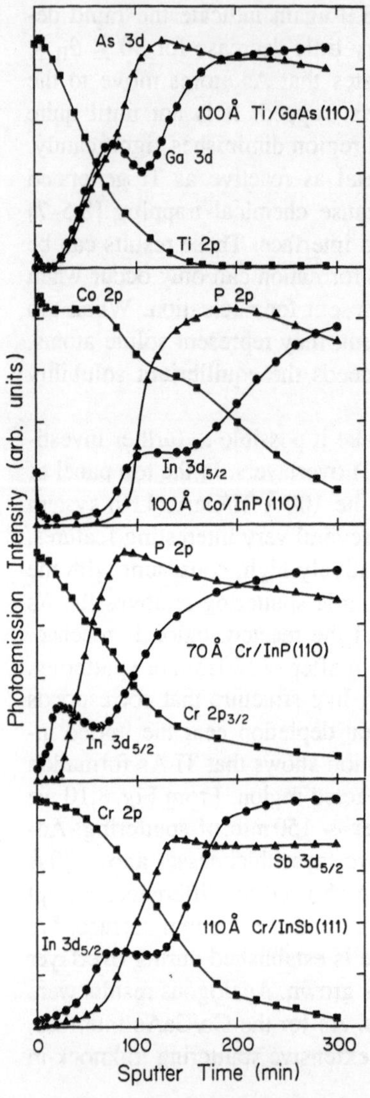

Fig. 8.10. Sputter profiles for reactive overlayers on GaAs(110), InP(110), and InSb(111). The characteristic valley-like structure in the cation profile is indicative of cation expulsion from the region near the buried interface where metall–anion reaction products form. (From [8.56])

The valley-like structure shown for Ti/GaAs is common for other reactive metal overlayers but it is not always as pronounced. Examination of the results for 70 Å Cr/InP(110) in Fig. 8.10 reveals an equivalent profile with an even lower amount of In in the region where Cr-P bonding has occurred and much greater amounts of In near the surface. We can speculate that formation of thicker films would allow the In concentration to diminish to nearly zero in the Cr film because of the strong tendency of In to surface segregate. Sputter profiles for 100 Å Co/InP(110) showed that the valley-like structure is reduced to a plateau and, further, that the ratio of In to P is small. For Cr/InSb(111), we find intermediate behavior and a weak minimum near the buried interface. The similarity in cation depletion and the variability in the amount of cation accumulation suggests the generality of the expulsion mechanism discussed above. The differences are likely to be related to the amount of surface segregation and interface solubilities and are then specific to the cations and metals under study. Although annealing studies have not been reported for the other systems shown in Fig 8.10, we expect that additional compound formation at the buried interface would result in sharpening of the cation profiles and greater segregation, provided ternary phase formation is not favored.

Comparison of the sputter profiling results for Co/GaAs in Fig. 8.6 to those for Co/InP in Fig.8.10 emphasizes the importance of interface reaction in determining the ultimate concentration profiles in the overlayer. For both GaAs and InP, there is a good lattice match with bcc Co ($\sim$ 1.8% for GaAs and 3.9% for InP). However, the heats of formation for Co-P compounds favor substantial chemical reactivity while those for Co-As are much weaker. As a result, Co-P nuclei form readily and reaction expands both laterally and vertically with increasing Co coverage. For Co/GaAs, domains of epitaxial Co can grow directly on GaAs, despite the release of Ga and As during the early stages of deposition.

Synchrotron radiation photoemission studies were undertaken for Co/InP(110), Cr/GaAs(110), and Au/GaAs(110) to further investigate cation expulsion from the buried interface region [8.64, 65]. In Fig. 8.11 we show the core-level attenuation curves obtained in those studies. For Co/InP(110), the In 4*d* intensity decreases as the Co deposition increases until a critical coverage of 1.6 Å but it then *increases* steadily until $\theta_{Co} = 7$ Å. Thereafter, the In 4*d* emission attenuates very slowly. In this range, the P 2*p* signal drops rapidly, excluding the possibility of cluster formation. These results indicate that there is a coverage at which In surface segregation is triggered and In atoms are driven to the vacuum surface. The sputter profiles for 100 Å Co/InP(110) then show the consequences of this ejection of the cation from the reacted region. Such onsets for cation surface segregation onsets have now been observed for other systems as well [8.64, 65].

For Cr/GaAs(110), the Ga 3*d* and As 3*d* attenuation curves of Fig. 8.11 are smooth and structureless with the rate of Ga burial greatly exceeding that of As because As atoms surface segregate. Several authors [8.66, 67] have demonstrated that Cr does not react strongly with As to form a stable Cr-As compound at 300 K because the Cr-As bonds are relatively weak. During the early stages of formation of this system, the Cr atoms and the released anions and cations intermix in the region in the interface region. Since Cr-As compound formation is not as compelling as that observed for Ti-As, there is much less cation expulsion than for Ti/GaAs. In order to determine whether these anion and cation profiles are metastable with respect to temperature, we formed 100 Å Cr/GaAs(110) interfaces at 300 K and compared the profiles to those obtained for identical interfaces after annealing at 360°C for 75 min. The results of Fig. 8.12 show that the Cr/GaAs interaction does not produce the Ga valley structure (from [8.56]). Instead, there is a monotonic decrease in both Ga and As intensities away from the substrate, more like the behavior observed for Co/GaAs. Annealing at 360°C increased the Ga and As content of the surface and the near-surface region, and

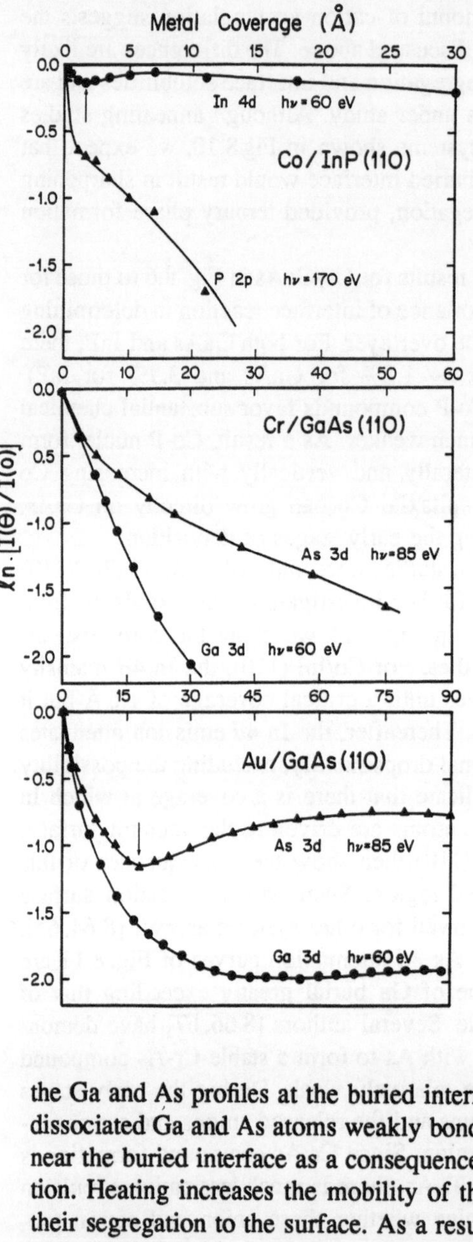

Fig. 8.11. Core-level attenuation curves for Co/InP(110) that show the onset of expulsion for In due to the nucleation and growth of Co-P bonding configurations. For Cr/GaAs(110), the Cr-Ga and Cr-As interactions are weaker and both Ga and As are present in the thickening layer in metastable bonding configurations. For Au/GaAs(110), both Ga and As are expelled above a critical thickness as the overlayer seeks to form a minimum-energy Au film. (From [8.56])

the Ga and As profiles at the buried interface became sharper. We conclude that dissociated Ga and As atoms weakly bonded to Cr exist in the intermixed region near the buried interface as a consequence of room temperature substrate disruption. Heating increases the mobility of these impurity-like atoms and promotes their segregation to the surface. As a result, the Cr layer becomes more ordered and of higher purity so that the free energy of the system is lowered.

For these reactive systems, atoms deposited from the vapor phase onto clean III-V semiconductor surfaces change the overall energy configuration of the near-surface bonds and can weaken or break anion–cation covalent bonds. Similar substrate modifications can occur for nonreactive adatoms, and the latter induce limited amounts of substrate disruption. For reactive systems, disruption continues until the reaction products that form become a self-limiting diffusion barrier.

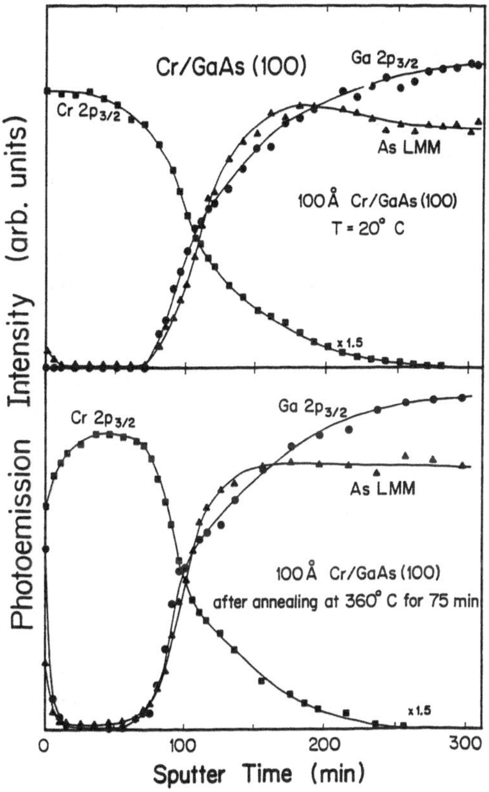

Fig. 8.12. Sputter profiles for Cr/GaAs(100) showing Ga and As intermixing in the film at room temperature but the expulsion of both when annealing makes it possible for the solute atoms to diffuse to the surface. (From [8.56])

Thereafter, the metal nucleates, incorporating some semiconductor atoms or forcing their segregation to its surface. Factors which control the concentration profile include the extent of substrate disruption (and thus the number of released anions and cations), the chemical reactivities of the metal with the anions or cations, the stability of the reaction products (if there are any), and the semiconductor atom solubilities in the metal and in the reaction products. The concentration profiles differ from those described for Co/GaAs because of the presence of the reacted region. However, many of the observations made in Sect. 8.2 are still applicable for reactive systems, including the mechanism of segregation and expulsion.

8.2.3 Au III-V Interfaces

A large number of Au/GaAs interface studies have been reported, and those by *Spicer* and coworkers [8.68–70], *Xu* et al. [8.71], *Grioni* et al. [8.54, 72] and *Andersson* et al. [8.73] are particularly relevant to our discussion here. It is generally agreed that Au adatoms induce disruption of the substrate but do not form large Au islands or compounds with Ga or As at room temperature. Instead, there are high concentrations of Ga and As atoms in the intermixed region during the early stages of formation and these semiconductor atoms are in a metastable, supersaturated-like solution.

Substrate disruption is completed by coverages of 10–15 Å. The synchrotron-radiation attenuation curves for Ga and As shown in Fig. 8.11 show the gradual decrease in Ga and As emission for $\theta_{Au} \leq 18$ Å. Above ~ 18 Å, there is a very surprising increase in the As intensity. This onset of *anion* surface segregation is of interest here because the interface reactions are different from those discussed above for reactive interfaces where cation segregation is observed. In particular, the onset of As segregation toward the surface cannot be due to expulsion from a reacted Au-Ga alloy region, as was the case when In was expelled from Co-P regions. Instead, increasingly pure Au aggregates form and the formation expels the As to the surface. This observation is then analogous to that found in Fig. 8.12, where annealing was needed to expel Ga and As from supersaturated solution in Cr. These expelled As atoms then float on the Au film because of their low solubility in Au once the Au layer has reached a critical thickness. This is reflected in the attenuation curves, the magnitude of the segregated As, and the absence of a valley structure in the As distribution profile of Fig. 8.13. Since Au-Ga bonding is more favorable, there is a greater chance of forming nuclei of Au-Ga or a higher solubility in Au. Hence, there is a weaker expulsion and a much lower concentration of segregated Ga. Analogous behavior can be seen in Fig. 8.13 for Au/InSb(110) although there is greater substrate disruption and an even higher Sb concentration on the Au surface.

Support for this anion segregation onset model comes from a photoemission study of *Ruckman* et al. [8.74] of the binary Au/Ge system where a similar rise

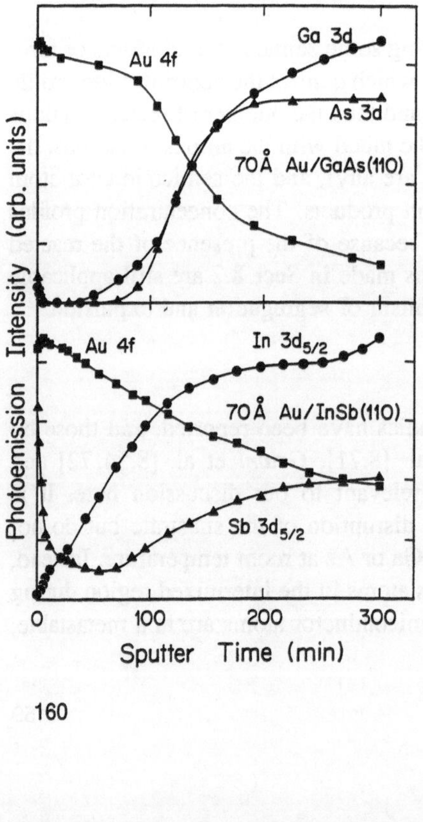

Fig. 8.13. Sputter profiles for Au/GaAs(110) and Au/InSb(110) showing surface segregation and the absence of any structure near the buried interface that would indicate preferential bonding of the semiconductor atoms with Au. (From [8.56])

in the attenuation curves for Ge was observed at a nominal Au coverage of 15–20 Å. In this case it was obvious that Ge expulsion toward the free surface cannot be caused by the formation of an Au-Ge phase. Instead, those results can be interpreted in terms of the lowering in energy of the interior of the film as pure Au forms and Ge is driven to the surface. It appears that an Au film of high purity decorated by a skin with a high concentration of semiconductor atoms is thermodynamically favored over the solution or any compound at room temperature. This is no longer the case, of course, when Au-Ga compound formation is made possible by thermal processing [8.75–78].

In conclusion, we have demonstrated that cations can be expelled from the region where the metal-anion compound occurs as the small nuclei evolve into microcrystallites with appreciable dimension. Analogous behavior is observed for noble-metal/semiconductor systems but the evolution of Au nuclei themselves is critical since they are the lowest-energy configuration. We have shown that the onset of surface segregation, the extent of the cation-deficient region, and the concentration of expelled cations depend on factors that include the amount of compound formation, the number of dissociated cations, and the solubility of the cations in the growing compounds.

8.3 Low-Temperature Interface Formation

The discussions of Sect. 8.2 focused on the properties of interfaces formed at 300 K. In this section, we emphasize changes that are a consequence of overlayer formation at low temperature, with special attention paid to Ti/GaAs, Cr/GaAs, and Ag/GaAs because these three overlayers are representative of reactive, disruptive, and nondisruptive systems. As we will show, chemical bonding is not quenched at low temperature and the amount of disruption induced by the atoms is rather insensitive to temperature for $T \leq 300$ K. However, processes related to diffusion and atom segregation are suppressed, and kinetic trapping gives rise to quite different concentration profiles across the interface.

8.3.1 Ti/GaAs(110)

The studies discussed above for Ti/GaAs(110) interface formation at 300 K showed strong reaction that was present from the lowest coverage, as was evident from the Ga and As $3d$ core-level evolution, and core-level intensity variations as a function of metal coverage indicated intermixing and the disruption of several monolayers of the substrate. There were then significant changes in atom concentrations as the amount of Ti increased and the film thickened. In the following, these results are compared with those observed for low-temperature atom deposition so as to correlate the fundamental properties of the interface with the temperature and formation process.

In Fig. 8.14 we show the Ga $3d$ core level emission from [8.79] for representative Ti coverages for atom-by-atom deposition at 60 K on n-GaAs(110) as

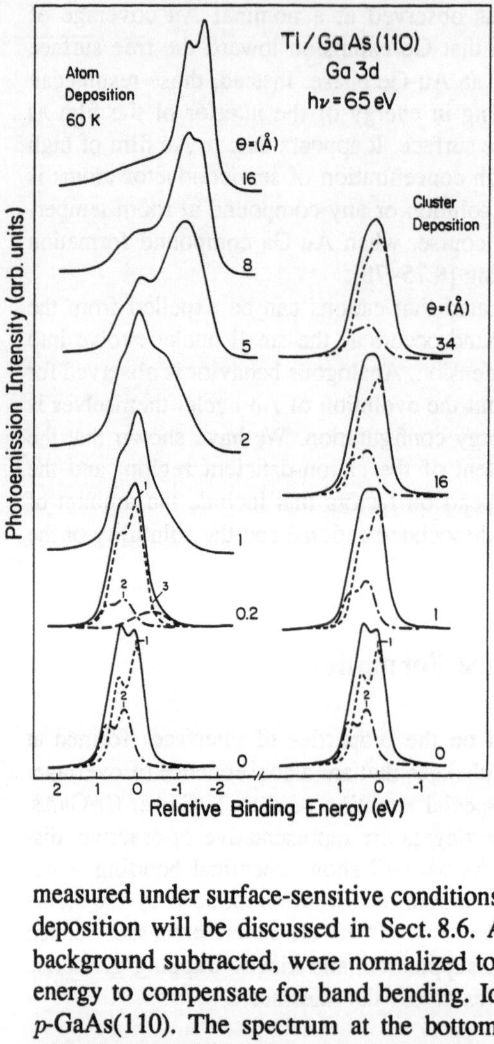

Fig. 8.14. Ga 3d core level evolution for Ti atom deposition onto GaAs(110) at 60 K. These results show that a reduction in temperature does not quench Ti-GaAs reactivity. The results for cluster deposition show that there are no reaction products that can be seen. (From [8.79])

measured under surface-sensitive conditions ($h\nu = 65$ eV). The results for cluster deposition will be discussed in Sect. 8.6. As before, core-level spectra had the background subtracted, were normalized to constant height, and were shifted in energy to compensate for band bending. Identical lineshapes were obtained for p-GaAs(110). The spectrum at the bottom corresponds to the cleaved surface with decompositions that show bulk and surface atom contributions. Disruptive interaction between the Ti adatoms and the substrate is evident for coverages as low as 0.02 Å (not shown). The Ga 3d spectrum for $\theta_{Ti} = 0.2$ Å clearly shows a broad new feature, component 3, at ~ 0.5 eV lower binding energy than the bulk peak. Simultaneously, emission from the surface-shifted component, number 2, diminishes rapidly and was not needed to fit the core level lineshape by 1 Å Ti deposition. Component 3 exhibits increasing broadening and it is rather featureless by $\theta = 2$ Å because of a wide range of inequivalent configurations for Ga atoms dissociated from the substrate. At higher coverage, component 3 sharpens as the environments of the released Ga atoms become better defined within the probed region as in the $\theta_{Ti} = 16$ Å spectrum. The total Ga 3d emission shift is 1.67 eV, in agreement with the final shift observed for atom deposition at 300 K. The increasing shift of component 3 to lower binding energy makes it possible to determine with confidence the intensity and binding energy of the attenuated substrate doublet, as required for quantitative band bending studies (see following section).

Analysis of the As 3*d* spectra (not shown) reveals a more complex behavior. At low coverages ($\theta < 1$ Å), the lineshape evolution for 60 K is similar to that at 300 K, where a Ti-induced doublet overlaps the surface component of the clean substrate. For $\theta_{Ti} > 5$ Å, the persistence of the high binding energy doublet can only be explained by postulating overlap of the substrate and a Ti-induced feature. At intermediate coverages, lineshape decompositions require three doublets to obtain acceptable results, i.e. statistically neutral residuals when fits were compared with data points. These findings suggest the existence of a range of bonding configurations. Significantly, comparison with the 300 K results reveals the absence of a doublet corresponding to segregated As atoms at 60 K because As outdiffusion is kinetically limited.

In Fig. 8.15 we show the total and component-specific Ga 3*d* attenuation curves. The total attenuation curve for interface development at 60 K has a slope with 1/e decay length of ~ 3.8 Å, and the even more rapid decay of the substrate component indicates surface disruption. For coverages above ~ 5 Å, the rates of attenuation of the substrate and component 3 are quite similar, indicating uniform coverage of substrate and disrupted regions. This is quite different from the behavior at 300 K, where Ga atoms intermix with the thickening Ti film over a more extended scale. Low temperature thereby provides kinetic trapping near the buried interface of atoms released from the substrate.

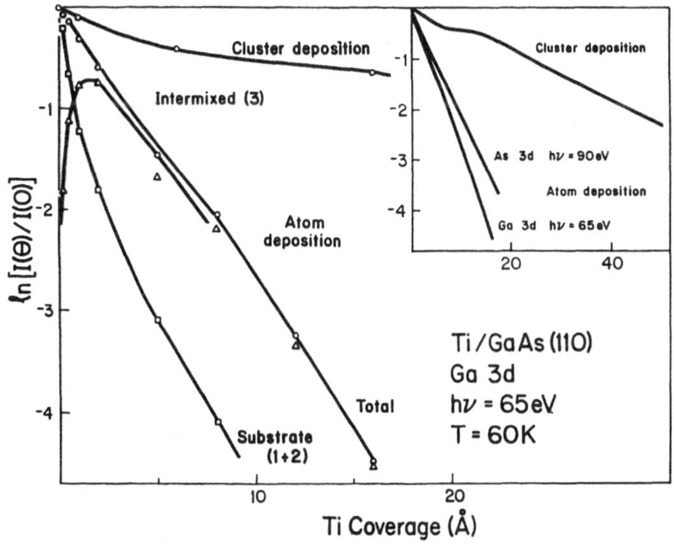

Fig. 8.15. Attenuation curves For Ti atom deposition onto GaAs(110) at 60 K showing that outdiffusion of Ga and As into the Ti overlayer is suppressed at low temperature. The results for cluster deposition show that the morphology is very different and large portions of the surface remain exposed to high coverage. (From [8.79])

From the attenuation curves, we can estimate the amount of disruption to be equivalent to ~ 3 ML for both 300 K and 60 K atom deposition. There is no evidence for differences that might reflect thermal quenching of chemical reactions that are responsible for substrate disruption. The fact that component 3 appears at the same final binding energy at 60 and 300 K indicates that Ga atoms released from the substrate are present in an equivalent dilute form in the intermixed region. This is confirmed by the well-defined Ga 3*d* lineshape for both cases. Kinetic trapping is particularly evident from the attenuation curves because the rate of total attenuation at 300 K is much slower than for 60 K, and the final Ga configuration is not reached until ~ 40 Å at 300 K, compared to ~ 12 Å for 60 K.

The inset in Fig. 8.15 shows total attenuation curves for Ga and As for Ti atom deposition at 60 K. While both substrate species are rapidly attenuated, the As intensity is higher than that of Ga at all coverages, even though the photoelectron mean free paths are the same. This shows that the As atoms have a greater tendency to outdiffuse when not bound in Ti-As bonds. This tendency for As expulsion is more clearly manifested at 300 K and is reflected by the presence of As on the surface of relatively thick films. As discussed in Sect. 8.2, such atom distributions are metastable, and the amount of As in the overlayer is probably greater than would be predicted from an equilibrium phase diagram, even at 300 K. Low-temperature formation creates an even less stable buried region because atom movement is kinetically limited.

Figure 8.16 summarizes schematically the morphologies produced by Ti atom deposition at 60 and 300 K. For atom deposition at 60 K, (a) shows that released As atoms are kinetically trapped near the buried interface, but that disruption

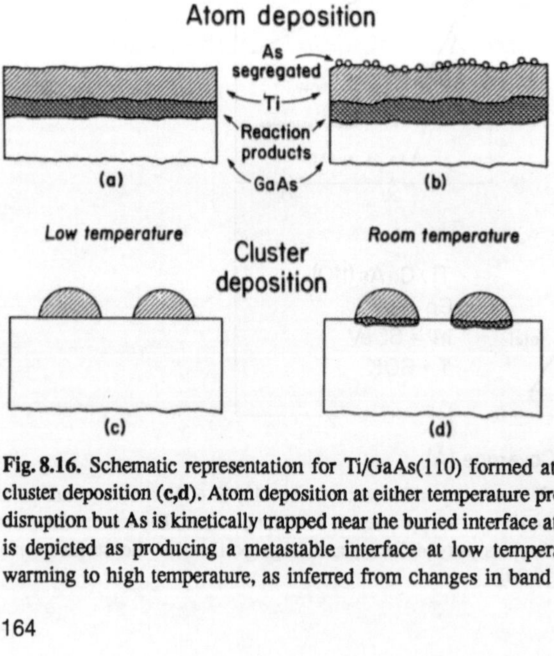

Fig. 8.16. Schematic representation for Ti/GaAs(110) formed at (a) 60 K and (b) 300 K and for Ti cluster deposition (c,d). Atom deposition at either temperature produces the same amount of substrate disruption but As is kinetically trapped near the buried interface at low temperature. Cluster deposition is depicted as producing a metastable interface at low temperature because reaction occurs upon warming to high temperature, as inferred from changes in band bending

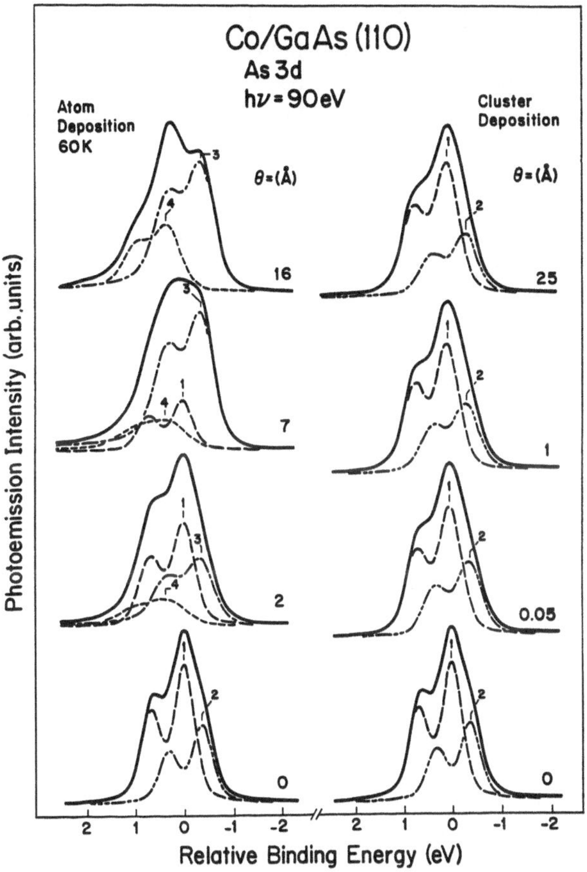

Fig. 8.17. Comparison of As 3*d* emission for Co atom deposition onto GaAs(110) at 60 K and for Co cluster deposition. The former shows substrate disruption and the appearance of Co-induced features equivalent to those of Fig. 8.2. Cluster deposition leads to a reduction in the surface core-level component due to surface unrelaxation around the clusters. (From [8.80])

and reaction occur to the same extent as at 300 K, sketched in (b). However, As surface segregation is observed at 300 K. (The results for Ti cluster deposition shown at the bottom of Fig. 8.16 will be discussed in Sect. 8.6).

8.3.2 Co/GaAs(110)

Figure 8.17 (from [8.80]) shows representative As 3*d* core level EDCs for Co/GaAs(110) following atom deposition at 60 K (*left*) and the deposition of preformed clusters (*right*, to be discussed in Sect. 8.6). In both cases, the clean surface spectra were recorded at 60 K. Features labeled 1 and 2 again arise from emission from bulk and surface As atoms, respectively. Alignment of the bulk emission provides a direct measure of changes in the surface Fermi level position, to be discussed in Sect. 8.4. The results of Fig. 8.17 allow direct comparison for atom deposition at 60 K and at 300 K (Fig. 8.2). (Note that the spectral resolution of Fig. 8.17 is substantially better than in Fig. 8.2.)

Co atom deposition at 60 K leads to reduced As 3*d* emission from the surface component and the appearance of a Co-induced feature (labeled 3) at −0.4 eV relative binding energy. For submonolayer coverages, it is difficult to distinguish between component 3 and the surface shifted component for As atoms at the relaxed (110) surface because the features are nearly degenerate in binding energy. (The rate of loss of the surface component can be determined, however, by analysis of the Ga 3*d* EDCs, since there were no such accidental degeneracies.) For depositions above ~ 1 Å, there is a second Co-induced feature (labeled 4) at +0.25 eV relative binding energy. Feature 3 dominates the spectra for $\theta_{Co} \geq 2$ Å. The binding energies of components 3 and 4 are very similar to those observed at low coverage for Co atom deposition at 300 K. These As 3*d* features persist to much higher coverage for 300 K deposition and exhibit coverage-dependent binding energy shifts that are indicative of atomic environment changes. Likewise, these Co-induced As features develop asymmetric lineshapes indicative of atoms in metallic environments and their incorporation in the evolving overlayer. (Such complications were readily handled by introducing a Doniach-Sunjic asymmetry, as discussed in Sect. 8.1.1.) It is clear from Figs. 8.17 and 2 that Co deposition disrupts the substrate at both 60 and 300 K.

In Fig. 8.18, we show the total integrated emission from the As 3*d* core levels for Co/GaAs obtained with $h\nu = 90$ eV for atom deposition at 300 K (*top* and 60 K (*bottom*). At both temperatures, the substrate attenuation representing the sum of the bulk and surface components is very rapid (1/e length ~ 1.5 Å) because of substantial substrate disruption. For $\theta_{Co} > 2$ Å the substrate attenuation slows to 1/e ~ 4 Å as disruption is reduced and more uniform growth is initiated. As discussed in Sect. 8.2.1, feature 3 is due to emission from As atoms segregated to the surface while component 4 is derived from As atoms intermixed in the growing Co overlayer near the surface. We note that the As attenuation rate at high coverage is much faster at 60 K than at 300 K (1/e ~ 12 Å) as outdiffusion of released As is kinetically inhibited. Moreover, the As 3*d* EDCs show coverage-dependent binding energies that reflect changes in atomic environment during 300 K overlayer development, but not at 60 K.

In Fig. 8.19 we show the total Ga 3*d* intensity for atom deposition at 60 K as well as the contributions from the substrate and intermixed Ga atoms (from [8.80]). Again, we observe rapid initial attenuation of substrate features below ~ 2 Å due to conversion of substrate atoms into intermixed atoms, i.e., Ga atoms in chemically distinguishable environments. The Ga 3*d* EDCs reveal a Co-induced feature at 60 K even for only 0.05 Å of Co. This feature shifts in binding energy with increased deposition and exhibits the asymmetric lineshape of Ga in a metallic environment. The continuous binding energy shift indicates increasing Co coordination, and the total attenuation rate for $\theta > 6$ Å (1/e ~ 4 Å) suggests that most of the disrupted Ga atoms remain near the buried interface.

It is important to emphasize that the Ga 3*d* EDCs and attenuation curves were indistinguishable for atom deposition at 60 and 300 K. This indicates that the redistribution of released Ga atoms during interface formation is not kinetically constrained, i.e., these atoms remain near the buried interface at both 60 and 300 K. In contrast, the As distribution is temperature dependent because surface segregation is clearly evident at 300 K but not at 60 K. A simple calculation

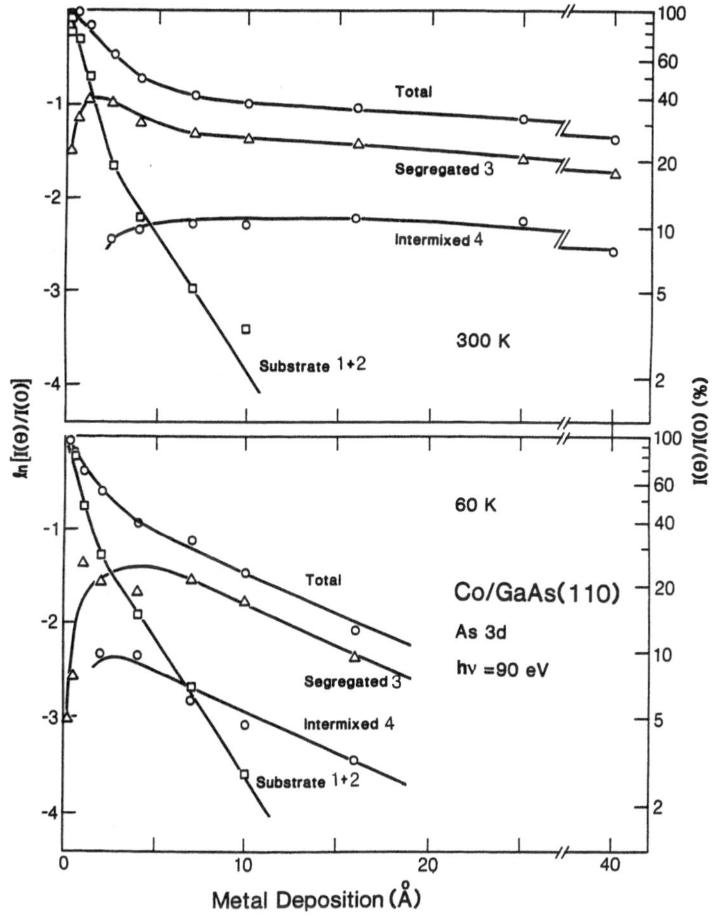

Fig. 8.18. Attenuation curves for Co deposition at 300 K and 60 K showing kinetic trapping of As at low temperature. (From [8.80])

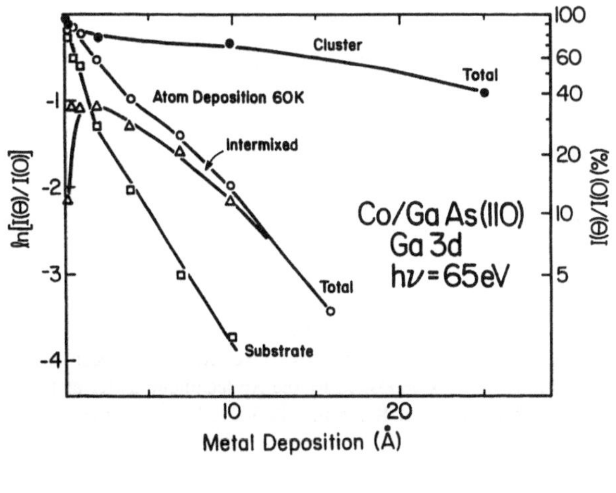

Fig. 8.19. Attenuation curves for Co atom deposition on GaAs(110) at 60 K and Co cluster deposition. The low-temperature results for Ga are equivalent to those at 300 K because Ga is not expelled from the interface region, i.e., there is no kinetically limited atom redistribution. The cluster results show inefficient coverage of the surface by the metal clusters. (From [8.80])

leads to an estimate that ~ 1–2 monolayer-equivalents of substrate are disrupted at both temperatures.

8.3.3 Ag/GaAs(110)

Previous studies of Ag/GaAs(110) interface formation at 300 K have shown weak substrate interaction and the formation of Ag clusters [8.81, 82]. These clusters assume metallic character at low deposition, but cover the surface rather inefficiently. This tendency to cluster is sharply reduced at 60 K because surface mobility is severely suppressed. In Fig. 8.20 we show Ga and As 3*d* total emission intensities as a function of Ag coverage for atom deposition at 60 and 300 K and for cluster deposition (Sect. 8.6). As demonstrated in [8.81, 82], the slow attenuation and the absence of Ag-induced spectral features are consistent with Ag cluster growth at 300 K with no detectable metal-induced substrate disruption. (The region beneath these Ag clusters cannot be probed with photoemission but, based on measurements for 60 K atom deposition, surface modification is likely, as will be discussed below.) Significantly, from Fig. 8.20 it can be seen that substrate attenuation is identical for cluster deposition (*triangles*) and 300 K atom deposition (*circles*). A rough approximation based on a hemispherical cluster model is able to fit the attenuation results with a cluster radius of ~ 30 Å for $\theta \geq 2$ Å for 300 K atom deposition [8.80]. The relatively rapid rate of attenuation during the earliest stages of growth is likely to be due to the formation of smaller clusters, while the increasing attenuation rate at higher coverage reflects overlap of large clusters to form a continuous film.

As can be seen from Fig. 8.20, the interface morphology for Ag/GaAs(110) for atom deposition at 60 K is very different from that at 300 K. The exponential

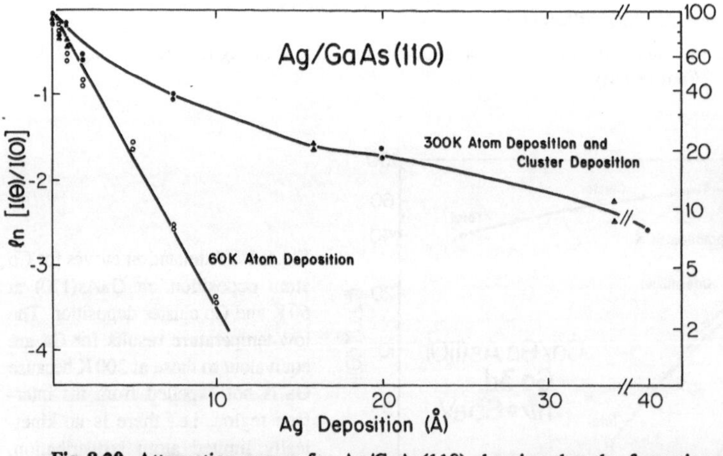

Fig. 8.20. Attenuation curves for Ag/GaAs(110) showing that the formation of a Ag layer at 60 K produces an abrupt interface that grows in an approximately layer-by-layer fashion. Deposition at 300 K produces clusters that are equivalent to clusters deposited by the Xe technique as far as their attenuation of the substrate is concerned. Warming the 60 K Ag film to 300 K leads to spontaneous cluster formation. (From [8.80])

decrease in the total substrate emission intensity ($1/e \sim 3$ Å) indicates nearly uniform layer-by-layer growth consistent with reduced Ag adatom mobility and the inhibition of Ag cluster formation. In this case, lineshape analysis of the Ga $3d$ EDCs reveals the growth of a Ag-induced interface component at -0.2 eV relative binding energy as the surface-shifted component is reduced. In particular, the surface component is converted to this interface component after $\theta_{Ag} = 2$ Å. The presence of this interface component is a consequence of atom deposition and is not correlated with the onset of the metallic character in the overlayer. A distinct interface component is not observed for As, but the bulk component is attenuated at the same rate as for Ga. We therefore infer that surface As atoms are either less sensitive to their local environment or that the As interface component lies too close in energy to the surface-shifted As feature to be resolvable. While the EDCs obtained at 300 K do not provide direct evidence for these components beneath the clusters (these regions are inaccessible to photoemission), their presence cannot be excluded.

In summary, these low temperature results for representative interfaces show that the chemical reactivity is not significantly altered at low temperature. They show, however, that the redistribution of atoms released from the substrate *is* altered, and kinetic trapping leads to the formation of a very metastable interface layer. Annealing to temperatures as low as 300 K results in the expulsion of trapped atoms, in effect analogous to that discussed above for interfaces formed at 300 K and annealed to higher temperature. Finally, these results show that clustering can be inhibited by reducing surface mobility. Interestingly, if Ag/GaAs(110) interfaces formed at 60 K are annealed by warming to 300 K, there is a morphology change and the Ag layer agglomerates into islands. In the following section we examine the consequences of these changes in the context of band bending.

8.4 Surface Photovoltaic Effects

The rectifying behavior associated with metal-semiconductor contacts was explained many decades ago by the formation of a depletion region on the semiconductor side of the interface. The potential barrier related to that layer, the Schottky barrier, is the most important parameter controlling the transport properties of such a junction, and a great many studies have focused on understanding the physical phenomena responsible for the formation of that barrier [8.1–24, 83–88].

Despite extensive study, the fundamental mechanisms involved in Schottky barrier formation remain controversial. In part, this is because metal–semiconductor interface formation involves simultaneous processes of the sort discussed above, including substrate disruption, nucleation of reacted species, outdiffusion of substrate atoms, and adatom clustering. It should be clear by now that these metal–semiconductor interfaces are generally far from ideal in terms of their atom arrangement and, therefore, their electronic structure.

Recent interest in Schottky barrier formation has involved effects related to low temperature interface evolution. For those interested in Schottky barrier formation, the original purpose for such low-temperature experiments was to simplify some of the intricate phenomena that take place at 300 K. In this way, it was hoped that the various interface phenomena could be isolated and their individual roles in barrier formation could be identified. This was partially successful because, for example, nucleation and growth kinetics depend on the temperature of the substrate during metal deposition (Sect. 8.3). At the same time, the results discussed above demonstrate that the amount of substrate disruption is rather insensitive to the substrate temperature during deposition.

8.4.1 Dependence of Band Bending on Temperature and Bulk Dopant Concentration

In Fig. 8.21 we schematically show how the surface Fermi level moves in the band gap of GaAs as a function of the amount of metal deposited at low temperature, LT. This behavior was observed by several groups for a significant number of metals. In particular, *Kahn* and coworkers [8.16–18] and *Spicer* and coworkers [8.3] have reported them for a variety of systems. Subsequent calculations by *Klepeis* and *Harrison* [8.86] and by *Allan* and coworkers [8.87] reproduced these Fermi level movements, achieving reasonable agreement with the experimental results sketched in Fig. 8.21. They showed that metals deposited on n-GaAs(110) at low temperature (taken to be 200 K or less in the above experiments) induced little change until a few angstroms had been deposited but that E_F moved toward midgap at higher coverages. This step has been attributed to the metallization of the overlayer. The behavior at low coverage for n-GaAs was taken to mean that the adatoms acted like donors, adding charge to the conduction band. Since electrons were already the majority carriers, there was little change in Fermi level position. Results for GaAs in the low coverage regime showed rapid E_F movement into the gap, consistent with the apparent donor-like character of the adatoms. *Spicer* et al. [8.3] sought to distinguish between reactive adatoms that

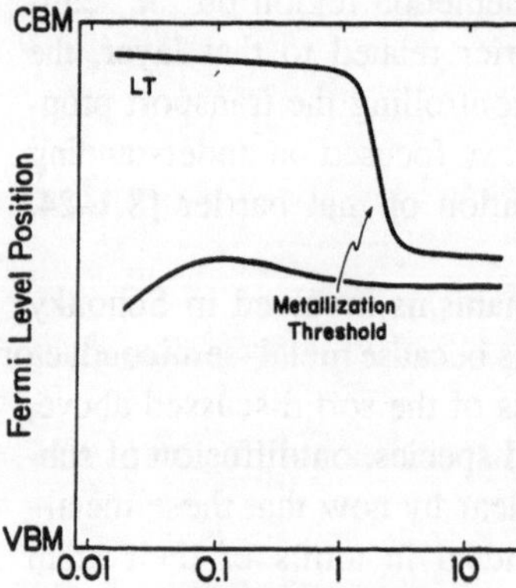

Fig. 8.21. Schematic summary of the results obtained recently for metal atom deposition onto GaAs(110) at low temperature. For n-GaAs, the surface Fermi level does not move deep into the gap until the metallization threshold. For p-GaAs, the adatoms induce rapid movement into the gap and an overshoot is often observed

would lead to defect formation upon condensation and nonreactive metals that would lead to pinning that could be understood in terms of metal-induced gap states.

In the following, we describe how these low-temperature studies present only part of an even more interesting story. By changing the dopant concentration and type, we show that Fermi level movement depicted in Fig. 8.21 is actually symmetric for n- and p-type samples of comparable concentration. This demonstrates that band bending is a sensitive function of the bulk dopant concentration. In all cases, the adatom-induced changes in band bending are relatively small in the coverage region below 1–2 ML for low dopant concentrations but significant changes are observed for the same number of adatoms deposited onto heavily doped samples. This behavior is observed for unreactive Ag, disruptive Au and Co, and reactive Ti, and there is analogous behavior for InP(110). At high coverage, E_F moves into the gap as overlayer metallization occurs. The low-coverage results can be understood in terms of nonequilibrium surface photovoltaic effects that give rise to band flattening, and the step into the gap reflects the formation of a conducting layer on the surface that shorts the surface photovoltage (SPV).

The procedures for the high-resolution synchrotron radiation photoemission experiments were analogous to those described above but special precautions were taken to eliminate spurious effects due to partially pinned clean surfaces. For the results discussed here (taken from [8.89–94], only those cleaves with E_F within 60 meV of the conduction band minimum, CBM, or valence band maximum, VBM, were used. The n-type (p-type) samples were Si (Zn) doped, and the doping concentrations were $2 \times 10^{18}\,\mathrm{cm}^{-3}$ (high doping, HD) and $1 \times 10^{17}\,\mathrm{cm}^{-3}$ (low doping, LD). Results for p-GaAs with a doping concentration of $4 \times 10^{19}\,\mathrm{cm}^{-3}$ (very high doping, VHD) were also obtained, as were results for n-type samples doped at $5 \times 10^{16}\,\mathrm{cm}^{-3}$ (very low doping, VLD). The measurements were conducted at 60 K, and sets of spectra were obtained under identical conditions for LD- and HD-samples of both n- and p-type. The points identifying the position of E_F in the gap were derived from analysis of bulk- and surface-sensitive photoemission spectra for Ga $3d$ and As $3d$ core levels. Line-shape analysis of core level emission made it possible to separate the substrate component from adatom-induced components and to measure changes in band bending with a precision of 30 meV [8.43].

The results discussed in the two previous sections have shown that Ti disrupts ~ 6 Å of the GaAs(110) substrate at 300 K, that Ti-As bonding configurations form, that the intermixed region close to the buried interface is Ga deficient, and that there is As atom segregation to the free surface [8.58]. The temperature-dependent studies have shown that the same amount of disruption occurs at 60 K as at 300 K but that kinetic processes involved with redistribution of the released semiconductor atoms are greatly reduced at 60 K [8.79]. They have also shown that adatom-induced chemical changes are indistinguishable for n- and p-samples, regardless of dopant concentration.

In the top panel of Fig. 8.22 (from [8.89]) we show E_F movement in the surface band gap upon Ti deposition at 60 K onto high dopant (HD, $2 \times 10^{18}\,\mathrm{cm}^{-3}$)

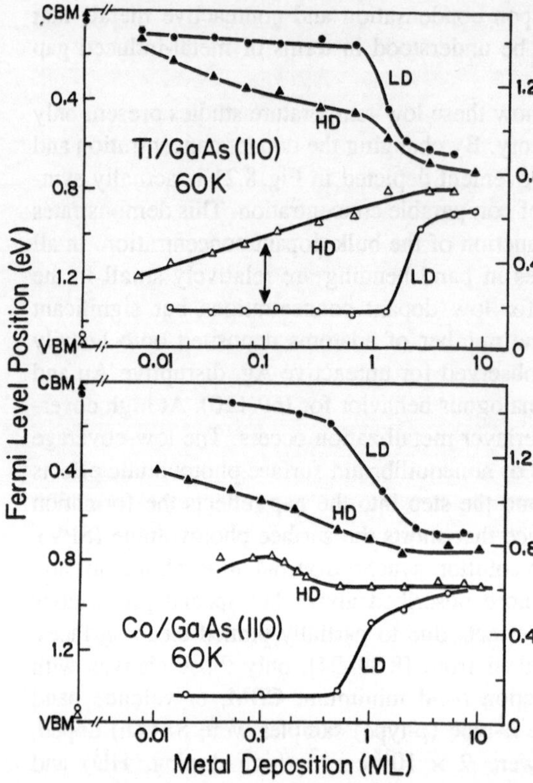

Fig. 8.22. Quantitative analysis of band bending as a function of Ti and Co atom deposition onto GaAs(110) at 60 K when samples of matched bulk dopant concentration are examined. As shown, the asymmetry sketched in Fig. 8.21 is an artifact of the dopant concentration. Samples of matched dopant concentration exhibit more symmetric behavior. (From [8.89])

and low dopant (LD, 1×10^{17} cm^{-3}) GaAs(110). There are almost no Ti-induced changes in E_F from 0.006 to $\sim$ 1 ML for LD n-GaAs. Between 1 and 4 ML, E_F moves to its final position 0.67 eV below the CBM. For LD p-GaAs, the movement of E_F is highly symmetric with that for LD n-GaAs, with almost no change below $\sim$ 1 ML and then a step toward midgap. Although this step behavior has been observed previously for n-GaAs, it has not been seen for p-GaAs, and the remarkable symmetry had gone unnoticed (compare to Fig. 8.21). For HD Ti/GaAs, E_F moves monotonically toward its final position with adatom deposition for both n- and p-GaAs, starting at the lowest coverage. For HD n-GaAs, there is a vestigial step at the same coverage as for LD GaAs, $\sim$ 1 ML. We stress that core-level and valence-band analyses indicate equivalent substrate disruption for LD and HD n- and p-GaAs. Indeed, while Ti/GaAs is the most reactive system studied to date, the symmetry is clearly independent of surface reactivity and disruption.

To demonstrate the generality of this symmetric Fermi level movement, we investigated Co [8.89], Ag [8.89], Au [8.92], Bi [8.95,96], and Sm [8.97] atom deposition on GaAs(110) at 60 K. The studies discussed above have shown that Co induces limited disruption and forms an epitaxial bcc overlayer at 300 K. At 60 K, an equivalent amount of disruption is observed, but liberated As atoms are kinetically trapped near the buried interface. For Ag, atom deposition leads

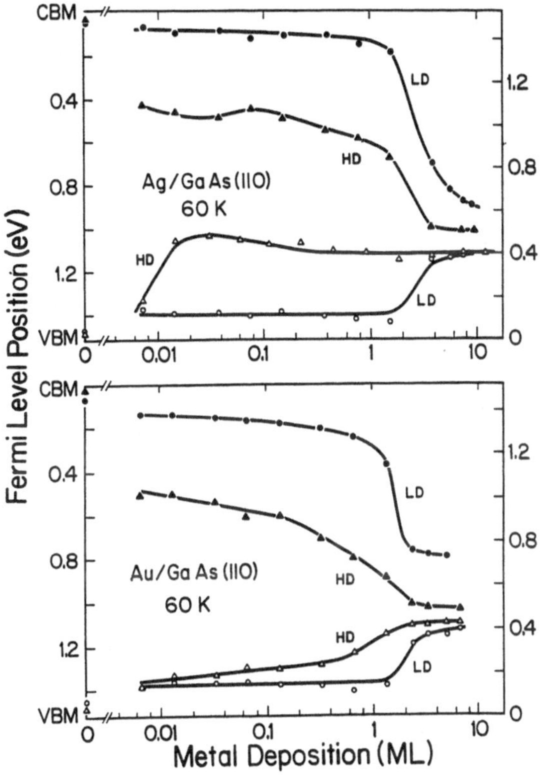

Fig. 8.23. Band bending changes for Ag/GaAs(110) and Au/GaAs(110) analogous to those of Fig. 8.22. For Au/GaAs, there is no p-type overshoot and the symmetry of E_F movement is particularly obvious. (From [8.92])

to spontaneous island formation at 300 K with no evident surface disruption. Thermally activated Ag surface diffusion is negligible at 60 K, and approximately layer-by-layer growth is observed [8.3, 4, 80]. Au deposition leads to substrate disruption with anion segregation. Bi deposition leads to an ordered array that is semiconducting at 1 ML coverage but then Stranski-Krastanov growth sets in and metallization occurs at higher coverage [8.95, 96]. Bi deposition induces no substrate disruption. Sm overlayers are highly reactive and exhibit valence changes as a function of temperature [8.97].

The lower panel of Fig. 8.22 shows the E_F movement induced by Co adatoms and Fig. 8.23 shows the behavior for Ag and Au. Analogous results were obtained for Bi. There is little change at low coverage for LD n- and p-GaAs(110), but E_F shifts rapidly into the gap after ~ 1 ML. For Co, Ag, Au, and Bi, the transition is completed by ~ 6 ML, ~ 10 ML, ~ 3 ML, and ~ 6 ML, respectively. As for Ti/GaAs, the half-way point occurs at approximately the same coverage for LD n- and p-GaAs, emphasizing a common mechanism for each metal. In contrast, movement is evident by 0.01 ML for these metals for HD n-GaAs, and like Ti, they exhibit a vestigial step as E_F drops in the gap. The vestigial step for HD n-GaAs suggests that the mechanism responsible for movement into the gap at lower coverage is supplemented by that which causes the step for LD samples. For HD p-GaAs, the previouslys observed overshoot of the Fermi level position is seen at low coverage for Co and Ag, followed by relatively little change [8.3,4,16–18]. HD p-GaAs results do not show a step near ~ 1 ML because of the overshoot induced at lower coverage. Of these metals, the results for Au/GaAs(110) show most dramatically the symmetry in E_F movement. For Au, the rise into the gap for LD p-GaAs at low coverages is relatively weak and the step observed at higher coverage is clearly evident. There is no evidence for an overshoot for Au/GaAs.

Since the results of Figs. 8.22 and 23 show dramatically different behavior for 60 K and 300 K, we examined the temperature dependent evolution for Al/GaAs(110) for a variety of temperatures. Figure 8.24 (from [8.91]) summarizes E_F movement for Al deposition on LD n-GaAs for substrates held at 60, 100, 150, 200, and 300 K. Substrate temperature variations produce a wide range of band bending below ~ 8 ML. In particular, the amount of induced band bending for 0.01 ML of Al more than doubles from 60 K to 200 K despite the fact that the Al atoms are likely to be highly dispersed and noninteracting on the surface. Moreover, the plateau region below ~ 2 ML is diminished as T increases, and it is completely washed out at 300 K. In this temperature range, the step toward midgap occurs at progressively lower coverage with midstep coverages of ~ 7 ML at 60 K, ~ 4 ML at 200 K and is not discernible at 300 K.

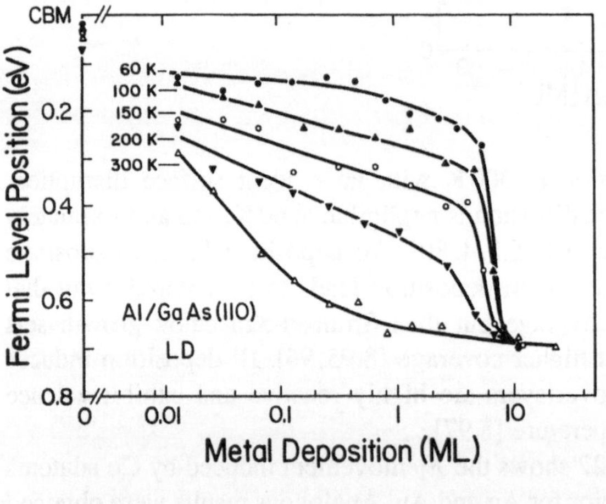

Fig. 8.24. Temperature-dependent results for Al/GaAs(110) interface evolution showing the gradual change from the band bending pattern observed at low temperature to that at room temperature. The step near 6 ML corresponds to the metallization threshold. (From [8.91])

The results of Fig. 8.24 raise fundamental questions about the temperature-dependent movement of E_F for interfaces formed at a given low temperature and heated or cooled to another. In particular, it might be argued that deposition at 60 K "freezes in" a different interface morphology or bonding than observed upon deposition at 200 K or 300 K because of kinetic inhibitions. In this case, warming might produce E_F positions equivalent to those characteristic of the higher temperature. If this were true, then temperature would play a critical role in dictating surface energy levels and E_F pinning. If such temperature-induced bonding changes were responsible for E_F movement, then a critical question would be whether E_F would move back toward the band extrema upon recooling.

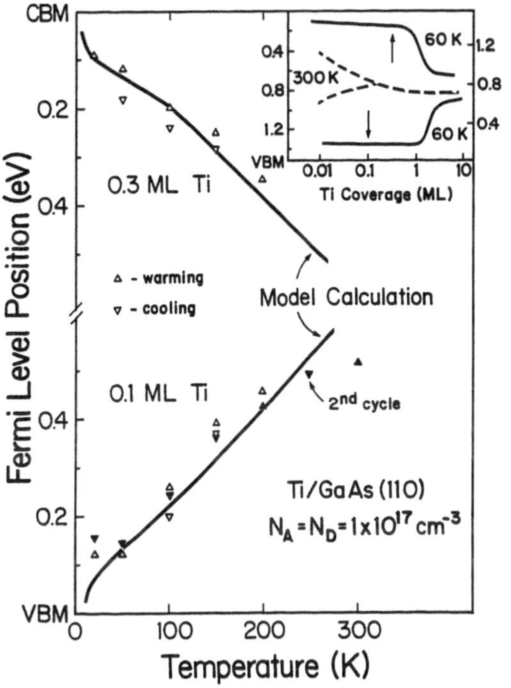

Fig. 8.25. Movement of the surface Fermi level induced by submonolayer amounts of Ti on GaAs(110) as the temperature of measurement increased from 20 K to 200 or 350 K and then returned to 20 K. As shown, the position of E_F in the gap moves in a fully reversible fasion with T. The solid lines correspond to the maximum amount of band bending predicted. (From [8.92])

In order to investigate the reversibility of E_F movement for metal–GaAs interfaces, we examined n-and p-GaAs(110) onto which submonolayer amounts of Ti, Co, and Ag atoms had been deposited. For these measurements, the sample temperature was variable from 20 to 350 K. Note that for a good cleave, temperature reduction from 300 K to 20 K moved E_F by only 50–60 meV due to the temperature dependence of E_F relative to the band extrema in the bulk and the changing band gap. Greater movement indicated that E_F was partially pinned.

Figure 8.25 (from [8.90]) summarizes E_F variations with temperature for Ti adatoms on n- and p-GaAs(110). The arrows in the inset denote coverages for which the measurements were done. While the selection of lightly doped substrates and low metal coverages was made to emphasize effects of temperature on band bending, the conclusions are quite general. The solid lines of Fig. 8.25 are based on model calculations that will be discussed presently.

The experiments involved the deposition of Ti at 20 K followed by warming and then recooling to 20 K (upward- and downward-pointing open triangles, respectively). A second cycle for p-GaAs produced the solid triangles. Deposition and starting the cycle at 300 K would produce equivalent results because Ti is immobilized on the surface by substrate reaction. As shown, the effects of temperature are completely reversible. In the 20–200 K range, band bending varies almost linearly with temperature. For 0.1 ML Ti/p-GaAs(110), however, the E_F position saturates $\sim$ 0.52 eV above the VBM at 300 K. This indicates convergence to an energy determined by the adatom-induced levels at this metal coverage (see 300 K results in inset), without thermal constraint. No changes were observed in the substrate core-level emission except for reversible phonon broadening, and we conclude that there were no changes in surface morphology. This is reasonable because Ti is reactive and chemical changes that immobilize the adatoms occur upon deposition at any temperature. Again, we stress that equivalent disruption has been seen for n- and p-GaAs at all temperatures so that the large difference in E_F position is not related to defect formation. The Fermi level reversibility demonstrates that warming does not introduce new defects of the sort proposed to account for E_F pinning since such defects could hardly be expected to disappear upon recooling.

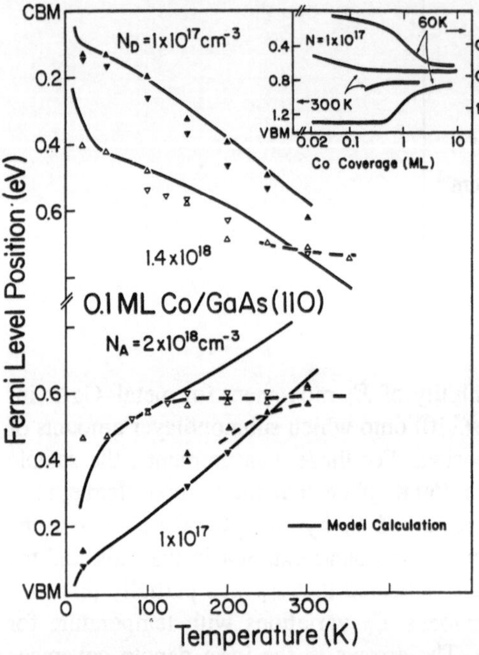

Fig. 8.26. Movement of the surface Fermi level for 0.1 ML Co/GaAs for n- and p-GaAs samples having matched high and low dopant concentrations. The solid lines again represent the maximum amount of band bending predicted as a function of T and N. The inset shows this behavior for overlayer growth on lightly doped n- and p-GaAs(110) at 60 K and 300 K

Additional insight into the effects of temperature on Fermi level movement can be gained by comparing results for different bulk dopant concentrations. Figure 8.26 summarizes the results for Co/GaAs(110). The inset shows the movement of E_F for 60 K and 300 K formation as a function of Co deposition for $N = 1\times10^{17}\,\mathrm{cm}^{-3}$. Again, triangles pointing Upward (downward) correspond to data measured during the warming (cooling) phase of the cycle. Filled symbols designate Fermi level positions for lightly doped substrates ($1\times10^{17}\,\mathrm{cm}^{-3}$) and the empty ones are for substrates doped at $2\times10^{18}\,\mathrm{cm}^{-3}$. For lightly doped p-GaAs the temperature was cycled from 20 to 300 to 20 to 300 K, and the second warming cycle produced Fermi level movement along the line established during the first cycle. For heavily doped n-GaAs, the cycle was extended to 350 K. The dashed lines in Fig. 8.26 guide the eye and emphasize that E_F movement saturates upon warming, as was the case in Fig. 8.25. They show that saturation occurs at lower T for greater N and that saturation occurs at lower T for p-GaAs than n-GaAs. Again, the solid lines are the result of model calculations.

Temperature cycling studies were carried out in the same manner used for Ti overlayers. The 0.1 ML Co coverage was chosen to minimize metallization effects and thus emphasize thermally induced E_F movement. For substrates doped to $1 \times 10^{17}\,cm^{-3}$, the Schottky barrier height increased linearly with temperature for both n- and p-GaAs. Indeed, the results for Co/GaAs and more reactive Ti/GaAs can be superimposed when the dopant concentration is the same. Hence, whatever is responsible for the observed behavior is rather insensitive to the chemical identity of adatoms and the detailed reaction products that they can induce, provided that overlayer morphologies are similar and stable with respect to temperature. Furthermore, the Ti and Co results indicate equally reversible and predictable changes of E_F position with thermal cycling.

The experimental results for lightly doped Co/p-GaAs show E_F saturation near 300 K, analogous to that for Ti/p-GaAs. This saturation is dramatically emphasized for heavily doped p-GaAs ($2 \times 10^{18}\,cm^{-3}$) in the lower portion of Fig. 8.26. As shown, E_F moves with T only for $20 \leq T \leq 120\,K$ and then saturates at 0.59 eV above the VBM. For heavily doped Co/n-GaAs, the corresponding saturation occurs at considerably higher temperature than for p-GaAs for comparable N, namely $\sim$ 250 K for n-type vs $\sim$ 120 K for p-type. This is expected because of differences in the maximum amount of band bending that these coverages can induce at 300 K or below. Prior to saturation, there are similar rates of E_F movement with temperature.

In the above we emphasized reversibility of E_F movement with temperature for interfaces whose morphology was stabilized by adatom–substrate chemical reaction. Ag atoms cluster spontaneously when deposited at 300 K, but the surface is more uniformly covered when adatoms are deposited at low temperature. The results in the inset of Fig. 8.27 show that band bending increases until $\sim$ 100 K, but that the rate of movement is accelerated at higher temperature. At 300 K, E_F was 720 meV from the CBM, in excellent agreement with 300 K results for this interface. Cooling reduced the amount of band bending, but there was an offset between the warming and cooling results that amounts to $\sim$ 140 meV at 20 K. For the first thermal cycle for Ag/GaAs, then, the reversibility was not complete. Subsequent warming and cooling showed movement into the gap and back along the lines established during the cooling phase of the first 300 K cycle.

To correlate these changes in E_F movement with the distribution of Ag atoms on the surface, we examined the valence bands as a function of temperature. The results shown in Fig. 8.27 indicate Ag d-band broadening and a shift toward higher kinetic energy (lower binding energy) with temperature increasing from 20 K to 300 K. The direction and magnitude of this shift followed the band bending (*inset*) and is related to cluster formation and surface electrostatics. This indicates that adatom-related states in the low coverage regime are related to intrinsic semiconductor energy levels rather than E_F [8.91]. Recooling to 20 K did not change the d band width, and the Ag 4d features did not return to their initial energy position. This reflects the formation of metallic states at the surface

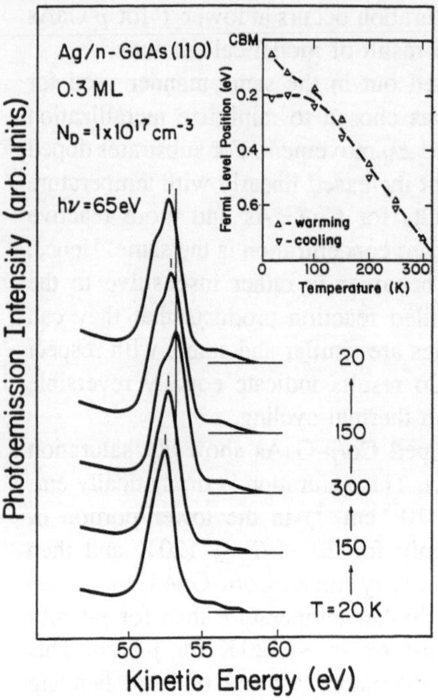

Fig. 8.27. Movement of the surface Fermi level for 0.3 ML Ag/*n*-GaAs as a function of temperature showing that thermal cycling does not produce complete reversibility (*inset*). The results that show the *d*-band evolution with temperature demonstrate that the metastable Ag layer formed at 20 K spontaneously forms clusters upon warming so that the nonreversible E_F movement is a consequence of morphology changes on the surface. (From [8.93])

due to changes in morphology. Note, however, that the morphology of interfaces formed at 20 K and annealed at 300 K to induce clustering was not identical to that produced by cluster deposition at 300 K (Sect. 8.5), because of differences in the details of the adatom clustering/growth process.

The experimental data in Figs. 8.22–27 demonstrate the following that must be explained by any successful model of interactions at semiconductor surfaces:

1) E_F movement depends critically on bulk dopant concentration N;
2) equal E_F movement requires fewer adatoms for HD than for LD samples;
3) adatom-induced states can act as either donors or acceptors;
4) E_F movement is similar for reactive and nonreactive metal adatoms;
5) E_F position in the gap depends on the temperature at which measurements are carried out, not the substrate temperature during deposition;
6) E_F movement in the gap is reversible with temperature provided that there are no changes in interface morphology;
7) metallization is important at high coverage, regardless of the low coverage behavior;
8) final E_F positions in the gap are dependent on T and N and deviate increasingly from the I-V values at low temperature and low doping.

Spicer et al. [8.3] recently argued that delayed E_F movement for n-GaAs was due to reduced defect generation at low temperature. This disagrees with the findings for Ti/GaAs, Co/GaAs, and Au/GaAs because the apparent electrical properties were analogous to those for nondisruptive Ag/GaAs and Bi/GaAs. Moreover, E_F evolution was sensitive to dopant concentration, but there were no differences in disruption for HD and LD substrates. The cyclic movement of E_F in the gap defies description within a defect model for a system in equilibrium.

A model that has gained considerable attention describes high coverage pinning in terms of metal induced gap states (MIGs) [8.83, 84]. These states result from the exponential decay of metal wavefunctions into the semiconductor, and the pinning position is determined largely by the charge neutrality point of the semiconductor. They are relatively unimportant in the low coverage regime because adatom–adatom interaction is insufficient to provide the required charge delocalization for overlayer metallicity. However, the abrupt step in E_F position after a few monolayers for LD n-GaAs certainly confirms the importance of overlayer metallization and charge delocalization. We now see that such delocalization is important for HD n-GaAs because the results of Figs. 8.22 and 23 for Ti, Co, Ag, and Au show a vestigial step that is evident even though E_F has been drawn into the gap at lower coverage. Likewise, the observation of the corresponding step for LD p-GaAs demonstrates that the effect of delocalization is general and symmetric for n- and p-type substrates. Finally, the results for Au/GaAs reveal the symmetric metallization step for p-GaAs (Fig. 8.23). While supporting a delocalization-based behavior at high coverage, the results of Figs. 8.22 and 23 also show an N-dependent high coverage E_F position which will be explained later in this section.

The band bending behavior for Al/n-GaAs (Fig. 8.24) can be partially understood in terms of overlayer metallization. Rapid E_F movement toward its final position at 4–8 ML correlates with the onset of overlayer metallicity [8.3, 4, 16–18, 91]. With increasing T, the step shifts to lower θ until no inflection is detected at 300 K. This suggests that the overlayer acquires metallic character at successively higher θ as T is reduced. Surprisingly, this progression is evident even below 200 K, although thermally activated Al surface diffusion on GaAs should be negligible below 200 K. This points to the importance of non-thermally-activated diffusion processes associated with adatom "cooling" upon deposition, as proposed by *Egelhoff* and *Jacob* [8.98]. It is interesting that the E_F position for Al/GaAs converges to 0.70 eV below the CBM for $60 \leq T \leq 300$ K. Within the context of MIGS, this suggests that the charge neutrality position is pinned relative to the CBM since the band gap changes by 0.1 eV in this temperature range.

The points summarized above demonstrate the importance of T- and N-dependences and force a reevaluation of low-temperature photoemission studies of metal–semiconductor interfaces as far as Schottky barrier evolution is concerned. Before proceeding, it should be noted that the width of the depletion region is related to N (for a given barrier height), and charge exchange between surface and bulk is controlled by this barrier. At low temperature and/or for a highly doped semiconductor, tunneling is the dominant mechanism of charge exchange. The probability for an electron to pass through the barrier created by band bending can be calculated with the WKB approximation. If we consider only electrons with energies corresponding to the conduction band mmimum of GaAs, then the probability of tunneling through a barrier with height of 0.2 eV is $\sim 10^{-4}$ in a HD sample but reduces sharply to $\sim 10^{-14}$ for a LD sample. Since the experiments show remarkable band bending differences for HD and LD samples at a given θ and T, it is reasonable that differences in tunneling probability are important for adatoms on the surface.

Aldao et al. [8.23] recently described in detail the mathematical formalism under which the maximum barrier height can be predicted at any N and T under the assumption that a minimum current exists. Figures 8.25–27 offer comparison between the calculated barrier heights (solid lines) and experimental values (symbols). Figures 8.25–27 also show E_F movement as T was cycled, demonstrating that the barrier height reversibility follows the predicted temperature dependence. Deviations occur above ~ 200 K because $q\phi_b$ has reached a maximum value that is dictated by the adatom-induced levels which it cannot exceed. Results for VHD show that the depletion width is narrow enough to pose no problem for tunneling, even at 20 K, and the predicted maximum barrier is always higher than the observed value. Hence, band bending is not controlled by charge exchange for VHD and should be almost independent of T, as observed.

Most recently, *Hecht* [8.93] and *Tersoff* [8.99] suggested that the same mathematical formulism as that used by *Aldao* et al. [8.23] could be used to describe a nonequilibrium configuration in which the barrier height is determined by the formation of states in the surface band gap so that the photoabsorption inherent in photoelectron spectroscopy introduces a surface photovoltage (SPV). The electrons and holes created by photoabsorption would be accelerated in opposite directions by the electric field of the depletion region, and minority carrier accumulation would occur at the surface and would tend to flatten the bands for both n- and p-type semiconductors. This is sketched in Fig. 8.28. This SPV would keep the interface under non-equilibrium steady-state conditions. Since the only carriers that can contribute to the SPV are those produced in or near the depletion region, the SPV should depend on the amount of band bending and the depletion width. If the injection current is large, the bands can approach the flat band condition so that the SPV would approach the surface barrier height. Such a saturation has been observed for Si at low temperature [8.100, 101] but there were only preliminary investigations of the effect for GaAs [8.102] and no investigations for any semiconductor as a function of N, T, and coverage.

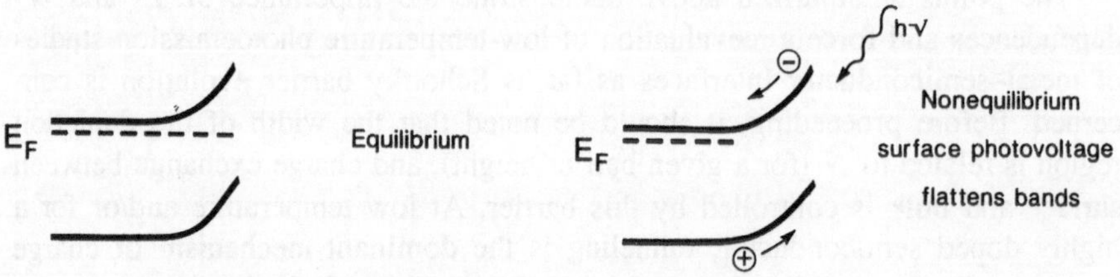

Fig. 8.28. Schematic of the effect of photoabsorption and electron–hole separation in the depletion region. The resulting surface photovoltage leads to band flattening

In the following, we consider the dependence of band bending on photon flux for n- and p-GaAs(110) of different N for temperatures $20 \leq T \leq 300\,\mathrm{K}$. The goal of the investigation was to determine how well the approximation of equilibrium implicit in the DCM model could be realized for a system under photon (or electron) illumination. The results indicate that the surface photovoltage can account for core level shifts of $\sim 100\,\mathrm{meV}$ for changes in the typical photon flux of 10^4. While seemingly small, recent analysis by *Hecht* shows that much of the temperature-dependent band bending observed for submonolayers of metal on LD substrates is due to this nonequilibrium SPV.

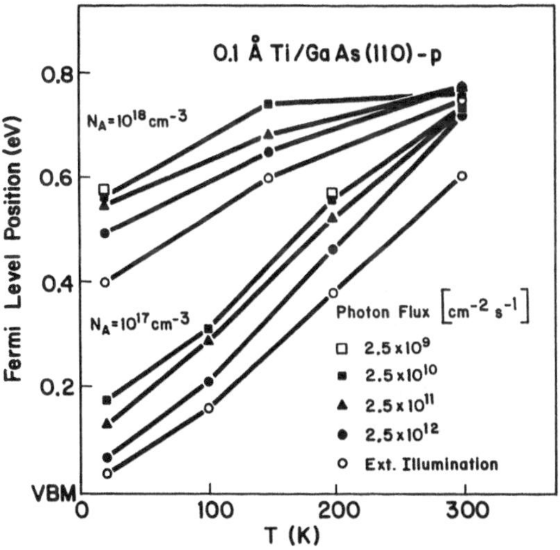

Fig. 8.29. Measured surface E_F movement as a function of temperature for 0.1 Å Ti deposited on low doped p-GaAs(110). Results show the effect of changing the photon flux as well as that of additional illumination with an external lamp. (From [8.94])

Figure 8.29 (from [8.94] summarizes the variation in E_F position for $20 \leq T \leq 300\,\mathrm{K}$ for LD p-type GaAs(110) after the deposition of 0.1 Å of Ti. Results are also shown for HD p-type GaAs(100). The experiment involved the deposition of the metal at 300 K followed by cooling to 20 K. No change in interface morphology was observed during cooling at this coverage, and the effects of temperature on band bending are completely reversible, as discussed above. The results of Fig. 8.29 represent movement along an iso-deposition line of the inset of Fig. 8.25 for ~ 0.1 ML and are analogous to those of Fig. 8.25. They differ, however, in that the core level spectra were collected with photon intensities ranging from $2.5 \times 10^9\,\mathrm{cm}^{-2}\,\mathrm{s}^{-1}$ to $2.5 \times 10^{12}\,\mathrm{cm}^{-2}\,\mathrm{s}^{-1}$ as well as under conditions with additional illumination from an intense external lamp. The flux of $2.5 \times 10^{12}\,\mathrm{cm}^{-1}\,\mathrm{s}^{-1}$ corresponds to the value used in previous studies and is defined here as the normal experimental flux I_N. (It is a convenient flux because it gives $\sim 6 \times 10^4$ counts/s for the Ga or As $3d$ core level under our high-resolution conditions.) The synchrotron photon flux was changed by adjusting the monochromator slits and by inserting an Al filter into the photon beam. These adjustments did not change the measured core level positions, a fact established by examining the behavior for clean, unpinned surfaces.

The results of Fig. 8.29 show that changes of two orders of magnitude in the primary flux do not influence the amount of band bending at 300 K, but that the additional flux from the external lamp induced a SPV of ~ 120 meV. At T = 200 K, band bending differences of ~ 100 meV were observed for spectra taken with I_N and $10^{-2} I_N$. Reduction to $10^{-3} I_N$ had no further effect, and we conclude that the SPV was negligible below $10^{-2} I_N$. For T = 100 K and 20 K, the measured Fermi level position moved toward the VBM (toward flat band conditions) as the flux was increased from $10^{-3} I_N$. The difference between measurements taken with I_N and $10^{-2} I_N$ was nearly constant at ~ 100 meV. This difference was reduced to ~ 50 meV when the experiment was carried out with n-type LD samples. These results demonstrate that the change in SPV of ~ 100 meV is produced when the flux changes by 10^2 for temperatures $\leq$ 200 K for low doped samples. This 100 meV contribution forms only a small portion of the overall T-dependent E_F movement (~ 600 meV) for this interface at low coverage.

An analogous set of experiments was undertaken for HD n- and p-type GaAs(110) after the evaporation of 0.1 and 5 Å of Ti, as shown in Figs. 8.29 and 30. For HD, the amount of band bending at any temperature changes relatively weakly with coverage, resembling the room-temperature results of the inset of Fig. 8.25 and not showing the metallization step. This shows that the depletion region is almost completely formed after the deposition of a small amount of Ti on HD GaAs, as previously reported. In this case, the SPV due to the primary flux were below 100 meV at all temperatures and nearly independent of coverage and temperature for $T \leq$ 200 K. The SPV induced by the external lamp was 170 ± 20 meV for n-type GaAs and 190 ± 20 meV for p-type GaAs. These results indicate that the SPV for a given flux is appreciably smaller for HD

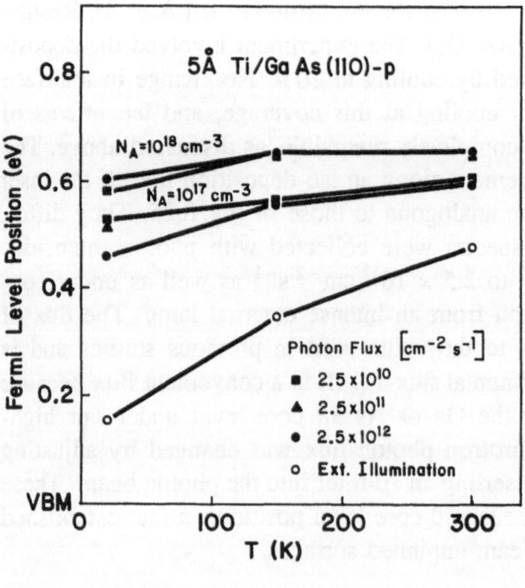

Fig. 8.30. Results analogous to those presented in Fig. 8.29 after deposition of 5 Å of Ti

samples than for LD samples although the changes in SPV per decade of flux are comparable at low temperature. These findings can be correlated with the dependence of band bending and depletion width on the dopant concentration.

We noted above that the high coverage positions of E_F in the gap for n- and p-GaAs were not identical for a given metal coverage, although they would be expected to be for bulk interfaces based on measurements of junctions formed by the deposition of thick metal layers. Moreover, they were dependent on the dopant concentration: the separation is 200–250 meV for LD GaAs and 100–160 meV for HD GaAs at 60 K. An important point of interest is the dependence of the SPV on the deposited metal because this nonequilibrium contribution could account for the high coverage difference measured in photoemission. Preliminary results for LD Ag/n-GaAs show rather remarkable findings. In particular, the effect of photon flux for 7 Å Ag deposited at 20 K was to change the measured band bending by $\sim$ 450 meV, from 850 meV at $10^{-2}I_N$ to 400 meV for I_N. The effect of external high intensity illumination for 7 Å Ag deposited at 20 K was to further flatten the bands by 150 meV relative to the I_N measurement, giving a total reduction in band bending of 600 meV. Note that 7 Å is above the metallization limit for Ag deposited at 20 K. SPV effects were also observed for 3 Å Ag deposited at 300 K when an increase in flux from $10^{-2}I_N$ to I_N induced a change in E_F position of $\sim$ 90 meV. This behavior for Ag is quite different from that for Ti, and we postulate that it can be understood by considering differences in photon absorption within the metal layers. In particular, the photon absorption coefficient for Ag is four times larger than for Ti for $h\nu$ = 65 eV. Absorption within the metal leads to the injection of photoexcited electrons from the metal into the depletion region of the semiconductor. These injected electrons then enhance electron–hole pair production and contribute to the SPV. Indeed, light absorption in the metal layer was proved to be important in CdS photocells long ago [8.103]. An alternative explanation would be that the extensive amount of substrate disruption that results from Ti deposition ($\sim$ 3 ML of substrate) increases dramatically the surface recombination rate relative to nondisruptive Ag overlayers. Although we cannot disregard the possible influence of surface recombination, the fact that an increase in Ag deposition from 7 Å to 8.5 Å enhanced the SPV indicates that photon absorption in the overlayer plays a definite role. Moreover, epitaxial Bi overlayers which do not induce disruption present small photovoltages [8.95, 96].

The results obtained for overlayers above the metallization threshold are of significance in the context of Schottky barrier models. We have observed that measurements based on core-level binding energies can yield incorrect E_F positions if the interface is assumed to be in equilibrium, i.e. E_F is taken to be the same for the sample surface and the spectrometer, for routinely used photon fluxes (on the order of 10^{12}–$10^{13}\,cm^{-2}\,s^{-1}$). As a consequence of SPV, these measurements would show a difference between final E_F positions at higher coverages for n- and p-type GaAs, an effect visible in the inset of Figs. 8.22, 23, 25, and 26. For Ti overlayers, the splitting is 100–200 meV for low temperature studies, consistent with the SPV measured for high coverage in this study. The difference is not significant for Ti overlayers at 300 K (< 30 meV), but it is still important for Ag overlayers. It is likely that the observed differences in final Fermi level position for n- and -type substrates, particularly at low temperature, can then be related to SPV effects and the absorption coefficients of the metals themselves.

In the above, we have shown that changes in the SPV correspond to $\sim$ 100 meV for $\sim 10^2$ changes in primary flux. These results are critically dependent on bulk dopant concentration and temperature. Previously, we have discussed the dependence of apparent Fermi-level position on N and T, assuming that the coupling of the surface to the bulk represented an equilibrium process. From the results of Figs. 8.29 and 30, one might be inclined to conclude that the SPV changes do not introduce major deviations from equilibrium since the dependence on photon intensity is relatively small. However, *Hecht* has calculated the surface photovoltage as a function of flux, assuming that the barrier height was equal to that observed at high coverage. (He adopted a formalism equivalent to that of [8.23] to describe charge flow across the depletion region but he related the charge flow to electron–hole pair separation rather than equilibrium coupling.) The agreement between the predicted apparent barrier height and the results of Figs. 8.29 and 30 is very good. Indeed, his reevaluation of the N- and T-dependences indicates that surface photovoltages are largely responsible for the Fermi-level behavior reported from photoemission results for low temperature and low doping.

The generality of the SPV can be extended by considering recent results for surfaces other than GaAs(110) and for other semiconductors. First, we note that results for InP(110) produce behavior analogous to that reported above, although the range of N values investigated has not yet been as great as for GaAs [8.104]. Second, for cleaves which are not quite ideal there is partial pinning due to defect levels at the surface, presumably due to steps. For these not-quite-perfect surfaces, it was possible to move the surface Fermi level from deep in the gap to close to the band extrema and this movement was cyclic with temperature. Again, this can be understood within the SPV because levels established by cleaving are analogous to those created by metal atom deposition in that they induce band bending at the surface. Third, *Brillson* and coworkers [8.6, 7] have recently reported results for MBE-grown GaAs(100) surfaces onto which adatoms were deposited at 300 K and at low temperature. For the surface that was prepared by thermally removing a protective As cap, they found that E_F was pinned near the center of the gap at 300 K, but that it was close to the CBM at low temperature. The effect was explained as due to detailed differences in surface preparation. We propose that it can be better understood within the context of the SPV. If the number of defects at the uncapped surface is relatively low, then the coupling between them and the bulk should exhibit the same temperature dependence as discussed for GaAs. Moreover, N was very low (10^{16} cm^{-3}) for those studies. Hence, the SPV would predict band flattening at low temperature. This is consistent with the nearly flat-band condition reported by *Brillson* and coworkers. We postulate that the density of surface levels for those uncapped

GaAs(100) samples was insufficient to produce defect-state delocalization [and pinning as for surface states on Si(111)].

8.4.2 Photoemission from Metallic Dots

In the previous section, we showed that the SPV can explain the N-dependent E_F movement at low T and the spread in E_F position above the metallization threshold. The SPV does not, however, establish the mechanism responsible for the metallization step itself. In the following, we demonstrate that the movement into the gap is related to the formation of a conducting surface layer that shorts out the SPV effect. These experiments were motivated by suggestions by *Hecht* [8.93] and *Tersoff* [8.99] that overlayer growth and metallization provide a conduction path from the illuminated (flat-band) region of the surface to a location on the surface or at its edge where more ohmic contact is made. To test this intriguing idea, we formed metallic dots on GaAs(110), illuminated those dots with synchrotron radiation, and used photoemission to measure the Fermi cutoff and the binding energies for Ga 3*d*, As 3*d*, and the metal core levels. For dots confined to pristine regions of the surface, the SPV prevailed to coverages above the metallization limit and the step was not observed. Temperature cycling demonstrated that flat-band conditions existed beneath the dots at 30 K but that approximate equilibrium pinning was established at 300 K. However, when conducting layers were established between the dots and the remainder of the sample, the Fermi cutoff returned to the equilibrium position, even at 30 K. We conclude that full surface coverage allows compensation of the SPV and that the step is a consequence of surface conduction.

The synchrotron radiation photoemission experiments were done with a fixed photon energy of $h\nu = 65$ eV. The photon flux of $\sim 10^{12}$ photons/cm^2 s did not vary enough to affect the measurements. Overlayers of Bi and Ti were deposited onto cold substrates through a $\sim$ 2 mm diameter aperture placed $<$ 1 mm from the surface (the cleaved surface was typically 4 mm$\times$4 mm). In this way, a dot isolated from the sample edges was formed, and its character could be identified by observation of a Fermi cutoff and the binding energies of the core levels. The inset of Fig. 8.31 schematically shows the experimental arrangement. The dots produced with this sample masking technique were not uniform in thickness around the perimeter, but they sufficed very well for our measurements. Indeed, the low density of atoms that migrated away from the dots provided a way of directly comparing very low coverage results with those for metallized regions by moving the sample in the beam.

Figure 8.31 shows photoemission spectra near E_F for a Bi dot of 10 Å thickness on *p*-type GaAs(110). The spectra at different temperatures are offset vertically for clarity, and they are referenced in energy to the equilibrium E_F position determined from a $\sim$ 100 Å thick Ti film. These results clearly demonstrate nonequilibrium T-dependent movement of the emission cutoff of semi-metallic Bi that is consistent with the SPV model and charging of the dot. This charging is due to the open circuit voltage at the interface resulting from accumulation of

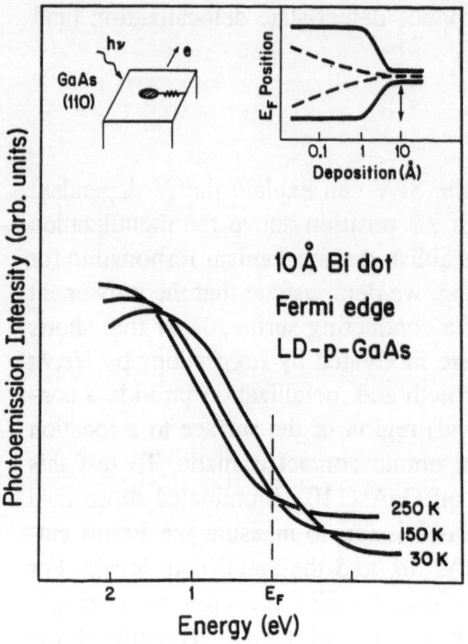

Fig. 8.31. Photoemission results showing spectra near the Bi Fermi cutoff as a function of substrate temperature for a 10 Å Bi dot deposited near the center of a lightly doped (LD) *p*-type GaAs(110) surface, as sketsched in the left inset. E_F is the equilibrium Fermi level of the spectrometer. The right inset shows qualitatively the band bending behavior as a function of deposition onto LD GaAs at 300 K (*dashed line*) and 50 K (*solid line*). The photoemission spectra show emission from above the E_F at low temperature that is due to a nonequilibrium SPV that flattens the bands beneath the Bi dot. The temperature-dependent movement of the Bi cutoff follows the arrow at 10 Å in the right inset (From [8.105])

minority carriers at the surface. It serves to oppose band bending for both *n*- and *p*-type substrates. It is important to note that 10 Å of Bi would result in near-midgap pinning at 30 K for conventional experiments where the surface is fully exposed to the evaporant (see arrow in inset of Fig. 8.31 at 10 Å). The fact that this is not observed for the dot indicates that a unique nonequilibrium configuration has been achieved and the bands become increasingly flat upon cooling despite the coverage being above the "metallization" step. The T-dependent movement of E_F (or the metal core levels) is a measure of band flattening beneath the dot.

In the previous section we discussed reversible E_F movement in the band gap as a function of T for a given amount of metal coverage across the surface. Figure 8.32 gives the corresponding movement of E_F for metallic dots. In the upper panel of Fig. 8.32 we show T-dependent results for a 10 Å Bi dot on LD *p*-GaAs(110). The solid line represents the change in band bending derived from the Bi Fermi level emission and the dashed line corresponds to the shift in the Ga and As core levels. Both reflect the same trends, but the magnitude of the shift is greater for the Bi dot. The smaller shift for Ga and As emission can be understood by recognizing that the Bi dot is opaque to photoelectrons emitted beneath it and that the Ga and As 3*d* photoelectrons were emitted from the surface region near the dot where the Bi coverage is very small. In this case, the Bi-induced equilibrium band bending would be much less and the temperature dependence would correspond to that for ~ 0.1 Å rather than 10 Å (see inset to Fig. 8.31).

We stress that when 10 Å of Bi is deposited across the entire surface, the Fermi level is pinned near midgap, even at 30 K, and no SPV effect is observed (Fig. 8.31, inset). In the absence of such a conducting layer, our results clearly

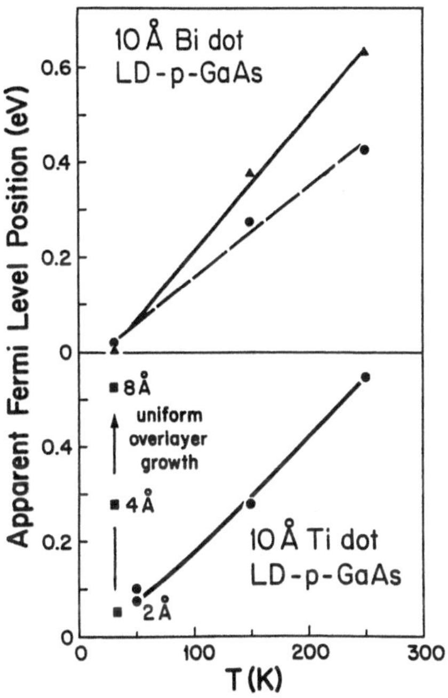

Fig. 8.32. Apparent Fermi level position in the gap for 10 Å Bi and 10 Å Ti dots on LD *p*-GaAs(110). The energy zero represents flat bands with E_F within ~ 30 meV of the valence band maximum. The dashed line was obtained from shifts of the Ga and As 3*d* emission; it is indicative of the SPV in regions near the dot covered by ~ 0.1 Å of Bi where the amount of band bending was less. The lower panel shows results for the Ti dot derived from Ga and As 3*d* spectra. The squares show that Ti deposition at 30 K over the entire surface induces movement of the apparent Fermi level to its final position by 8 Å. Hence, the SPV is large for isolated metallic dots at low temperature, but similar depositions across the full surface produced a shunted diode with no appreciable SPV. (From [8.105])

demonstrate a substantial SPV at coverages well above the metallization threshold. This suggests that the step for lightly doped samples at low temperature is a consequence of the changing conductivity of the surface.

To test the idea that surface conductance could alter the apparent Fermi level position, we first formed a metallic 10 Å Ti dot on *p*-type GaAs and then deposited Ti incrementally across the surface at 30 K. The bottom panel of Fig. 8.32 shows the change in band bending for the Ti dot upon cycling from 30 K to 250 K and back to 30 K. The results are again derived from Ga and As 3*d* core-level spectra and represent changes in SPV in the regions of the surface around the Ti dot, i.e., regions covered only by submonolayer amounts of Ti. As previously established, cycling between low and high T results in reversible E_F movement for submonolayer metal coverages. As for Bi, the Ti valence band features exhibit emission above the Fermi level at low T because of nonequilibrium band flattening beneath the Ti dot. Again, for uniform Ti deposition across a GaAs(110) surface, a 10 Å film is sufficient to pin E_F at its final position at 30 K because it exceeds the metallization threshold of Fig. 8.31 This establishes that the Ti dot is electrically isolated from the sample edge.

To understand more fully the differences between the dot experiment and the fully-exposed-surface experiments, we deposited 2 Å of Ti on the 10 Å dot surface at 30 K. From Fig. 8.31 (inset), the apparent E_F position for a uniform 2 Å film would be close to the band edge at 30 K and metallization should not have been observed. From Fig. 8.32 the effect of this uniform 2 Å deposition was negligible (square symbol labelled 2 Å). When 4 Å of Ti was deposited onto the dot and the surface, band bending changed by ~ 200 meV. Note that uniform coverage of 4 Å corresponds to partial movement over the metallization threshold of Fig. 8.31. When the uniform overlayer thickness was increased to 8 Å , the amount of band bending equalled that for a conventional metallized surface *and* for the 10 Å dot at 250 K (Fig. 8.32). No SPV was observed. While these measurements predominantly sample regions of the surface not directly beneath the dot (due to photoelectron attenuation by the dot), they establish that uniform metallic coverages are sufficient to pin E_F at low temperature but isolated metallic coverages are not.

The results of Fig. 8.32 make it clear that metallic dots of Bi and Ti are charged during the photoemission process at low temperature. This SPV is reduced for the isolated dot as the temperature increases because a compensating current is established through the depletion region beneath the dot. Remarkably, the SPV for the dot is shorted at low temperature when the extended surface exhibits metallicity. These results therefore demonstrate that the metallization threshold reflects the ability of the surface to pass a current able to compensate the SPV. Such shorting would occur for both n- and p-type samples, explaining the symmetric movement of E_F observed at low temperature. Morever, the SPV would account for the apparent spread in the pinning position above the metallization threshold and its dependence on the bulk dopant concentration. Thus, the metallizaton threshold of Fig. 8.31 is important, but not for the reasons previously thought [8.14–18, 23, 85–87]. It appears unrelated to an alteration of the fundamental interactions of the overlayer with the substrate as the overlayer becomes metallic.

It is interesting to speculate about the nature of the contact that is established upon uniform surface metallization because this contact is clearly different from that beneath an isolated metallic dot. The metallization threshold has now been observed at low temperature for metals ranging from the semimetals like Bi (which does not disrupt the surface) to metals like Ag (which does not disrupt the surface) and Sm and Ti (where disruption and intermixing is pronounced). Despite differences in surface morphology, reactivity, and the conductivity of the species that evolve, the half-way point for the step falls between ~ 0.8 ML for Co and ~ 4 ML for Cr (or ~ 6.8 ML for Al) and the step width is generally less than ~ 4 ML. At this point, we see no clear correlation between surface reactivity and the metallization parameters. Moreover, the SPV is able to maintain essentially flat-band conditions during overlayer growth, starting at very low coverage and extending beyond the metallization limit, indicating that direct recombination rates are very small.

Hecht recently postulated that a uniform film could establish a shunt resistance such that the metallized surface represents a leaky diode with a shorting of the open circuit voltage that leads to band flattening [8.39]. The results of Figs. 8.31 and 32 establish the importance of such a conducting layer. One must assume, therefore, that there are sites on the cleaved surface or at its edge where an ohmic contact is established. Presumably, these contacts occur at heavily stepped regions of the surface where the density of states in the gap is large, but the details remain unknown.

8.5 Interface Formations with Metal Ions

In the preceding sections, we discussed interface evolution as a function of temperature, and we investigated the influence of surface chemistry on band bending. Here, we focus on the preparation of the interface by the deposition of metal ions. The purpose of this section is to examine the consequences when the parameters of interface formation are changed. Whereas the low-temperature studies of Sect. 8.4 and the cluster deposition studies of Sect. 8.6 were intended to reduce adatom interactions with the substrate, ion deposition experiments seek to increase that interaction. With a partially ionized beam of atoms accelerated to several hundred electron volts, it is possible to change the energetics and alter surface morphologies and defect densities. Significantly, studies of band bending for interfaces produced by this process at 300 K show that surface disruption is relatively unimportant in terms of the final band bending [8.106, 107]. This conclusion is particularly important insofar as Schottky barrier models are concerned because several of them require that pinning be related to defects in the near-surface region.

The ion beams used in thin film deposition processes can be divided into two basic configurations [8.108]. In one, the desired material is ionized and deposited at low energy ($\sim$ 100 eV) directly onto a substrate (primary ion beam deposition). In the other, the desired material is sputtered from a target by an energetic beam of inert or reactive gases. The deposition of atoms from a charged particle beam to form a contact presents attractive features compared to neutral deposition because it affords control over deposition energy and gives the ability to direct and mass-analyze the beam.

The reason for using ions for overlayer growth is to impart kinetic energy that can affect film kinetics and growth processes [8.109, 110]. Ion–surface interactions can include enhancement of adatom migration, intermixing, defect creation, sputtering, and implantation. Thus, it is of interest to investigate the possibility of using ions to change film growth characteristics and to correlate surface effects with the electronic properties, i.e., the Schottky barrier.

For our studies, we constructed a compact single-grid, ultrahigh-vacuum-compatible, low-energy ion gun capable of utilizing solid source materials. The ion source is compact, built with ultrahigh-vacuum-compatible components, bakeable, able to produce ions of high-boiling-point solids, able to yield ion fluxes of $> 20\,\mathrm{nA/cm^2}$ with an efficiency greater than 1% at deposition rates of $\sim$ 1 ML/min, and able to produce uniform ion beam intensities over a 10 cm diameter at a distance of $\sim$ 40 cm from the source. The design allows the temperature of the entire ion source to be maintained sufficiently high to avoid condensation of the ions or evaporated atoms within the source. During operation, the pressure in the experimental chamber increases from 5×10^{-11} torr to 8×10^{-10} torr for In and $\sim 2 \times 10^{-9}$ torr for Ag.

The ionization mechanism used in our ion source is electron impact ionization in a low voltage gas discharge. The neutral atom density can be several times the ion density, but the ionization efficiency can be high enough to produce

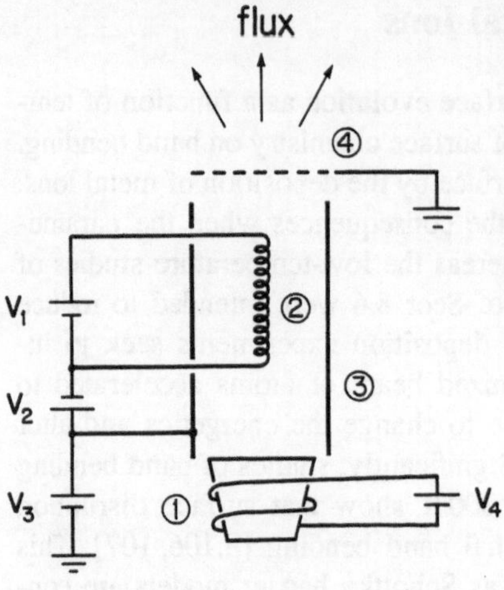

Fig. 8.33. Schematic of the ion source. The metal to be evaporated is heated by a tungsten coil around the alumina crucible (1). A tungsten filament (2) provides energetic electrons to ionize the metal atoms contained within the anode (3). A grounded grid (4) extracts the ions from the plasma established in the source. V_1 is the filament power supply, V_2 is the discharge power supply, V_3 controls the ion energy, and V_4 is the power source used to evaporate the metals from the crucible. (From [8.107])

rates of 4% for Ag and 7% for In since extraction is more efficient for ions. The fraction of ionized metal atoms is determined by comparing the amount of metal deposited (measured with a crystal thickness monitor) with the ion current (measured with a Faraday cup).

Figure 8.33 shows a schematic cross-section of the cylindrically symmetric ion source. The evaporants are heated in an alumina crucible, and the evaporated atoms are injected into the discharge chamber or anode (a stainless steel cylinder 14 mm in diameter and 30 mm long). The discharge is sustained by thermionic emission of electrons from a spiral tungsten filament. In operation, the anode is maintained at a positive potential, and the target is grounded. The ion density in the discharge chamber depends on the electron mean free path (via the pressure and the ionization cross section), the electron emission current, and the total electron path length (which depends on the chamber geometry and any external magnetic fields). A fraction of the ions accelerated by the extraction grid voltage passes through the grid apertures and forms the beam. During source operation, heating of the evaporant was provided by radiation from the hot filament and bombardment of the anode walls by electrons thermionically emitted from the filament. This was enough to mantain the evaporation for Bi, but we needed to heat the alumina crucible for In and Ag.

8.5.1 Ag/ZnSe(100)

The following studies of ZnSe(100) interface formation focus on Ag neutrals and ions because Ag neutrals are nondisruptive. The accelerated ions had sufficient energy to overcome kinetic barriers that might prevent reaction and defect formation. At the same time, the energies were small enough to ensure that the ions were confined to a region near the interface (~ 4 Å average range for 400 eV Ag ions in ZnSe [8.111]).

In these experiments, undoped n-type ZnSe(100) samples were grown on GaAs(100) substrates using molecular beam epitaxy (MBE). Clean surfaces were

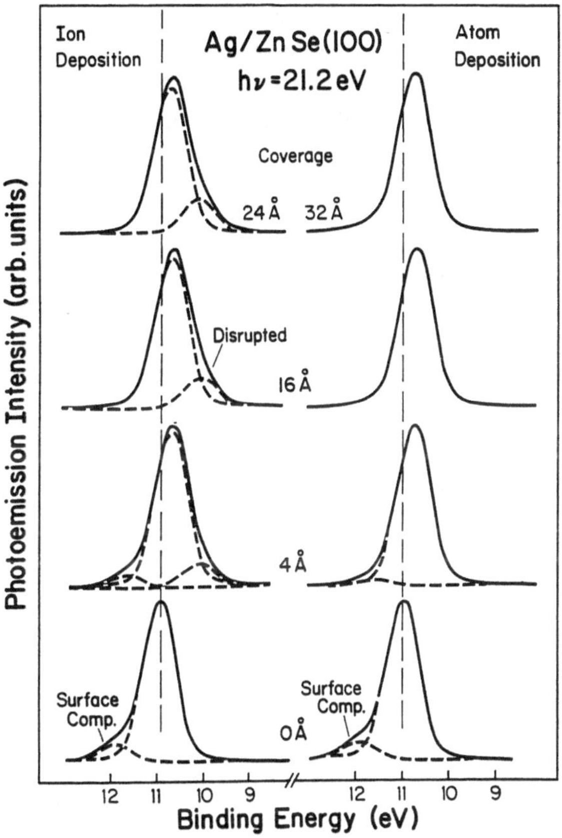

Fig. 8.34. Zn 3*d* core level emission for the deposition of Ag ions (*left*) and Ag neutrals (*right*). Ion deposition induces disruption and the dissolving of Zn in the overlayer, whereas Ag atoms induce no such disruption. (From [8.106])

prepared in the measurement chamber by Ar^+ sputtering at 500 eV, followed by annealing at 400°C for 20 min. The surface structure was determined with low energy electron diffraction (LEED), and Auger electron spectroscopy was used to confirm the cleanliness of the sample. The reconstructed $c(2 \times 2)$ structure corresponds to a Zn-terminated surface.

Figure 8.34 shows the Zn 3*d* core level emission as a function of Ag deposition. In contrast to those presented earlier, these results were not corrected for band bending. During interface formation by Ag atom deposition (right panel), the emission from the substrate shifts to lower binding energy because of band bending and is attenuated slowly. The only change in lineshape is the gradual disappearance of the surface-shifted component. The spectra for deposition using a partially ionized beam are shown in the left panel. Inspection shows that a shoulder develops on the low binding energy side of the bulk Zn 3*d* emission. These ion-deposition spectra could be fit with a component that corresponds to the substrate and another shifted by 0.60 eV. From studies of numerous metal overlayers on ZnSe(100), and those above for GaAs, the second component stands as clear indication of Zn dissolved in the metal overlayer [8.112]. We conclude that the effect of the partially ionized Ag beam is to break ZnSe bonds in such a way that Zn atoms are released and dissolved in the Ag overlayer. As a consequence, the interface morphology is more complex for ion deposition than for thermal deposition. Examination of the Ag $4d$ valence band emission shows equivalent evolution for ion and atom deposition because the shape of the Ag valence band is not very sensitive to the presence of small amounts of Zn in solution.

Quantitative assessment of the attenuation rate of the substrate Zn $3d$ emission shows very similar behavior for ionized- and neutral-Ag deposition. Here, the kinetic energy of the Zn photoelectrons is $\sim 10\,eV$ relative to E_F. For both forms of deposition, the substrate attenuation rate is slower than expected for layer-by-layer growth, and the substrate emission decay rate is even slower at higher coverage. For a nonreacted, nondisrupted interface, this indicates that three-dimensional Ag-island growth occurs, as expected. However, the fact that the attenuation is the same for ion deposition is surprising because Ag ions create defects at the ZnSe surface. This increases the density of Ag nucleation sites, and a more uniform coverage would be expected. That this is not observed indicates that the Ag nucleation density is not very dependent on the perfection of the ZnSe(100) surface.

For Ag atom deposition, the quality of the LEED pattern deteriorated gradually, and the reconstructed spots were too weak to be detected by 2 Å. Although spots of the primary lattice were visible to $\sim$ 16 Å, they were weak and broadened. This is consistent with Ag cluster formation and the continued exposure of the clean surface to the electron beam. For ion deposition, the reconstructed spots were not visible after 1 Å, and no LEED spots could be observed above 8 Å . Hence, LEED indicates island formation for both forms of deposition, but the long range order of the ZnSe(100) surface degrades faster under ion beam depositions.

To correlate these surface structural changes with band bending, we measured the changes in Zn $3d$ core level energies as a function of Ag deposition, as summarized in Fig. 8.35. The accuracy in measuring changes is $\pm 30\,meV$

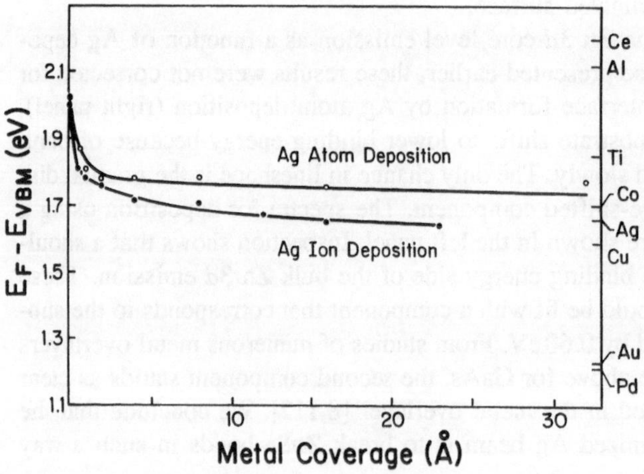

Fig. 8.35. Movement of the Fermi level in the gap of ZnSe(100) induced by Ag ions and Ag neutral atoms. These results are equivalent, to within experimental eror and run-to-run reproducibility, indicating that surface disruption amounting to at least one monolayer has little effect on Schottky barrier formation. (From [8.106])

and the run-to-run reproducibility is 0.1 eV. This reproducibility for ZnSe(100) wafers reflects variations in the quality of sputter-annealed surfaces. For Ag ion deposition, we now see that band bending is 60–100 meV larger at any given coverage than for neutral Ag deposition. The final position of E_F (1.05 eV below the CBM, as noted at the right of Fig. 8.35) is effectively identical to the value of 1.06 reported for evaporated Ag [8.12, 13]. Hence, the effect of the ion-induced disruption is not more than 100 meV. This is a conservative upper bound since a smaller value would have been deduced from [8.112].

A recent systematic investigation of metals on ZnSe(100) showed a wide range of Fermi level pinning positions, as summarized on the right axis of Fig. 8.35 for Ce, Al, Ti, Co, Ag, Cu, Au and Pd [8.112]. As can be seen, the metal-to-metal spread is larger than 1 eV. A clear correlation was found between metal work function and the observed barrier height, in a way similar to the Schottky model. At the same time, there was also a correlation between barrier height and reactivity of the metal overlayer [8.113, 114] because metals with low work functions are usually more reactive. Hence, there was concern that the observed relationship between Schottky barrier height and metal work function might have been accidental, and the true driving force might involve the chemistry at the interface. In that context, the results presented here are particularly important because the surface chemistry could be modified with energetic ion deposition. Significantly, the Schottky barrier varied less than 100 meV, a value which is much less than the variation among different metals. We conclude that the properties of the metal overlayer, rather than details of the interface morphology, determine the Schottky barrier height for ZnSe(100).

Comparison of the present results with the predictions of the defect model [8.115], where there would be similar pinning position for all metals, shows that the model does not provide an accurate description of Schottky barrier formation for ZnSe. One might argue that the formation of defects by metal deposition is suppressed for ZnSe compared to GaAs because the cohesive energy of ZnSe is much larger (−163 vs −71.1 kJ/mole) and simple condensation cannot create defects. However, the present results show that ions deposited with 300–400 eV are able to create defects at and near the surface. Even for partially ionized beams, the number of defects should be more than sufficient to pin E_F. The absence of significant differences in Schottky barrier height between atom and ion deposition argues that defects do not play an important role in pinning for ZnSe.

8.5.2 In/GaAs(110) and Ag/InP(110)

The ability to directly compare the effects of neutral-atom and ionized-atom deposition on clean semiconductor surfaces makes it possible to determine changes in the electrical properties while altering the chemistry and morphology of the surface. For Ag/ZnSe(100), we have shown that the electrical properties are remarkably insensitive to the structural properties. In the following, we demonstrate that the conclusions concerning interface formation, defect formation, and Schottky barrier formation are general.

Previous studies of interface development for In/GaAs(110) at 300 K have shown no evidence of an exchange reaction, and overlayer growth has been characterized by the formation of In clusters from submonolayer coverages. The movement of the Fermi level has also been reported to be slow compared to other metals [8.17, 116, 117], and recent studies have reported E_F positions close to midgap (0.70–0.75 eV above the VBM) [8.17, 117].

Figure 8.36 shows a series of Ga 3*d* and In4*d* core level spectra taken after the deposition of neutral In atoms (left) and In ions (right). These spectra have been normalized and shifted to compensate for band bending. Indium deposition leads to the appearance of new structures that correspond to the spin-orbit-split In4*d* core levels. Detailed lineshape analysis shows that only one new doublet is needed to achieve a good fit for the full spectrum at all coverages for atom deposition (two components for Ga 3*d* to represent bulk and surface atoms and one for In4*d*). No changes were observed in the substrate emission except for the loss of intensity caused by covering the substrate. These findings reflect the development of an abrupt, nonreactive interface from which there is no release of Ga or As atoms due to adatom-induced disruption. This nonreactive character is in agreement with previous reports ([8.117] and citations therein). The slow attenuation of the GaAs core level intensities can be understood by the formation of In clusters. Indeed, In clustering to form three-dimensional islands on GaAs(110) has been reported by *Savage* and *Lagally* [8.118] from nominal coverages as low as 0.4 Å at 300 K.

The right portion of Fig. 8.35 summarizes results obtained after the deposition of a partially ionized beam of In on GaAs(110). Inspection quickly shows that the

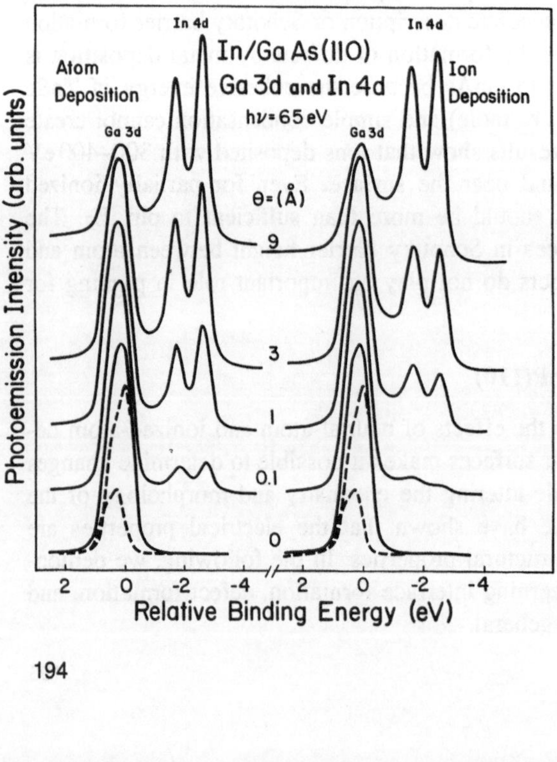

Fig. 8.36. Ga 3*d* and In 4*d* core level spectra for In atom and ion deposition on GaAs(110). While spectra for atom deposition can be fit by adding a single doublet for metallic In, spectra for ion deposition need two components, as shown in Fig. 8.37. (From [8.107])

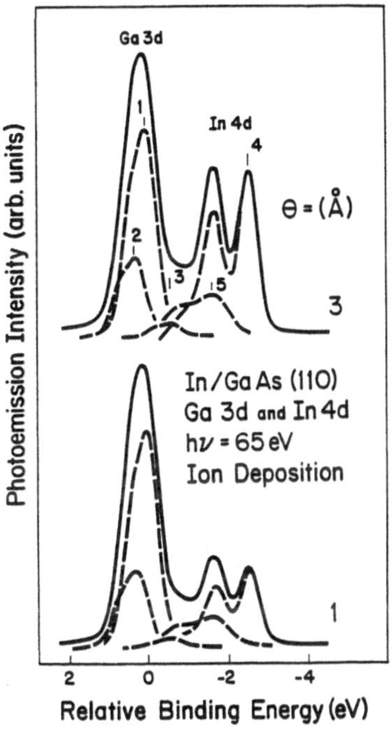

Fig. 8.37. Lineshape decomposition for In ion deposition on GaAs(110). Features resulting from bulk and surface Ga atoms are labeled 1 and 2. Feature 4 corresponds to metallic In. These features are enough to fit spectra for atom deposition. Two additional features, labeled 3 and 5, are needed for ion deposition. Feature 3 corresponds to Ga released from the substrate and feature 5 to covalently bonded In. (From [8.107])

In4*d* core-level lineshape differs from that obtained for neutral atom deposition. Attempts to fit these results in a manner similar to that for atom deposition failed: To obtain acceptable fittings, two new components were needed, as shown in Fig. 8.37. One corresponds to a new In doublet, labeled 5, and one to a new Ga doublet, labeled 3. Feature 5 is shifted 0.95 eV to higher binding energy relative to feature 4, i.e., relative to metallic In. This binding energy difference agrees with previous results observed for metallic and covalently bonded In. Indeed, studies of In deposition onto InP(110) showed the same energy difference [8.64, 65]. As observed for many disruptive metal–GaAs(110) interfaces, the Ga 3*d* core level shows a new doublet at lower binding energy as atoms released from the substrate intermix in the forming overlayer (see discussion above). This suggests an exchange reaction involving In and Ga that results from ion deposition but not neutral atom deposition; it appears to be similar to that for Al deposition onto GaAs(110) at 300 K [8.91]. A similar assignment to component 5 was proposed by *Chin* et al. [8.117] when chemical reactions were induced by heating 10 ML In/GaAs(110) above the melting point of bulk In. We conclude that In ion deposition leads to an overlayer growth mode that is similar to that observed for atom deposition but that clustering is accompanied by substrate disruption that amounts to ~ 0.33 ML after 15 Å deposition.

In order to correlate changes in surface atom distribution with changes in band bending, we show in Fig. 8.38 results for ion and neutral depositions for In/GaAs(110). Ion deposition accelerates band bending evolution slightly, as observed for Ag/ZnSe(100), but the final E_F positions in the gap differ by only

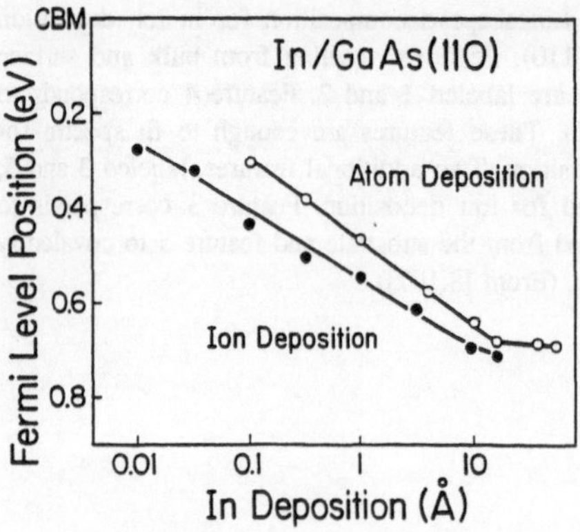

Fig. 8.38. Fermi level position movement for In/GaAs(110) for atom and ion deposition. Although ion deposition accelerates the band bending process slightly, the final position is reached slowly despite obvious substrate disruption. (From [8.107])

30 meV (0.69 eV below CBM for atom deposition and 0.72 eV for ion deposition), an amount that is not considered to be physically significant. Moreover, while ion deposition induces band bending by 0.01 Å , the magnitude is much smaller than fully occupied defect-induced states would produce. From the intensity ratio of components 4 and 5 for In (Fig. 8.37), the number of defects can be estimated and the expected band bending can be calculated. For example, 0.1 Å deposition produces $\sim 2 \times 10^{13}$ defects/cm^2. Since 10^{12} electrons/cm at the surface would move E_F to midgap, full pinning should be observed well below 0.1 Å if every induced state at the surface held an electron. Although ions do induce band bending at low coverages, E_F approaches midgap gradually and does not exhibit a quadratic dependence on the number of surface defects. This implies that the amount of charge per induced state is much less than one electron. Hence, slow movement of E_F toward midgap is not a consequence of the lack of disruption and the creation of surface defects. Instead, it is a fundamental characteristic of In-GaAs(110) interactions that control the occupancy of the adatom-induced states.

These results for In/GaAs(110) demonstrate that the Schottky barrier is *not* changed when disruption is induced for an interface which is normally free of disruption. To demonstrate that this is a general result, we can consider the Ag/InP(110) system where adatom deposition at 300 K is known to cause surface disruption [8.104]. In Fig. 8.39 we show representative In4*d* EDCs for neutral atom and ion deposition of Ag on InP(110). Ag neutral atom deposition induces a new feature in the lower binding energy side of the spectra at 300 K and 60 K, see [8.104]. At high coverages, ion deposition enhances this feature relative to that for atom deposition. The component corresponding to In atoms segregated to the (cluster) surface represents a larger portion of the total signal for ion deposition than for atom deposition. Interestingly, it is more intense for atom deposition at lower coverages. The clear persistence of the substrate feature after 20 Å of Ag indicates incomplete substrate coverage.

Ion bombardment has been observed to favor the growth of larger islands while inhibiting the rate of secondary nucleation and providing more uniform island size distribution [8.109]. These results can be understood in terms of dif-

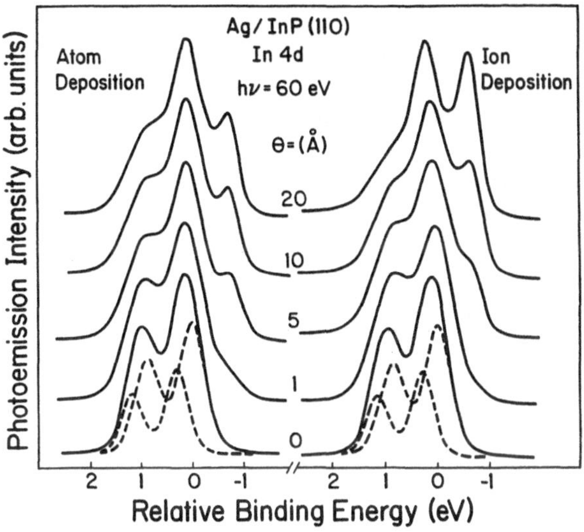

Fig. 8.39. Representative In 4*d* EDCs for Ag atom and ion deposition on InP(110). The EDCs are shifted to account for band bending. Comparison shows clear enhancement of substrate disruption at higher coverage for ion deposition. (From [8.107])

fusion enhancement, together with sputter-induced disssociation of small islands and subsequent surface diffusion that feeds larger, more stable islands. Substrate heating increases adatom migration during standard vapor deposition, and similar effects can be expected as a consequence of using accelerated ion beams. Indeed, this more rapid attenuation of the substrate emission intensity indicates the same trend in our studies. These studies have also shown, however, that the substrate is disrupted and that the atom distribution depends on the technique used to deposit the atoms.

Figure 8.40 shows the Fermi level evolution for Ag/InP(110). As for In/GaAs(110) and Ag/ZnSe(100), ion deposition induces larger band bendings for the same coverages, but the E_F positions differ by less than 120 meV at low coverage and by only 50 meV after 20 Å deposition, despite interface morphology differences. For both atom and ion deposition, E_F movement into the gap is gradual, and final positions had not been reached for 20 Å depositions. This slow movement to-

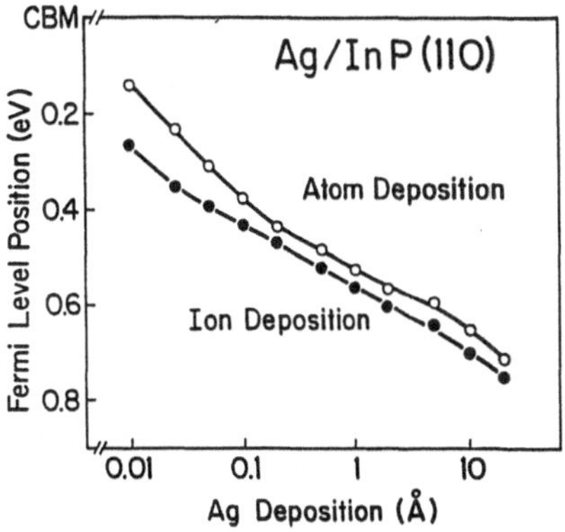

Fig. 8.40. Fermi level position evolution for Ag/InP(110) for atom and ion deposition. Ion depositions accelerate the band bending process slightly, but defects related to substrate disruption do not play an important role in Fermi level pinning. (From [8.107])

ward midgap is not consistent with the idea of fully occupied disruption-induced defect states near the surface since the final pinning position would be reached at submonolayer coverages.

Slow movement of E_F into the gap has been observed before for nonreactive overlayers, and it has been attributed to the lower rate of defect formation [8.116]. *Zunger* and coworkers [8.119, 120] suggested that the pinning mechanism for Al/GaAs(110) could be the transition from isolated Al atoms into Al clusters such that cluster formation would release sufficient energy to trigger Al-Ga exchange reactions, $\sim 3\,eV$. In turn, this would produce defects that would pin E_F. In contrast, for In/GaAs(110) it was proposed that clusters grow from very low coverage without a transition that could release enough energy to form defects. The present ion deposition results challenge ideas related to threshold energies: with ion deposition we have imparted almost two orders of magnitude larger energy than the $\sim 3\,eV$ needed to induce disruption, and substantial disruption did not change E_F evolution. Even for partially ionized beams, the number of defects should be enough to pin the Fermi level at very low coverages for both GaAs(110) and InP(110).

Since the observed band bending behavior *cannot* be understood in terms of fully charged adsorbate-induced defect (acceptor) levels, we must consider how these levels couple to bulk states near the CBM through the depletion layer. As discussed above, this coupling depends on temperature, dopant concentration, and coverage, and experiments have established that the fractional occupation of these adsorbate-induced states (and hence the amount of band bending at the interface) can be controlled by varying these parameters. The relatively slow E_F movement for In/GaAs(110) and Ag/InP(110) compared to other metal/III-V interfaces can be understood as arising from specific differences in the nature of the adsorbate-induced gap states at the surface. Specifically, broader, more delocalized states in the gap would be expected to be filled more slowly as a function of Fermi level position than more-well-defined states. As the density of surface adatom-induced states increases with metal deposition, broader states result in gradual E_F movement since the states are distribtuted over a wider range of energies. For sharper adsorbate-induced states, E_F movement should be much faster and, in the limit of complete localization, the final E_F position should be reached at submonolayer coverage. The results presented here indicate that there is coupling to broad adsorbate-induced levels at the surface.

8.6 Interfaces Formed by Metal Cluster Deposition

The results of the previous section have shown that atom and ion deposition of metal atoms onto semiconductor surfaces produces complex interface morphologies, including substrate disruption, atomic interdiffusion, alloy or compound formation, and structural changes of the substrate surface.

To create "ideal" boundary regions, we have developed a method of bringing preformed metal clusters into contact with atomically clean semiconductor surfaces [8.121, 122]. Our approach has been to isolate the substrate from the impinging metal atom flux to avoid the complex processes at the surface that are due to adatom impact and bonding. Thus, isolated adatoms are prevented from interacting with the surface until they have agglomerated into clusters. To achieve this, we first condense a thin layer of solid Xe on the semiconductor surface at 60 K and deposit metal atoms onto this buffer layer. Adatom mobility on Xe is sufficient to assure the formation of metallic clusters. These clusters come into contact with the clean, relaxed GaAs(110) surface when the Xe buffer is desorbed.

In this section, we show that metal–GaAs(110) interfaces formed by cluster deposition display unique band bending, with E_F positions that are largely metal-independent for n-GaAs but more dependent on the metal for p-GaAs. The metals Ag, Al, Au, Co, Ga, and Ti were chosen because they were known to exhibit a variety of interfacial interactions for atom deposition. In particular, Ag and Ga atoms cluster spontaneously at 300 K, and it was possible to directly compare band bending changes for systems which were expected to have the same final morphology. Titanium was chosen because of its highly reactive refractory character. Cobalt induces disruption, and Al induces exchange reactions at 300 K. These metals also exhibit a very wide range of melting temperatures, so that effects related to sintering or wetting could be expected.

8.6.1 Cluster Morphology

For our studies of metal cluster deposition, we chose a Xe coverage of 200 L. This is well into the Xe multilayer regime, and estimates based on GaAs core-level attenuation by the Xe multilayers [8.122] indicate the Xe layer to be ~ 30 Å thick. This was sufficient to minimize interaction of the deposited atoms with the substrate prior to Xe desorption. In preparation for the metal cluster experiments, we confirmed that the properties of the GaAs substrate were not modified by Xe adsorption at 60 K and desorption during warming to 300 K. Substrate Ga and As 3d core level EDCs taken before and after this process differed only in the Gaussian width of the surface and bulk features, i.e., phonon broadening. Most importantly, the binding energy of the substrate core levels changed by less than 30 meV. This shift is consistent with the temperature-dependent position of E_F in the semiconductor gap, since EDCs for the clean surface were collected at 60 K and those following the desorption were generally collected at 300 K.

The preformed clusters are brought into contact with the GaAs(110) surface when the Xe buffer layer is desorbed. A crude estimate of the size of the clusters can be obtained by assuming them to be hemispheres distributed across the surface. In this case, the normalized substrate emission can be written

$$\frac{I(\theta)}{I(0)} = 1 - \frac{3\theta}{2R}\left\{1 + \frac{2}{R^2}\left[\lambda(R+\lambda)\exp\left(-\frac{R}{\lambda}\right) - \lambda^2\right]\right\} ,$$

where λ is the photoelectron mean free path, R is the cluster radius, and θ the amount of metal deposited, in angstroms. Estimates based on the rate of attenuation by these clusters give $R \cong 10$ Å for $\theta = 1$ Å, corresponding to ~ 100 atoms in the cluster. Significantly, attenuation of the substrate emission is nearly identical for each of the metals investigated. Attenuation results for Ti, Co, and Ag clusters deposited onto GaAs(110) are shown in Figs. 8.19 and 20; points obtained with the other metals fall on the same line as that established by Ti and Co. For depositions of 5–15 Å on Xe, R varies between 30 and 40 Å (3000–6000 atoms per hemisphere). This large estimated cluster size is consistent with the high metal mobility on Xe surfaces because of the weak metal–Xe interactions.

To gain direct information regarding the structure and morphology of the metal cluster/GaAs(110) interface, scanning and transmission electron microscopy studies were undertaken. In Fig. 8.41 we show results for 7 Å Au(cluster)/GaAs(110) As can be seen from the plan view at the top of the figure, there has been considerable sintering and network formation, but large portions of the surface are exposed. The size of the large central aggregate is approximately 1000 Å. Figure 8.41b shows a transmission electron micrograph of a cross section through the Au(cluster)/GaAs interface. For this section, the lateral spacing between clusters along the [110] direction is 200–250 Å, and there is intimate contact between the clusters and the substrate. Figure 8.41c shows a high-resolution lattice image of one of the clusters. Inspection of the left portion of this cluster shows lattice images of (111) Au planes separated by 2.5 Å. The region at the right side of this cluster also appears to be well defined, but the orientation of the Au(111) planes is different. This indicates that the Au cluster consists of several grains of Au. From this image, we can see an abrupt interface without preferential lattice orientation between the metal cluster and the semiconductor surface. The thickness of the clusters is ~ 60 Å. There is no evidence for reaction between Au and the GaAs substrate. The slightly fuzzy image in the central part of the metal cluster close to the substrate (indicated by the arrow) is probably due to differently oriented small grains (or distortion at the grain boundary) where the lattice spacing between planes is smaller than the resolution of the microscope (below 1.4 Å).

8.6.2 Cluster Metallicity and Substrate Modification

From the above, it is clear that the metal aggregates formed by cluster deposition onto GaAs(110) are rather extended, even for 7 Å nominal coverage. In order to investigate the onset of metallic character, we have studied the valence band evolution for clusters corresponding to depositions between 0.1 and 34 Å. Repre-

Fig. 8.41a–c. Transmission electron micrographs for 7 Å (cluster) deposition onto GaAs(110). The plan view (**a**) reveals the formation of a connected network of Au with large regions of surface that are exposed. A transverse slice (**b, c**) reveals the profile of the clusters. The results in (**c**) reveal lattice images for Ag and GaAs that show that sintering of at least two crystallites of Au have occurred to produce the single observed island. (From [8.122])

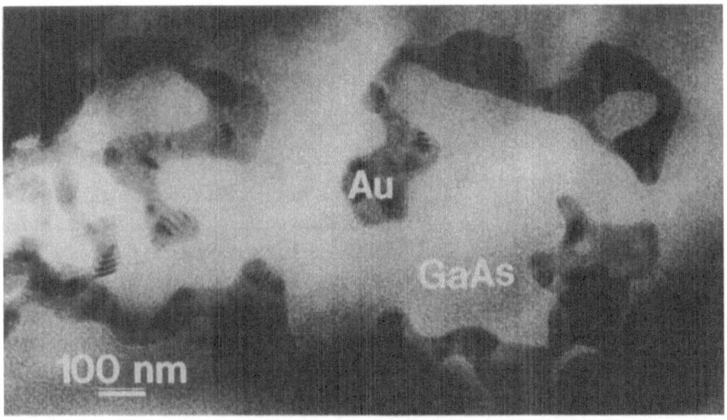

a

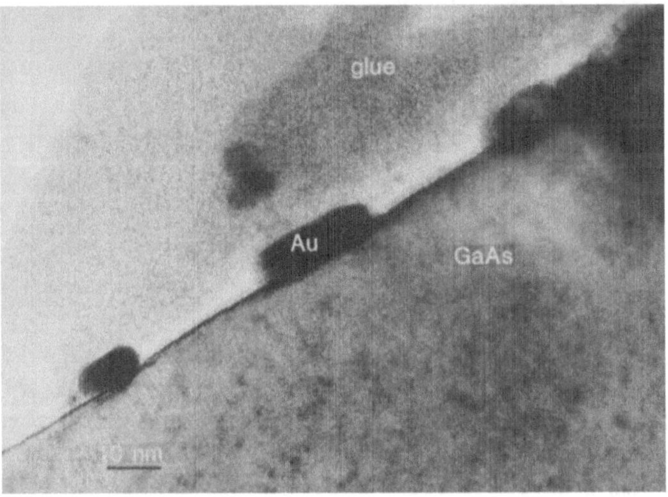

b

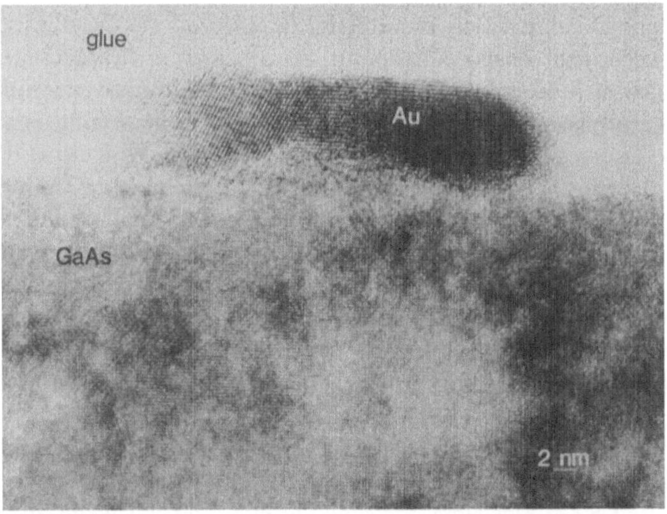

c

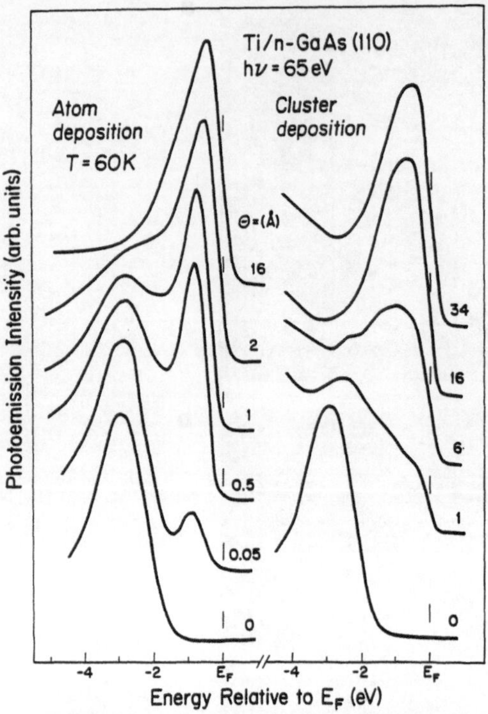

Fig. 8.42. Valence band results for Ti/n-GaAs(110) for atom deposition at 60 K and cluster deposition. Atom deposition leads to the gradual formation of a metallic overlayer, as evidenced by the appearance of a Fermi level by 2 Å. For cluster deposition, there is a distinct Fermi level cutoff by 1 Å because the clusters are metallic, but emission from exposed GaAs is also very high. (From [8.122])

sentative results are shown at the right of Fig. 8.42 for Ti clusters and at the right of Fig. 8.43 for Ag clusters, with energies referenced to the Fermi level, E_F. For comparison, the left side of each figure shows results obtained by conventional atom deposition at 60 K. For 0.05 Å atom deposition onto n-GaAs at 60 K, the Ti d-derived states appear in the gap centered ~ 1 eV below E_F. For 0.5 and 1 Å Ti atom deposition at 60 K, the overlayer is still below the metallization threshold, and there is no emission at E_F. Between 1 and 2 Å, however, the metallization threshold is exceeded, and d-band emission at E_F is observed. Broadening of the d-band emission also occurs and by 16 Å deposition the overlayer is fully metallic. In contrast, cluster deposition of 1 Å gives rise to a metallic Fermi level cutoff. This is expected because the attenuation measurements indicate that the clusters are 10 Å in radius, and such a cluster should be metallic. The very different lineshapes for 1 Å (cluster) and 1 Å (atom) deposition demonstrate that the clusters cover a relatively small portion of the surface and emission from the GaAs substrate is still strong. In contrast, atom deposition leads to a more uniform coverage of the GaAs surface, and substrate emission is much lower for like metal coverages. At higher coverage, the spectra are more similar to those for atom deposition because almost all of the surface is covered by Ti clusters and the substrate emission is strongly attenuated.

Examination of the changing Ag d-band width for Ag cluster and atom deposition onto GaAs(110) makes it straightforward to follow the developing overlayer metallicity (Fig. 8.43). For atom deposition at 60 K, the full width at half

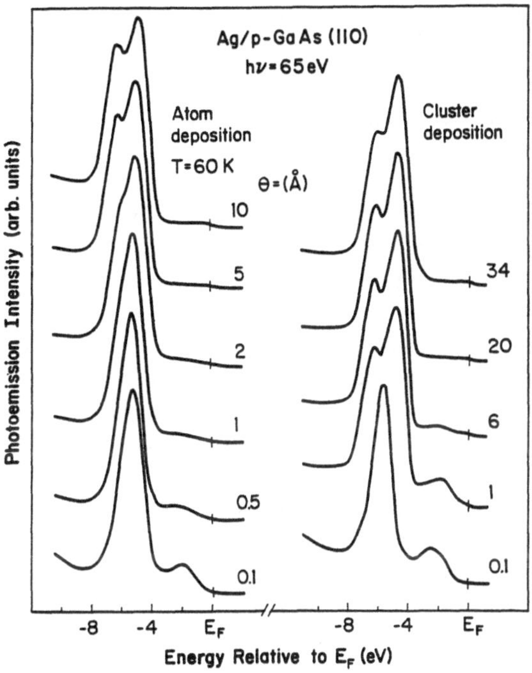

Fig. 8.43. Valence band evolution for Ag/*p*-GaAs for atom and cluster deposition analogous to that of Fig. 8.42. (From [8.122])

maximum increases gradually until it reaches its fully metallic value at ~ 5 Å [8.82, 123]. In contrast, the Ag d-band emission for 1 Å cluster deposition exhibits fully metallic width and shape. Close inspection of the emission near E_F shows a distinct metallic cutoff for all cluster coverages ≥ 1 Å. The step is difficult to see in Fig. 8.43 because of the much stronger d-band emission relative to the sp-states at E_F, but it is clear when the scale is expanded. In contrast, a metallic Fermi edge is not observed for atom deposition until $\theta \geq 5$ Å. These results indicate that cluster deposition produces metallic clusters, even for $\theta = 1$ Å.

In order to determine the morphology of the contacts between clusters and semiconductor surfaces, we have examined the Ga and As $3d$ core level emission as a function of deposition. For direct comparison, we have also examined the effects of atom deposition at 60 and 300 K. Figure 8.17 compares As $3d$ EDCs taken at $h\nu = 90$ eV for Co cluster and atom deposition while those in Fig. 8.14 consider Ti deposition.

The clean cleaved spectra of Fig. 8.17 were collected at 60 K, but the spectra following cluster deposition and Xe desorption were collected at temperatures between 100 K and 300 K. While this made it possible to investigate the temperature stability of the interface, it introduced changes in the core-level lineshapes associated with thermal broadening. These changes are reflected in the increase in Gaussian width for the core level spectra following cluster deposition shown in Fig. 8.17. These spectra reveal a decrease in the emission intensity from surface atoms (feature 2) relative to that from bulk atoms (feature 1) as the nominal coverage is increased. Most important, *no adsorbate-induced features* are required to fit the core-level spectra. The stability of these interfaces was demonstrated by the absence of any band bending or lineshape changes while warming to 300 K over a period of ~ 3 h. In contrast to the cluster deposition results, examination of the As core level spectra for atom deposition at 60 K shows two Co-induced features that dominate for $\theta \geq 2$ Å, as discussed in Sect. 8.2.1.

Figure 8.14 shows Ga 3*d* core level EDCs following Ti cluster and atom deposition. For atom deposition at 60 K, substantial changes in the Ga 3*d* lineshape occur, as discussed in Sect. 8.2.2. Inspection of the Ti cluster results shows that there are no Ti-cluster-induced features in the Ga 3*d* EDCs. Indeed, the only changes following cluster deposition reflect an increase in phonon broadening and a decrease in the relative intensity of emission from surface-shifted Ga atoms compared to bulk Ga atoms. The large size of the clusters compared to the photoemission probe depth ($\sim$ 80 Å cluster thickness vs $\sim$ 10–20 Å probe depth) makes it impossible to exclude the possibility of metal–substrate interaction beneath the clusters. As we will discuss shortly, the very different Fermi level positions for cluster and atom deposition on n-GaAs(110) suggest that different mechanisms determine E_F position for the two deposition techniques. Any deviation from the cluster-characteristic E_F position that occurs following cluster deposition can be taken as an indication of metal–substrate interaction caused either by the increase in system temperature or by the inherent instability of the particular metal cluster with respect to metal–GaAs interaction.

Additional studies with Al, Ag, Au, and Ga clusters show that the only lineshape changes observed in both the As 3*d* and Ga 3*d* EDCs involve phonon broadening and the decrease in the surface to bulk (S/B) emission intensity ratio as the nominal cluster coverage is increased. The absence of cluster-induced features suggests that this new deposition process is not disruptive and produces abrupt junctions. In contrast, there are radical lineshape changes following atom deposition, and various amounts of substrate disruption, atomic intermixing, reaction, and clustering are revealed by careful lineshape analysis (Sect. 8.3).

It is particularly instructive to consider differences between Ag cluster and atom deposition. Analysis of the core-level spectra shows distinctly different S/B ratios for atom deposition at 60 K and 300 K, and for direct Ag cluster deposition. For 300 K atom deposition, the ratio remains equal to the clean surface value of ~ 0.5 even at $\theta = 20$ Å, corresponding to $\sim 60\%$ surface coverage. This indicates that the exposed portions of the substrate are largely unperturbed by the Ag clusters. In contrast, atom deposition at 60 K leads to complete loss of emission from the surface-shifted atoms for $\theta \geq 2$ Å as the uniform overlayer forms and the surface atoms interact with the Ag overlayer. The S/B ratio decreases gradually with increasing coverage for preformed cluster deposition. This indicates that the clusters modify the surface structure in the region surrounding them and induce the loss of the surface component. (We expect that the metal overlayer leads to partial or complete surface unrelaxation beneath the clusters, but this area cannot be probed by photoemission.) We shall return to this shortly in discussions involving Schottky barrier formation since this process affects Fermi level movement at the surface.

8.6.3 Cluster-Induced Band Bending

One of the remarkable properties of interfaces formed by cluster deposition is that the Fermi level position in the gap is almost independent of the amount of metal deposited. To demonstrate this, we show in Fig. 8.44 the position of E_F in the gap as a function of metal coverage for Ag, Al, Au, Co, Ga, and Ti clusters deposited onto n-GaAs(110) doped at $1\times10^{17}\,\mathrm{cm}^3$. Each point was obtained from a complete cluster deposition experiment, i.e., obtaining a new GaAs(110) surface by cleaving, determining that E_F was within 60 meV of the CBM (unpinned), condensing a 30 Å Xe buffer layer, depositing the metal onto the buffer layer, desorbing the Xe, and, finally, measuring the Fermi level movement by analysis of changes in the Ga and As core-level energies using two photon energies for each core level. As shown, E_F moved to a position $\sim$ 260 meV below the CBM for 0.02 Å cluster deposition and then gradually moved to $\sim$ 320 meV below the CBM by 35 Å. From the attenuation curves, we know that $\sim$ 90% of the substrate is covered by metal clusters after 30–40 Å, and we expect no E_F movement at higher θ. We emphasize that E_F moved by only $\sim$ 60 meV over three orders of magnitude of metal coverage, and that the changes are entirely metal-independent, based on approximately 30 cluster experiments summarized in Fig. 8.44. Such coverage-independence and lack of metal specificity has not been found in atom deposition experiments and cannot be described with existing models of Schottky barrier formation. Likewise, the position of E_F high in the gap is contrary to that expected for pinning positions associated with deep levels or defects, or metal-induced gap states.

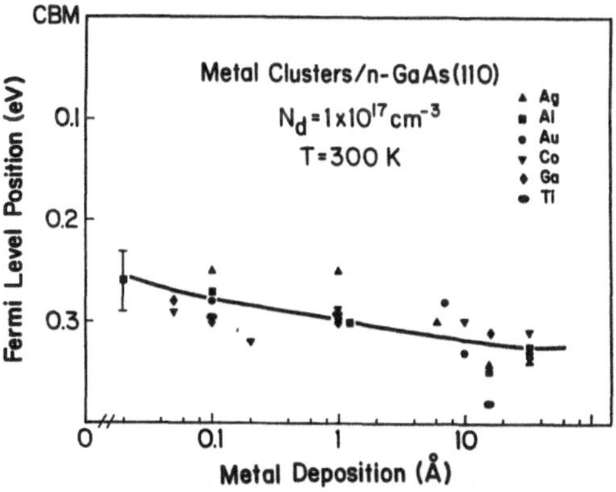

Fig. 8.44. Fermi level positions for n-GaAs(110) produced by cluster deposition of Ag, Al, Au, Co, Ga, and Ti. Each point represents a complete experiment starting with a fresh cleave. Within experimental error, all of the points fall along a single line, indicating the metal-insensitivity of the pinning position for these metals. (From [8.122])

Figure 8.45 shows the very different evolution of the surface Fermi level for metal cluster deposition on p-GaAs(110) doped at $2\times10^{18}\,\mathrm{cm}^{-3}$. The results show the movement to 100–200 meV above the VBM by 0.1 Å, and essentially no change until 6–8 Å, where E_F rises to its final position in the gap. For p-GaAs, the Fermi level evolution is somewhat metal-dependent with energies ranging from $\sim$ 370 meV above the VBM for Al to $\sim$ 650 meV above the VBM for

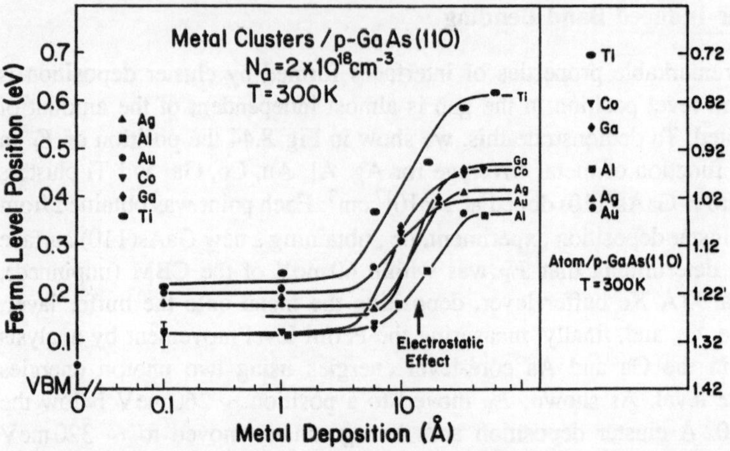

Fig. 8.45. Fermi level positions for p-GaAs showing much greater metal-specificity than for n-GaAs. The step near 10 Å occurs because the spacing between clusters becomes comparable to the depletion width parallel. The surface is homogeneously pinned above 10 Å but not below 10 Å for this high dopant concentration. Results for low dopant concentrations show no such step. (From [8.122])

Ti after 25–30 Å deposition. It is noteworthy that the ordering observed at low coverage (Ti, Au, Ag, Ga, Al, Co) is not the same as that at high coverage (Ti, Ga, Co, Ag, Au, Al), so that the step height is also metal-specific.

The very obvious difference between the results of Figs. 8.44 and 45 is the step for p-GaAs that occurs at 6–8 Å. We note that the p-GaAs samples were doped at $2\times10^{18}\,cm^{-3}$ while the n-GaAs samples were doped at $1\times10^{17}\,cm^{-3}$. This suggests that the step is due to the coverage-related changes in cluster separation with the final E_F position achieved when cluster spacing becomes comparable to, and ultimately less than, the substrate depletion width. This transition from isolated clusters on a largely exposed surface to clusters that are close enough to produce uniform surface pinning occurs near $\sim$ 10 Å. For 0.4 eV band bending, corresponding to an average final E_F position in Fig. 8.44, the depletion width is 170 Å for $N = 2\times10^{18}\,cm^{-3}$ and 760 Å for $N = 1\times10^{17}\,cm^{-3}$. The simple hemispherical model described above gives an estimate of R = 30 Å for θ = 6 Å so that the spacing between uniformly distributed clusters would be $\sim$ 40 Å. Plan-view TEM results for 7 Å Au clusters (Fig. 8.41) show that clusters are not distributed uniformly. The cross-sectional view along the [110] direction shows an average island separation of $\sim$ 250 Å. The fact that a step is observed in every case points to common morphological evolution of the clusters on the surface, consistent with the attenuation results, and equivalent modification of surface electrostatics. The low coverage regime corresponds to dispersed clusters on n- and p-GaAs, but the shorter depletion width for more heavily doped p-type samples requires higher coverages to realize uniform surface pinning by the clusters.

To prove that the step near 10 Å depends on the bulk dopant concentration and is electrostatic in origin, we investigated Ag metal cluster deposition on p-

GaAs doped at $1 \times 10^{17}\,cm^{-3}$, i.e., the same dopant concentration as that for the n-type samples. As discussed in [8.80], E_F moved substantially into the gap for 0.02 Å of Ag deposition onto lightly doped samples and it moved only gradually for higher depositions. The step seen for the more heavily doped p-type samples was not observed. This confirms that the change from the low coverage region to the final E_F position for the higher doped samples is electrostatic in origin. We conclude that the gradual changes for lightly doped n- and p-GaAs are a consequence of the cluster spacing being slightly greater than the depletion width. The Fermi level position measured above the step for p-GaAs should then be taken as representative of the average surface effect of the metal clusters.

In the discussion above, we concluded that cluster deposition produced stable interfaces for the weakly reactive metals as well as for Co and Al. This conclusion is supported by following the movement of E_F in the gap as the cluster/GaAs interfaces are warmed from $\sim$ 100 K to 300 K. These studies found the E_F position to be independent of temperature for Al, Ga, Co, Ag, and Au, provided account is taken of the slight temperature dependence of the position of E_F and the value of the band gap of GaAs. For highly reactive Ti, however, reaction was not completely suppressed upon warming, and investigations of E_F movement showed temperature dependence. The results of Figs. 8.44 and 45 give the Fermi level position for Ti cluster deposition for the lowest temperature measured, but E_F moved toward midgap as the temperature was increased to 300 K. Immediately following Xe desorption at $\sim$ 100 K, E_F was 0.38 eV below the CBM. After $\sim$ 120 min, E_F had reached its final position 0.67 eV below the CBM. This final position is in good agreement with the value observed for atom deposition. Warming experiments for HD p-GaAs at high coverage showed no change because the final E_F position for cluster and atom deposition is the same. At lower coverage, warming of an interface formed with 0.2 Å Ti on the highly doped p-GaAs showed no change in E_F since the clusters were widely separated and reaction beneath them would not alter E_F across the uncovered surface. During these warming experiments, the changes that were observed in the Ga and As core levels can be attributed to phonon broadening. As previously noted, due to the movement of E_F during warming we infer that the Ti cluster is so unstable when in contact with GaAs that thermodynamically driven reaction takes place despite kinetic constraints. Some reaction by $\sim$ 100 K cannot be excluded. Since E_F remained far from midgap for temperatures up to 300 K for the other metal clusters, we conclude that those interface regions are free of disruptive metal–substrate interaction.

In Table 8.1 we compare the highest coverage E_F positions for cluster deposition for Al, Ag, Au, Co, Ga, and Ti with those for atom deposition at both 60 and 300 K. These comparisons are based on measurements with the same bulk dopant concentration N. The energy shifts were obtained by lineshape analysis of the substrate core levels, as discussed in Sect. 8.1.4. Table 8.1 references all energies to the CBM. When account is taken of the temperature dependence of the band gap of GaAs, we see that the barrier heights for n-GaAs and p-GaAs do not change with temperature. Significantly, the spread in the values for the E_F

Table 8.1. Fermi level positions below CBM [cV] for metal cluster deposition and for metal atom deposition at 60 K and 300 K. Values for cluster deposition are from this study. Values for atom deposition are for the highest coverage reported in [8.16–18, 79, 80, 91]. All energies are referenced to the CBM, E_{CBM}, with error bars of ±30 meV. The Schottky barrier, $E_F - E_{VBM}$, would also be nearly independent of temperature if the *p*-GaAs results for atom deposition were referenced to the VBM

Metal	*n*-GaAs(110)			*p*-GaAs(110)			Ref.
	Cluster deposition	Atom deposition 300 K	Atom deposition 60 K	Cluster deposition	Atom deposition 300 K	Atom deposition 60 K	
Ag	0.32	0.89	0.89	1.00	1.02	1.12	8.80
Al	0.33	0.70	0.70	1.05	0.96	1.04	8.91
Au	0.33	0.90	0.87	1.02	1.04	1.10	8.122
Co	0.32	0.67	0.67	0.96	0.82	0.92	8.80
Ga	0.34	0.60	0.84	0.95	0.87	0.87	8.16–18
Ti	0.38	0.73	0.67	0.79	0.72	0.79	8.79

position for clusters is only 60 meV for *n*-GaAs and the barrier height is about half of that for atom deposition. For *p*-GaAs, the barrier heights are the same to within 30 meV for atom and cluster deposition for Ag and Au, deviate by less than 100 meV for Al, Ga, and Ti, and by 140 meV for Co. Atom deposition always gives the position farther from the VBM when there is deviation.

Models that address temperature-dependent E_F evolution [8.3, 4, 16–19] in the high-coverage regime are particularly interesting in view of the cluster deposition results because E_F movement toward midgap is believed to be related to the onset of metallicity and wavefunction delocalization in the overlayer. It is intriguing, then, that deposition of fully metallic clusters on *n*-GaAs results in E_F positions so different from those obtained after metallization by atom deposition. The behavior for *n*-GaAs indicates equivalent barrier heights for all stable clusters, implying that the Fermi energy position is related to the GaAs(110) surface rather than the chemical identity or any other property of the overlying cluster. For clusters on *p*-GaAs, E_F is more metal-dependent and moves farther into the gap. This asymmetric behavior suggests that these clusters induce electrically active donor and acceptor states but that the acceptor states lie deeper in the gap.

According to the MIGS model, metallic overlayers should pin the Fermi level at the charge neutrality point in the semiconductor band gap, and this pinning positions should show little metal specificity. For GaAs, this point is estimated to be 0.7 eV above the VBM [8.83, 84, 124]. Our results show that none of the final E_F positions are close to this predicted MIGS value for *n*- or *p*-GaAs for cluster deposition. Likewise, from Table 8.1, the results for atom deposition on *n*-GaAs are far from midgap while those for *p*-GaAs are closer but exhibit a metal-dependent spread of 260 meV. These facts suggest that MIGS are not very important in determining the Fermi level position following cluster deposition.

Another model of Schottky barrier formation relates pinning to antisite defects created by adatom condensation. Such defects produce distinct pinning positions that are intrinsic to the semiconductor. For cluster deposition, we see that Fermi level movement is unrelated to adatom-induced defects created in the GaAs. This is reasonable because there is no detectable cluster-induced substrate disruption. We also see that the final E_F positions for atom deposition are inconsistent with energies (0.52 and 0.75 eV above the VBM) predicted by defect models [8.1].

We propose that cluster deposition leads to surface unrelaxation and, therefore, bond modification around and beneath the clusters. This hypothesis is supported by results of detailed analysis of the ratio of the surface to bulk core level intensities, S/B, for Ga and As atoms. In the discussion of core level lineshapes it was shown that S/B decreased from the clean surface value as the cluster coverage increased. Those changes in S/B are to be expected if the clusters modify the GaAs(110) surface relaxation in the regions around them. The relaxation that is produced by cleaving gives rise to the surface component observed in the clean cleave spectra, and, as is well known, sweeps the surface states out of the gap. Unrelaxation would reintroduce surface states into the band gap of GaAs. The deposition of thick clusters could also lead to unrelaxation beneath them, but this effect would not be observed because the photoemission signal from these regions is attenuated by the thick metal overlayer. On the other hand, this unrelaxation must extend beyond the cluster perimeter as our photoemission probe does indeed detect changes in S/B. The loss or partial loss of surface relaxation would result in a large number of intrinsic surface states in the band gap. Calculations of such intrinsic states for unrelaxed GaAs(110) have shown surface levels within $\sim$ 150 meV of the CBM [8.125] and states at $\sim$ 450 meV above the VBM. Hence, unrelaxation for p-GaAs should lead to E_F movement deeper into the gap while E_F would remain close to the CBM for unrelaxed n-GaAs. The fact that the E_F position for n-GaAs is independent of metal is consistent with this model. The cluster deposition results on p-GaAs suggest, however, that there are additional factors affecting the positions of E_F in the gap which are reflected in the spread in final pinning values (260 meV).

It is very interesting that the sensitivity to the nature of metal clusters is lacking from the E_F results on the n-type substrates. A plausible scenario for this behavior would necessarily involve two parallel mechanisms for Schottky barrier formation at these interfaces. The first and dominant mechanism would be associated with surface unrelaxation and the reintroduction of intrinsic states in the gap. The calculations for these states agree with our results in the sense that the ultimate E_F positions for p-type GaAs lie deeper in the band gap than for the n-type. Also, the existence of these states in the gap is experimentally supported by recent STM measurements on GaAs(110) covered by Sb islands [8.25]. The second mechanism is associated with metal-specific modifications of the barrier height on p-type semiconductors, but apparently plays no role for the n-type. The precise nature of this mechanism is not currently understood.

It is particularly interesting to further examine unrelaxation for Ag/GaAs. As we noted previously, Ag atom deposition at 300 K results in the spontaneous formation of Ag clusters, but no change in S/B is observe This suggests that the uncovered substrate is unchanged by the presence of the Ag clusters. We speculate that under these conditions the Ag clusters grow in a manner which allows accommodation of the cluster to the substrate surface structure. Specifically, the sequential deposition of Ag atoms and their migration across the GaAs surface to nucleation sites for cluster growth may allow the Ag atoms to agglomerate without altering the surface structure. In contrast, deposition of large preformed Ag clusters allows no such metal accommodation, and the substrate areas around (and possibly beneath) the clusters are modified by the cluster presence. The similar final morphologies, but quite different final E_F positions (Table 8.1) for Ag/GaAs interfaces formed by these different techniques, represent a clear demonstration of the dependence of band bending upon the structural details of the intimate contacts and the energetics of the cluster formation process.

It is clear from the discussion above that the deposition of preformed metal clusters produces abrupt surfaces that are structurally very different from those produced by any other technique. The details of the Fermi level evolution for these interfaces are not completely understood at this time, but the abrupt disruption-free nature of the interfaces make them particularly attractive for metal-semiconductor junction modeling. Certainly, these results demonstrate the importance of the energetics associated with atom condensation and bond formation as far as E_F evolution is concerned. Consequently, comparison of results for Ag clusters grown spontaneously on GaAs during 300 K atom deposition and those deposited with the cluster technique shows dramatically different band bending.

8.7 Prospects and Future Developments

This chapter has sought to demonstrate that the number of parameters that can play a significant role in metal–semiconductor interface formation is quite large. Historically, their complex and interrelated nature has frustrated studies of the fundamental properties of interfaces. However, control over these parameters has become increasingly advanced. It is just this control which is making it possible to gain deeper insight into the role of each parameter. Ultimately, of course, the goal is to be able to predict the properties of an interface, including its chemistry, its morphology, and its electrical properties.

The parameter space emphasized in the above sections has included:

1) details of the semiconductor surface (surface plane exposed, surface reconstructions, surface quality, degree of bonding covalency and ionicity, etc.);
2) the chemical activity of the adatoms in terms of their bonding with the substrate and their bonding with other adatoms;
3) the number of adatoms deposited onto the surface;
4) the temperature of the surface when atoms are deposited;
5) the energy with which the atoms are deposited;
6) the temperature of the interface during measurements or associated with processing;

7) the bulk dopant concentration and dopant type;
8) the effect of external illumination with electrons or photons;
9) the way that surface processes are altered when atom deposition is replaced by preformed cluster deposition.

The influence of many of these parameters has just recently been recognized, and it is certain that their roles will be examined carefully in the next few years. Likewise, dynamic properties that occur at surfaces will also be examined in detail, including those related to surface chemical reaction that are induced by the probe. As an example, we note that the oxidation of GaAs(110) at 20 K is a direct consequence of the photon beam used in photoemission experiments.

Acknowledgements. It is a great pleasure to draw attention to the excellent work by C.M. Aldao, I.M. Vitomirov, G.D. Waddill, Steven G. Anderson, M. Vos, Xu Fang, S.A. Chambers, J.J. Joyce, M. Grioni, M. del Giudice, and M.W. Ruckman as reviewed in this chapter. Interactions with them, many other members of the Electronic Materials Group of the University of Minnesota, and the community at large have been stimulating and rewarding.

This work has been supported by the Office of Naval Research, the National Science Foundation, and the Army Research Office. The synchrotron radiation photoemission experiments were done at Aladdin, a user facility supported by the National Science Foundation and operated by the University of Wisconsin. The assistance of the staff of that laboratory is gratefully acknowledged. Data acquisition and data analysis has been enhanced immeasurably by access to IBM minicomputers obtained through a Materials Science and Processing Grant from IBM.

References

8.1 A great many important contributions have come from the application of the tools of surface science for interface research, including those listed below. Here, we emphasize references related to photoemission. Others would include scanning tunneling microscopy, LEED, medium-energy ion scattering, Rutherford backscattering, etc. See, for example:
J.M. Poate, K.N. Tu, J.W. Mayer (eds.): *Thin Films – Interdiffusion and Reaction* (Wiley Interscience, New York 1978)
L.C. Feldman, J.W. Mayer (eds.): *Fundamentals of Surface and Thin Film Analysis* (North-Holland, New York 1986)
8.2 W.E. Spicer, Z. Liliental-Weber, E. Weber, N. Newman, T. Kendelewicz, R. Cao, C. McCants, P. Mahowald, K. Miyano, I. Lindau: J. Vac. Sci. Technol. B **6**, 1245 (1988)
8.3 For a detailed discussion see: W.E. Spicer, R. Cao, K. Miyano: In *Metallization and Metal–Semiconductor Interfaces*, NATO Advanced Study Institute, Series B, Physics, Vol. 195, ed. by I.P. Batra (Plenum, New York 1989) and references therein
8.4 Z. Liliental-Weber: J. Vac. Sci. Technol. B **5**, 1007 (1987)
8.5 L.J. Brillson: Surf. Sci. Rep. **2**, 123 (1982) and references therein
8.6 R.E. Viturro, J.L. Shaw, C. Mailhiot, L.J. Brillson, N. Tache, J. McKinlay, G. Margaritondo: Appl. Phys. Lett. **52**, 2052 (1988)
8.7 L.J. Brillson et al.: J. Vac. Sci. Technol. B **6**, 1263 (1988)
8.8 J.R. Waldrop, S.P. Kowalczyk, R.W. Grant: J. Vac. Sci. Technol. **21**, 607 (1982)
8.9 J.H. Weaver: in *Analysis and Characterization of Thin Films*, ed. by K.N. Tu, R. Rosenberg (Academic, New York 1987)
8.10 E.H. Rhoderick, R.H. Williams: *Metal–Semiconductor Contacts*, 2nd ed. (Clarendon, Oxford 1988)
8.11 R. Ludeke: Surf. Sci. **168**, 290 (1986)

8.12 G. LeLay: Surf. Sci. **132**, 169 (1983)
8.13 G.W. Rubloff: Surf. Sci. **132**, 268 (1983)
8.14 M. Prietsch, M. Domke, C. Laubschat, G. Kaindl: Phys. Rev. Lett. **60**, 436 (1988)
8.15 C. Laubschat, M. Prietsch, M. Domke, E. Weschke, G. Remmers, T. Mandel, J.E. Ortega, G. Kaindl: Phys. Rev. Lett. **62**, 1306 (1989)
8.16 K. Stiles, A. Kahn: Phys. Rev. Lett. **60**, 440 (1988)
8.17 K. Stiles, A. Kahn, D.G. Kilday, G. Margaritondo: J. Vac. Sci. Technol. B **5**, 987 (1987)
8.18 K. Stiles, S.F. Horng, A. Kahn, J. McKinley, D.G. Kilday, G. Margaritondo: J. Vac. Sci. Technol. B **6**, 1392 (1988) and references therein
8.19 W. Mönch: Phys. Rev. Lett. **58**, 1260 (1987)
8.20 K.L.I. Kobayashi, N. Watanabe, T. Narasawa, H. Nakashima: J. Appl. Phys. **58**, 3758 (1985)
8.21 N. Watanabe, K.L.I. Kobayashi, T. Narasawa, H. Nakashima: J. Appl. Phys. **58**, 3766 (1985)
8.22 J.L. Freeouf, J.W. Woodall: Appl. Phys. Lett. **39**, 727 (1981)
8.23 C.M. Aldao, I.M. Vitomirov, G.D. Waddill, S.G. Anderson, J.H. Weaver: Phys. Rev. B **41**, 2800 (1990) and references therein
8.24 J.H. Weaver, Zhangda Lin, F. Xu: "Surface Segregation at Evolving Metal–Semiconductor Interfaces", in *Surface Segregation and Related Phenomena*, ed. by P.A. Dowben, A. Miller (CRC Press, Boca Raton, FL 1990) Chap. 10, and references therein
8.25 R.M. Feenstra, P. Mårtensson: Phys. Rev. Lett. **61**, 447 (1988)
8.26 P.N. First, J.A. Stroscio, R.A. Dragoset, D.T. Pierce, R.J. Celotta: Phys. Rev. Lett. **63**, 1416 (1989)
8.27 E.E. Koch, D.E. Eastman, Y. Farge: *Handbook on Synchrotron Radiation* (North-Holland, New York 1983)
8.28 H. Winick, S. Doniach (eds.): *Synchrotron Radiation Research* (Plenum, New York 1980)
8.29 G. Margaritondo, J.H. Weaver: "Photoemission Studies of Valence States" in *Methods in Experimental Physics and Surfaces*, ed. by M. Lagally, R.L. Park (Academic, Cambridge, MA 1985) Chap. 4
8.30 B. Feuerbacher, B. Fitton, R.F. Willis (eds.): *Photoemission and the Electronic Properties of Surfaces* (Wiley Interscience, New York 1978)
8.31 M. Cardona, L. Ley (eds.): *Photoemission in Solids I, General Principles*, Topics Appl. Phys., Vol. 26 (Springer, Berlin, Heidelberg 1978)
8.32 L. Ley, M. Cardona (eds.): *Photoemission in Solids II, Case Studies*, Topics Appl. Phys., Vol. 27 (Springer, Berlin, Heidelberg 1979)
8.33 C.N. Berglund, W.E. Spicer: Phys. Rev. **136**, 1030, 1044 (1964)
8.34 H.Y. Fan: Phys. Rev. **68**, 43 (1945)
8.35 W.E. Spicer: Phys. Rev. **112**, 114 (1958)
8.36 H. Mayer, H. Thomas: Z. Phys. **147**, 149 (1959)
8.37 D.A. Shirley: Phys. Rev. B **5**, 4709 (1972)
8.38 L. Kleinman: Phys. Rev. B **3**, 2982 (1971)
8.39 D.R. Penn: Phys. Rev. B **13**, 5248
8.40 L.I. Schiff: *Quantum Mechanis* (McGraw-Hill, New York 1955)
8.41 S. Doniach, M. Sunjic: J. Phys. E **3**, 265 (1970)
8.42 K. Siegbahn et al.: *ESCA Applied to Free Molecules* (North-Holland, Amsterdam 1959) p. 104
8.43 J.J. Joyce, M. del Giudice, J.H. Weaver: J. Electron Spectrosc. Relat. Phenom. **49**, 31 (1989)
8.44 G.K. Wertheim, P.H. Citrin: In [8.31]
8.45 J.B. Pendry: In [8.30]
8.46 G.K. Wertheim, S.B. Dicenzo: J. Electron Spectrosc. Relat. Phenom. **37**, 57 (1985)
8.47 G.A. Prinz, J.J. Krebs: Appl. Phys. Lett. **39**, 397 (1981)
8.48 C. Vittoria, F.J. Rachford, J.J. Krebs, G.A. Prinz: Phys. Rev. B **30**, 3903 (1984)
8.49 G.A. Prinz: Phys. Rev. Lett. **54**, 1051 (1985)
8.50 K. Schroder, G.A. Prinz, K.-H. Walker, E. Kisker: J. Appl. Phys. **57**, 3669 (1985)
8.51 G.A. Prinz, E. Kisker, H.B. Hathaway, K. Schroder, K.-H. Walker: J. Appl. Phys. **57**, 3024 (1985)

8.52 J.R. Waldrop, R.W. Grant: Appl. Phys. Lett. **34**, 630 (1979)
8.53 F. Xu, J.J. Joyce, M.W. Ruckman, H.-W. Chen, F. Boscherini, D.M. Hill, S.A. Chambers, J.H. Weaver: Phys. Rev. B **35**, 2375 (1987)
8.54 M. Grioni, J.J. Joyce, J.H. Weaver: J. Vac. Sci. Technol. A **4**, 965 (1986)
8.55 J. Nogami, T. Kendelewicz, I. Lindau, W.E. Spicer: Phys. Rev. B **34**, 669 (1986)
8.56 D.M. Hill, F. Xu, Zhangda Lin, J.H. Weaver: Phys. Rev. B **38**, 1893 (1988)
8.57 M.W. Ruckman, J.J. Joyce, J.H. Weaver: Phys. Rev. B **33**, 7029 (1986)
8.58 M.W. Ruckman, M. del Giudice, J.J. Joyce, J.H. Weaver: Phys. Rev. B **33**, 2191 (1986)
8.59 R. Ludeke, D. Straub, F.J. Himpsel, G. Landgren: J. Vac. Sci. Technol. A **4**, 874 (1986)
8.60 R. Ludeke, G. Landgren: Phys. Rev. B **33**, 5526 (1986)
8.61 F. Xu, D.M. Hill, Z. Lin, S.G. Anderson, Y. Shapira, J.H. Weaver: Phys. Rev. B **37**, 10295 (1988)
8.62 F. Xu, Z. Lin, D.M. Hill, J.H. Weaver: Phys. Rev. B **35**, 9353 (1987); ibid. **36**, 6624 (1987)
8.63 O. Wada, S. Yanagisawa, H. Takanashi: Appl. Phys. Lett. **j29**, 263 (1976)
8.64 F. Xu, C.M. Aldao, I.M. Vitomirov, Z. Lin, J.H. Weaver: Phys. Rev. B **36**, 3495 (1987)
8.65 C.M. Aldao, I.M. Vitomirov, F. Xu, J.H. Weaver: Phys. Rev. B **37**, 6019 (1988)
8.66 J.H. Weaver, M. Grioni, J.J. Joyce: Phys. Rev. B **31**, 5348 (1985)
8.67 M.D. Williams, T. Kendelewicz, R.S. List, N. Newman, C.E. McCants, I. Lindau, W.E. Spicer: J. Vac. Sci. Technol. B **3**, 1202 (1985)
8.68 P.W. Chye, I. Lindau, P. Pianetta, C.M. Garner, C.Y. Su, W.E. Spicer: Phys. Rev. B **18**, 5545 (1978)
8.69 W.P. Petro, T. Kendelewicz, I. Lindau, W.E. Spicer: Phys. Rev. B **34**, 7089 (1986)
8.70 D. Coulman, N. Newman, G.A. Reid, Z.Liliental–Weber, E.R. Weber, W.E. Spicer: J. Vac. Sci. Technol. A **5**, 1521 (1987)
8.71 F. Xu, Y. Shapira, D.M. Hill, J.H. Weaver: Phys. Rev. B **35**, 7417 (1987)
8.72 J.J. Joyce, M. Grioni, M. del Giudice, M.W. Ruckman, F. Boscherini, J.H. Weaver: J. Vac. Sci. Technol. A **5**, 2019 (1987)
8.73 T.G. Anderson, J. Kanski, G. Le Lay, S.P. Svensson: Surf. Sci. **168**, 301 (1986)
8.74 M.W. Ruckman, J.J. Joyce, J.H. Weaver: Phys. Rev. B **34**, 5118 (1986)
8.75 T. Yoshiie, C.L. Bauer: J. Vac. Sci. Technol. A **1**, 554 (1983)
8.76 T. Yoshiie, C.L. Bauer, A.G. Milnes: Thin Solid Films **111**, 149 (1984)
8.77 J.R. Lince, C.T. Tsai, R.S. Williams: J. Mater. Res. **1**, 537 (1986)
8.78 E. Beam, D.D.L. Chung: Mater. Res. Soc. Symp. Proc. **37**, 595 (1985)
8.79 C.M. Aldao, G.D. Waddill, S.G. Anderson, J.H. Weaver: Phys. Rev. B **40**, 2932 (1989); a study of Ti/GaAs(110) interfaces formed at 60 and 300 K
8.80 G.D. Waddill, C.M. Aldao, I.M. Vitomirov, S.G. Anderson, C. Capasso, J.H. Weaver: J. Vac. Sci. Technol. B **7**, 950 (1989)
8.81 R. Ludeke, T.-C. Chiang, T. Miller: J. Vac. Sci. Technol. B **1**, 581 (1983)
8.82 K.K. Chin, S.H. Pan, D. Mo, P. Mahowald, N. Newman, I. Lindau, W.E. Spicer: Phys. Rev. B **32**, 918 (1985)
8.83 J. Tersoff: Surf. Sci. **168**, 275 (1986); J. Vac. Sci. Technol. B **3**, 1157 (1985)
8.84 S.G. Louie, J.R. Chelikowsky, M.L. Cohen: Phys. Rev. B **15**, 2154 (1977)
8.85 W. Mönch: J. Vac. Sci. Technol. B **6**, 1270 (1988); Europhys. Lett. **7**, 275 (1988)
8.86 J.E. Klepeis, W.A. Harrison: J. Vac. Sci. Technol. B **7**, 964 (1989)
8.87 I. Lefevre, M. Lannoo, G. Allan: Europhys. Lett. **10**, 359 (1989) and private communication
8.88 R.E. Viturro, S. Chang, J.L. Shaw, C. Mailhiot, L.J. Brillson, A. Terrasi, Y. Hwu, G. Margaritondo, P.D. Kirchner, J.M. Woodall: J. Vac. Sci. Technol. B **7**, 1007 (1989) and private communication
8.89 C.M. Aldao, S.G. Anderson, C. Capasso, G.D. Waddill, I.M. Vitomirov, J.H. Weaver: Phys. Rev. B **39**, 12977 (1989)
8.90 I.M. Vitomirov, G.D. Waddill, C.M. Aldao, S.G. Anderson, J.H. Weaver: Phys. Rev. B **40**, 3483 (1989)

8.91 S.G. Anderson, C.M. Aldao, G.D. Waddill, I.M. Vitomirov, S.J. Severtson, J.H. Weaver: Phys. Rev. B **40**, 8305 (1989)
8.92 S.G. Anderson, C.M. Aldao, G.D. Waddill, I.M. Vitomirov, C. Capasso, J.H. Weaver: Appl. Phys. Lett. **55**, 2547 (1989)
8.93 M.H. Hecht: Phys. Rev. B Rapid Commun. **41**, 7918 (1990)
8.94 C.M. Aldao, G.D. Waddill, P.J. Benning, C. Capasso, J.H. Weaver: Phys. Rev. B Rapid Commun. **41**, 6092 (1990)
8.95 G.D. Waddill, C.M. Aldao, C. Capasso, P.J. Benning, Yongjun Hu, T.J. Wagener, M.B. Jost, J.H. Weaver: Phys. Rev. B **41**, 5960 (1990)
8.96 Yongjun Hu, T.J. Wagener, M.B. Jost, J.H. Weaver: Phys. Rev. B **40**, 1146 (1989)
8.97 T. Komeda, S.G. Anderson, J.M. Seo, M.C. Schabel, J.H. Weaver: J. Vac. Sci. Technol. (in press)
8.98 W.F. Egelhoff, Jr., I. Jacob: Phys. Rev. Lett. **62**, 921 (1989)
8.99 J. Tersoff: Private communication
8.100 J.E. Demuth, W.J. Thompson, N.J. DiNardo, R. Imbihl: Phys. Rev. Lett. **56**, 1408 (1986)
8.101 K. Markert, P. Pervan, W. Heichler, K. Wandelt: J. Vac. Sci. Technol. A **7**, 2873 (1989)
8.102 G. Margaritondo, L.J. Brillson, N.G. Stoffel: Solid State Commun. **35**, 277 (1980)
8.103 R. Williams, R.H. Bube: J. Appl. Phys. **31**, 968 (1960)
8.104 I.M. Vitomirov, C.M. Aldao, G.D. Waddill, C. Capasso, J.H. Weaver: Phys. Rev. B **41**, 8465 (1990)
8.105 G.D. Waddill, T. Komeda, Y.N. Yang, J.H. Weaver: Phys. Rev. B Rapid Commun. **41**, 10283 (1990)
8.106 M. Vos, C.M. Aldao, D.J.W. Aastuen, J.H. Weaver: Phys. Rev. B **41**, 991 (1990) [Ag/ZnSe(100)]
8.107 C.M. Aldao, D.J.W. Aastuen, M. Vos, I.M. Vitomirov, G.D. Waddill, P.J. Benning, J.H. Weaver: Phys. Rev. B **42**, 2878 (1990) [Ag/InP(110) and In/GaAs(110)]
8.108 J.M.E. Harper: In *Thin Film Processes*, ed. by J.L. Vossen, W. Kern (Academic, New York 1978)
8.109 M.-A. Hasan, S.A. Barnett, J.-E. Sundgren, J.E. Greene: J. Vac. Sci. Technol. A **5**, 1883 (1987)
8.110 S.-N. Mei, T.-M. Lu, S. Robert: IEEE Electron. Dev. Lett. **8**, 503 (1987)
8.111 J. Lindhard, M. Scharff, H.E. Schiott: Mat.-Fys. Medd. K. Dan. Vidensk. Selsk. **33**, No. 14 (1963)
This theory was developed for ions with an energy of many keV and the application to the current case should be considered only as an educated guess. For defect depth distributions, see:
J.Y. Tsai, E. Chason, K.M. Horn, D.K. Brice, S.T. Picraux: Nucl. Instrum. Methods B **39**, 72 (1989)
8.112 M. Vos, F. Xu, S.G. Anderson, J.H. Weaver, H. Cheng: Phys. Rev. B **39**, 10744 (1989)
8.113 L.J. Brillson: Phys. Rev. Lett. **40**, 260 (1978)
8.114 J.L Shaw, R.E. Viturro, L.J. Brillson, D. Kilday, M. Kelly, G. Margaritondo: J. Vac. Sci. Technol. A **6**, 1579 (1988)
8.115 I. Lindau, T. Kendelewicz: CRC Crit. Rev. Solid State Mat. Sci. **13**, 27 (1987)
8.116 R.R. Daniels, T.-X. Zhao, G. Margaritondo: J. Vac. Sci. Technol. A **2**, 831 (1984)
8.117 K.K. Chin, T. Kendelewicz, C. McCants, R. Cao, K. Miyano, I. Lindau, W.E. Spicer: J. Vac. Sci. Technol. A **4**, 969 (1986)
8.118 D.E. Savage, M.G. Lagally: J. Vac. Sci. Technol. B **4**, 943 (1985)
8.119 A. Zunger: Phys. Rev. B **24**, 4372 (1981)
8.120 R.R. Daniels, A.D. Katnani, T.-X. Zhao, G. Margaritondo, A. Zunger: Phys. Rev. Lett. **49**, 895 (1982)
8.121 G.D. Waddill, I.M. Vitomirov, C.M. Aldao, J.H. Weaver: Phys. Rev. Lett. **62**, 1568 (1989)
8.122 G.D. Waddill, I.M. Vitomirov, C.M. Aldao, S.G. Anderson, C. Capasso, J.H. Weaver, Z. Liliental-Weber: Phys. Rev. B **41**, 5293 (1990)
8.123 R. Ludeke, T.-C. Chiang, D.E. Eastman: J. Vac. Sci. Technol. **21**, 599 (1982)
8.124 V. Heine: Phys. Rev. A **138**, 1689 (1965)
8.125 D.J. Chadi: Phys. Rev. B **18**, 1800 (1978)

9. Electronic States in Semiconductor Superlattices and Quantum Wells: An Overview

Massimo Altarelli

With 3 Figures

Theoretical calculations of electronic subbands in two-dimensional semiconductor systems have been performed in Si metal-oxide-semiconductor (MOS) structures [9.1] and later in III-V heterostructures. In the latter case, the tight-binding method [9.2–4], the pseudopotential method [9.5, 6] or ab initio schemes based on the density functional formalism [9.7] were exploited, in addition to the envelope-function or effective-mass method.

Tight-binding methods have the remarkable advantage of treating the whole Brillouin zone on the same footing, and of being computationally simple. On the other hand, they have difficulty in incorporating the effect of external fields (magnetic, electric and strain fields), the Coulomb field of impurities, excitonic interactions, etc. The pseudopotential method is computationally more complex and again, external fields and excitons are not easily incorporated. Ab initio calculations are conceptually appealing, but limited by their computational complexity to short-period superlattices. In addition, the well-known difficulties [9.8] of the local density approximation in producing the correct value of the energy gaps are also present.

The envelope-function method (summarized in the next section) has enjoyed a large popularity in spite of its limitations (most notably, the difficulty in handling states resulting from the admixture of different k-points in the bulk band structure), because of its intuitive character and of its great flexibility. In its most popular application, it relates the conduction band levels of a complex system, such as a GaAs quantum well between $Al_xGa_{1-x}As$ barriers, to the "particle-in-a-box" quantization of elementary quantum mechanics. The envelope-function, or effective-mass method is, however, much more powerful and is able, in its most sophisticated multi-band version, to handle some common experimental situations in which elementary models provide no guidance. Examples of such situations for a variety of systems will be discussed in the next section. The straightforward extension of the envelope-function method to include the external fields is a most attractive feature, and we will present a brief discussion of the influence of external electric and magnetic fields on the electronic states. The results that we will present show that, in the appropriate limit, the effect of electric and magnetic fields on the electronic states can be predicted by using semiclassical Bloch dynamics for the zero-field energy bands, i.e., by applying simple textbook procedures to the superlattice band structure. The major conclusion is therefore that in superlattices the band structure concept has the same validity and significance as for ordinary crystals.

9.1 Envelope-Function Description of Electronic States

9.1.1 Generalities

In the envelope-function method (see e.g. [9.9, 10] and references therein) the wavefunction in each layer of a superlattice is written in terms of products of a $k = 0$ Bloch function of the corresponding bulk semiconductor and a slowly-varying envelope-function with wavelength of the order of the layer thickness. Suppose, for example, that the two materials have a parabolic conduction band edge at the Γ point, with effective masses m_A, m_B and energies $E_{\Gamma A}$, $E_{\Gamma B}$, respectively. Then we write the wavefunction of the conduction subbands as

$$\begin{aligned} \psi^A(\boldsymbol{r}) &= F^A(\boldsymbol{r}) u_{\Gamma A}(\boldsymbol{r}) \quad \text{in the } A \text{ layers} \,, \\ \psi^B(\boldsymbol{r}) &= F^B(\boldsymbol{r}) u_{\Gamma B}(\boldsymbol{r}) \quad \text{in the } B \text{ layers} \,, \end{aligned} \tag{9.1}$$

with F^A, F^B satisfying the equations

$$-\frac{\hbar^2}{2m_{A,B}} \nabla^2 F^{A,B} = \left(E - E_{\Gamma A,B}\right) F^{A,B} \,. \tag{9.2}$$

If the two materials are joined at an A-B interface, say the $z = 0$ plane, the boundary conditions are

$$\begin{aligned} &F^A(x, y, z = 0^-) = F^B(x, y, z = 0^+) \,, \\ &\frac{1}{m_A} \frac{\partial}{\partial z} F^A(x, y, z = 0^-) = \frac{1}{m_B} \frac{\partial}{\partial z} F^B(x, y, z = 0^+) \,. \end{aligned} \tag{9.3}$$

If the assumption $u_{\Gamma A}(r) \cong u_{\Gamma B}(r)$ holds, then the boundary conditions (9.3) ensure the conservation of the total probability current at the interface, and are therefore physically justified [9.11]. This assumption restricts the applicability of the method to pairs of materials with very similar chemical nature and band structure. In practice, it has been proven applicable to III-V compounds and alloys with a direct band gap, such as GaAs-$Al_xGa_{1-x}As$, for $x < 0.4$.

The interest of (9.1–3) lies in the fact that they are susceptible to far-reaching generalizations. First, any potential varying slowly on the scale of the lattice parameter of the constituents can be included in (9.2):

$$\frac{\hbar^2}{2m_{A,B}} \left[-\nabla^2 + V^{A,B}(\boldsymbol{r})\right] F^{AB}(\boldsymbol{r}) = \left(E - E_{\Gamma A,B}\right) F^{A,B} \,, \tag{9.4}$$

thus allowing consideration of space charge effects in doped superlattices, charge transfer across the interfaces, external fields, etc. A second very important generalization is the inclusion of band coupling, whenever it is impossible to construct the subband wavefunctions from a single, nondegenerate, nearly parabolic band edge as implied by (9.1). This happens frequently, for reasons that will become clear in the next sections. Specific examples include:

i) Band degeneracy near an extremum, as in the case of the valence band maximum at Γ in all cubic semiconductors.
ii) Deviations from parabolicity, as in the conduction band of direct-gap semiconductors. For narrow-gap materials, like InAs or InSb, the nonparabolicity is relatively large for energies near the band minimum [9.12], but even in GaAs has a sizable effect on levels higher than 0.1 eV above the band edge.
iii) Situations specific to heterostructures in which the single-band approach fails. If, for example, the two materials have a staggered band line-up, then, in a large and interesting energy region, the wavefunction has conduction band character on one side of the heterojunction, and valence band character on the other. InAs-GaSb superlattices provide an example of such a situation.

In all of these cases, it is necessary to describe more than one band, say n bands, at a time on the same footing. This is accomplished via the $k.p$ formalism [9.12], in which the n band energies $E_l(\boldsymbol{k})$, $l = 1,2,\ldots,n$ are obtained as eigenvalues of the $n \times n$ matrix $H_{lm}(\boldsymbol{k})$, expanded, for k near the origin, up to the second order in k:

$$H_{lm}(\boldsymbol{k}) = E_l(0)\delta_{lm} + \sum_{\alpha=1}^{3} P_{lm}^{\alpha} k_\alpha + \sum_{\alpha,\beta=1}^{3} D_{lm}^{\alpha,\beta} k_\alpha k_\beta \; , \tag{9.5}$$

where α and β run over the x, y and z directions. The P and D coefficients are matrices written in terms of momentum matrix elements between the Bloch functions at $k = 0$ of the n bands in question, and for each material they play the role of effective mass parameters.

In the general framework of the effective-mass theory in its many-band formulation, as adopted, for example, in the theory of acceptor impurities, we replace (9.1) with

$$\begin{aligned} \psi^A(\boldsymbol{r}) &\cong \sum_{l=1}^{n} F_l^A(\boldsymbol{r}) u_{\Gamma l A}(\boldsymbol{r}) + \ldots \; , \\ \psi^B(\boldsymbol{r}) &\cong \sum_{l=1}^{n} F_l^B(\boldsymbol{r}) u_{\Gamma l B}(\boldsymbol{r}) + \ldots \; , \end{aligned} \tag{9.6}$$

where the terms neglected are higher-order corrections proportional to ∇F_l [9.9]. The effective-mass equation (9.4) is now replaced by a set of n differential equations:

$$\sum_{m=1}^{n} \left(H_{lm}(-\mathrm{i}\nabla) + V(\boldsymbol{r})\delta_{lm}\right) F_m(\boldsymbol{r}) = E F_l(\boldsymbol{r}) \tag{9.7}$$

for each material, where, as in (9.4), $V(\boldsymbol{r})$ is a slowly varying potential. The boundary conditions which must complement (9.7) are

$$F_l^A(x,y,z=0^-) = F_l^B(x,y,z=0^+)\,,$$

$$\sum_{m=1}^{n}\left(\sum_{\alpha=x,y}(D_{lm}^{z\alpha}+D_{ml}^{\alpha z})\,k_\alpha - 2\mathrm{i}D_{lm}^{zz}\frac{\partial}{\partial z}\right)F_m \quad \text{continuous at } z=0\,. \tag{9.8}$$

The boundary conditions (9.8), in analogy to (9.3), ensure the continuity of the probability current, provided that one can assume

$$u_{\Gamma lA}(r) \cong u_{\Gamma lB}(r) \quad l=1,2,\dots n\,. \tag{9.9}$$

The boundary conditions are therefore valid only if the two materials have a set of $k=0$ band edges which can be grouped in pairs with similar symmetry and chemical origin. Given the success of the envelope-function method, it appears that (9.9) holds for pairs of lattice-matched III-V (or II-VI) compounds, as far as the lowest conduction band or the upper valence-band edge is concerned. It is important to notice that relation (9.9) implies the equality of the P coefficient matrix for the two materials A and B [9.12].

9.1.2 Discussion of the Envelope-Function Approximation

In deriving the equations for the envelope-function approximation, many assumptions, some of them quite restrictive, have to be made. It is quite clear that there are situations in which these assumptions are untenable. In addition to the cases of strong or rapidly varying potentials $U(\boldsymbol{r})$, which cause the breakdown of the effective-mass approximation also for impurity states, the condition (9.9) cannot be satisfied in many interesting heterostructures. This occurs in all of those cases in which the eigenstates to be computed are derived from the mixing of Bloch waves from different points of the Brillouin zone (e.g., the conduction-related states in Si-$\mathrm{Si}_x\mathrm{Ge}_{1-x}$ heterostructures, or in GaAs–$\mathrm{Al}_x\mathrm{Ga}_{1-x}\mathrm{As}$ with $x > 0.4$, when the alloy has an indirect gap, etc.). In a similar situation, one must resort to other methods, such as the empirical tight-binding scheme [9.2–4], or the pseudopotential method [9.5, 6, 13]. In either case, some of the attractive features of the envelope-function method are lost. Alternatively, one can devise an extension of the envelope-function method suitable for such heterostructures, but at the price of introducing somme additional parameters into the theory. *Akera* et al. [9.14] have applied such a method to GaAs–$\mathrm{Al}_x\mathrm{Ga}_{1-x}\mathrm{As}$ heterostructures with $x > 0.4$.

Also in those cases where the two semiconductors have the relevant band extremes which, because of their similarity in symmetry and chemical origin, can be paired to approximately satisfy (9.9), a few words of caution on the accuracy of the envelope-function calculations are necessary. Our discussion of boundary conditions is based on the assumption of perfect periodicity of both media up to a geometrical plane defining the interface. This is certainly an idealization and, in the important case in which one of the components is an alloy (e.g., GaAs–$\mathrm{Al}_x\mathrm{Ga}_{1-x}\mathrm{As}$) it is impossible even to define such a plane. It would be more realistic to talk about an interface layer, comprising several atomic planes, separating the two semiconductors, and characterized by given reflection and transmission coefficients. These coefficients would also be unknown input parameters of the model. In fact, when it comes to detailed quantitative comparisons, one should remember that (9.2, 3) and (9.5–8) already require a large number of input parameters, which we attempt to list here:

1) Electronic structure parameters of bulk compounds: gaps, effective masses, $k \cdot p$ coefficients, as well as dielectric constants, g-factors, deformation potentials, etc., in the presence of external fields.
2) Band offsets.
3) Structural and compositional parameters of the heterostructure: layer thicknesses, alloy compositions, doping levels, etc.
4) Additional parameters for special situations: e.g., Fermi energy pinning level for heterostructures with free surfaces.

The first group of parameters are not known with good accuracy, especially for ternary alloys, so that values for the parent binary compounds are often adopted instead. The band offsets are not easy to measure or calculate, and their values are often controversial. Notice also that sometimes one may even require their dependence on external perturbations (e.g., pressure). The third group of parameters are essential to any comparison of theory with experimental results for a given sample. As far as layer thicknesses are concerned, X-ray diffraction methods often provide much better values than estimates from growth rates [9.15–17]. The last group of parameters is of importance only in very specific situations, so that systematic comparisons of the predictions of envelope-function calculations and experiment are fortunately still possible for a wide range of heterostructures.

In spite of this, the envelope-function results are in overall good agreement with experiment, sometimes even more so than one would have expected. When comparisons with calculations that are free from the problem of boundary conditions, e.g., tight-binding calculations, are performed, the agreement is very good [9.2, 3, 14, 18, 19].

There is nevertheless a conceptual need to rigorously justify the envelope-function method and the boundary conditions. An attempt in this direction is being made by *Burt* [9.20, 21], but more work is clearly necessary.

9.1.3 Examples of Results

a) $GaAs–Al_xGa_{1-x}As$

The envelope-function method is applicable to the $GaAs–Al_xGa_{1-x}As$ system for all states deriving from Γ-point bulk Bloch functions. This excludes conduction subbands high enough in energy to approach the secondary minima at X and L, and therefore also the indirect alloys, for $x > 0.4$, at all energies. The valence subbands, on the other hand, are accessible to the method at all alloy compositions. The valence band structure near the Γ-point provides the first example of deviation from the "particle-in-a-box" model, because of the fourfold degeneracy of the bulk bands. We have therefore to consider, following (9.5), the 4×4 Luttinger Hamiltonian [9.22], which can be written in the form

$$\begin{array}{c|cccc|} & 3/2 & 1/2 & -1/2 & -3/2 \\ 3/2 & a_+ & b & c & 0 \\ 1/2 & b^* & a_- & 0 & c \\ -1/2 & c^* & 0 & a_- & -b \\ -3/2 & 0 & c & -b^* & a_+ \end{array} \tag{9.10}$$

with

$$\begin{aligned} a_\pm &= E_v - \tfrac{1}{2}(\gamma_1 \pm \gamma_2)(k_x^2 + k_y^2) - \tfrac{1}{2}(\gamma_1 \pm 2\gamma_2)k_z^2 \,, \\ b &= \sqrt{3}\gamma_3(k_x - \mathrm{i}k_y)k_z \,, \\ c &= \frac{\sqrt{3}}{2}\left(\gamma_2(k_x^2 - k_y^2) - 2\mathrm{i}\gamma_3 k_x k_z\right) . \end{aligned} \tag{9.11}$$

The bulk solutions of (9.10) give the spin-degenerate light- and heavy-hole bands. The degeneracy is actually lifted in non-inversion-symmetric materials, like GaAs, by terms linear in the k vector [9.23]. Such terms are, however, extremely small and we neglect them here. The valence bands are anisotropic, and therefore one has to expect that also in the superlattice the band dispersion will be different for different directions in the $k_x k_y$-plane. However, this warping of the bands is small and a good approximation ("axial model") is obtained by replacing [9.24, 25] γ_2 and γ_3 in the c matrix elements [see the last of (9.11)] by $\gamma = 1/2(\gamma_2 + \gamma_3)$. With this replacement, (9.10) acquires cylindrical symmetry about the z-axis, and the bands are isotropic in the $k_x k_y$-plane.

The solution of the four-component equation with the boundary conditions was originally pursued numerically [9.25] but, as shown by *Andreani* et al. [9.26], an analytic reduction to a search for the zeros of a small determinant is possible. At $k_x = k_y = 0$ there is no mixing of the light and heavy components. It is therefore possible to identify each subband as heavy or light. This is a consequence of the envelope-function approximation, and a more complete description is given in [9.13]. As one moves out of the $k = 0$ axis, the mixing grows rapidly and produces a nonparabolic behavior. This was first pointed out by *Nedorezov* [9.27], who solved the problem exactly in the limit of infinitely high barriers. A particularly striking feature of the results is the positive in-plane effective mass of the first light-hole subband, which changes sign and becomes hole-like only for $k \sim \pi/L$, where L is the well thickness.

A very straightforward demonstration of the heavy–light mixing effect is provided by calculations of resonant hole tunneling in p-type double barrier heterostructures [9.28, 29]. In Fig. 9.1, the transmission coefficient D for a heavy-hole impinging on a 50 Å GaAs quantum well confined by two 30 Å barriers and emerging again as a heavy hole is shown, for $k_x = k_y = 0.036(2\pi/a)$, where a

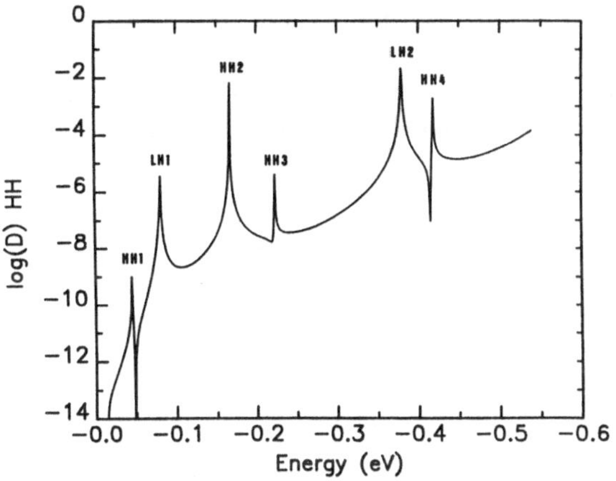

Fig. 9.1. Transmission coefficient of a heavy-hole impinging on a double barrier heterostructure, with barrier height $V_0 = -0.55$ eV, thickness 30 Å, and quantum well thickness 50 Å. The coefficient is computed for $k_x = k_y = 0.036(2\pi/a)$ (see text) and as a function of incoming energy (with respect to the top of the bulk valence band). Heavy- and light-hole quasi-bound resonances are labeled HH and LH, respectively. (From [9.29])

is the bulk lattice parameter of GaAs, as a function of energy. It is evident in Fig. 9.1 that sharp resonances appear also at the energies of *light-hole* quasi-bound states in the well. An even more striking feature of this result is the presence of strongly asymmetric resonance lineshapes, due to the interference of resonant heavy-hole tunneling with nonresonant light-hole channels. One can view the tunneling process as one in which the incoming heavy hole has, upon impinging on the heterostructure, a certain probability amplitude of being converted to a light hole, and of being converted again to a heavy hole at the opposite end of the heterostructure. Light holes tunnel much better than heavy ones, except when transmission in the heavy hole channel is resonant. In that case strong interference occurs between the two channels with comparable amplitudes.

From the experimental point of view, the complexity of the valence bands is revealed by intra- and interband optical experiments. Analysis [9.30–32] of Raman scattering [9.33] results from p-type modulation-doped wells, and polarized luminescence [9.34] in n-type structures confirms the strong light–heavy hole mixing, and the upward curvature of the light-hole first subband. Further confirmation is offered by magneto-optical experiments, to be discussed later.

b) InAs–GaSb Superlattices and Quantum Wells

The InAs–GaSb system is particularly interesting [9.35] because of the peculiar band line-up of these heterostructures. The top of the valence band of GaSb is estimated to lie ~ 0.15 eV above the bottom of the InAs conduction band [9.35, 36]. Electrons tend to be transferred to the InAs layers, unless the layers are so thin that the particle-in-a-box quantization energy reverses the order of the levels, locating empty InAs quantum well conduction states above the full GaSb valence states. Therefore, charge transfer occurs only for InAs layers thicker than ~ 90 Å, and in this case self-consistency plays a crucial role in determining the level structure. Such a calculation [9.37] was performed in the Hartree approximation and in the 6×6 band model, which neglects the split-off valence band. The results show strong valence–conduction subband mixing, with small (few meV) hybridization gaps opening up at subband crossing points. For a 120 Å InAs–80 Å GaSb superlattice, the density of transferred carriers is $\sim 0.9 \times 10^{12}\,\mathrm{cm}^{-2}$, thus providing good justification for the Hartree approximation. Magneto-optical experimental results, to be discussed later, show good agreement with the envelope-function calculations. Single quantum wells of InAs between GaSb barriers, however, show a concentration of electrons in InAs much larger than that of holes in the barriers [9.38]. The origin of these extrinsic charges is not clear, although a plausible explanation is that the well is in a strong band-bending region, deriving from the proximity of the GaSb free surface of the outer barrier [9.39].

c) CdTe–HgTe Superlattices

Superlattices of II-VI compounds have recently attracted much attention, and among them the CdTe–HgTe system is especially interesting. This is a consequence of the zero-gap character of HgTe and of the band line-up, which puts the HgTe Γ_8 edge above the corresponding edge of CdTe. The value Δ of this band offset is controversial, the estimate $\Delta = 0.04\,\mathrm{eV}$ from magneto-optics [9.40, 41] being confirmed by Raman measurements [9.42], but contradicted by photoemission results [9.43, 44] which suggest a much larger value $\Delta \approx 0.35\,\mathrm{eV}$.

These systems provide an excellent test for the validity of the boundary conditions (9.3) and (9.8). This is because the Γ_8 "light holes" are in fact Γ_8 "electrons" in HgTe, i.e., they form an unoccupied band with positive mass in this material, due to the interchange in the energy position of the Γ_6 and Γ_8 edges. There are therefore three types of Γ_8 states in these superlattices: Γ_8 "electrons" and Γ_8 heavy holes, both confined in the HgTe wells, and in addition unusual interface states [9.19, 45] arising in the gap between Γ_8 light holes in CdTe and Γ_8 electrons in HgTe and decaying exponentially on both sides of an interface. If one neglects the other bands for simplicity, it is easy to see that the boundary conditions (9.10) allow such states if m_A, m_B have opposite signs.

The good agreement [9.19] between the results of envelope-function calculations, in which the boundary conditions are crucial in predicting the interface states, and tight-binding results provides an a posteriori confirmation of the envelope-function method. On the other hand, the fact that the band offset value carefully measured by photoemission has to be drastically reduced to account for infrared and Raman results leads to the conclusion that some aspects of the physics of this system are not understood. *Jaros* et al. [9.6] went so far as to claim that the envelope-function method is not applicable to these systems, because of the strength of the Hg potential, and that pseudopotential calculations can reconcile the observed features of the electronic structure with a large band offset. This statement is however controversial [9.46, 47] and very recently it was shown, by an envelope-function calculation [9.48], that the properties of the electronic structure relevant to the interpretation of optical experiments (gap, effective mass, etc.) are strongly nonmonotonic functions of the band offset, with similar values for $\Delta = 0.04\,\text{eV}$ or $0.35\,\text{eV}$. This would provide a resolution of the paradox within the framework of few-band calculations.

9.2 External Fields

9.2.1 Generalities

As indicated in the discussion of the envelope-function method, any external field varying slowly on the scale of the lattice parameter of the constituents can be included in the envelope function equations. In particular:

- Electric fields [9.49], via the potential $\boldsymbol{F} \cdot \boldsymbol{r}$ [diagonal in (9.7)].
- Magnetic fields, by replacing ∇ with $\nabla - (e/c)\boldsymbol{A}$, where $\boldsymbol{A}$ is the vector potential [9.50].
- Strain fields [9.51, 52] via the potential $e \cdot d$, where e and d are the strain and deformation potential tensors, and via the strain dependence of the band gap discontinuities.
- Screened Coulomb fields of donor and acceptor impurities, via the potential $\pm e^2/\varepsilon|\boldsymbol{r} - \boldsymbol{R}|$, where $\boldsymbol{R}$ is the position of the impurity [9.53, 54].

9.2.2 Electric Fields

We shall start our discussion with electric fields and show how the application of standard solid state physics textbook procedures to this artifial "solid" leads to predictions in agreement with experiment, as well as with the results of more refined quantum mechanical calculations.

The effect of an external electric field on a superlattice is particularly interesting because it allows one to observe [9.55–57] some spectacular phenomena, such as Wannier-Stark quantization and Bloch oscillations, which are not accessible in ordinary solids. Let us consider the Hamiltonian for an electron subject to a periodic potential $V_{\text{per}}(z)$ and to an electric field F:

$$H = \frac{p^2}{2m} + V_{\text{per}}(z) - eF_z \,. \tag{9.12}$$

It is easy to see that if $\psi(z)$ is an eigenstate of (9.12) with energy E, then $\psi(z + \nu d)$, where ν is an integer and d is the period of V_{per}, is also an eigenstate, with energy $E - eF\nu d$. Eigenstates belong therefore to "ladders" of equidistant levels with spacing eFd. This characteristic energy eFd (or frequency $f = eFd/h$) is also found in the semiclassical derivation of Bloch oscillations [9.58]. Let $E_n(\boldsymbol{k})$ be the dispersion of the nth band (or subband) in the system, and let $F = (0, 0, F_z)$

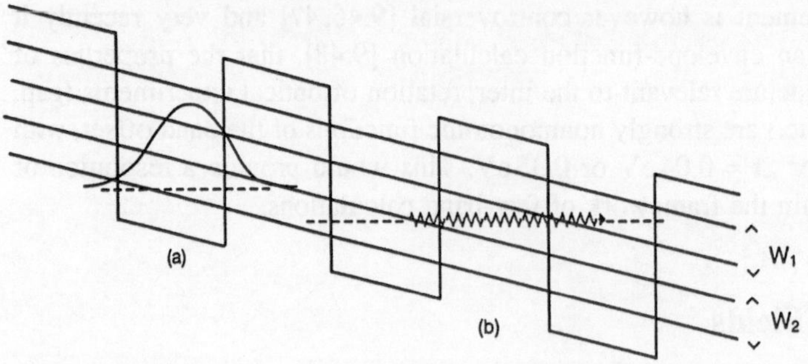

Fig. 9.2. Superlattice in an external electric field (schematic): (a) sketch of the wavefunction of an electron executing oscillatory motion in the semiclassical picture; (b) Zener tunneling process between two states as in (a) belonging to different bands W_1 and W_2, separated in space, but with the same energy

be the external field. If Zener tunneling (Fig. 9.2) between different bands is negligible, the semiclassical equation of motion of an electron in subband n is

$$\hbar \frac{dk_z}{dt} = eF_z \tag{9.13}$$

with solution

$$k_z(t) = k_z(0) + \frac{e}{\hbar} F_z t \, . \tag{9.14}$$

The periodicity of the band structure for $k_z \to k_z + 2\pi/d$ implies that the particle executes a periodic motion, with the characteristic frequency f discussed above.

In a superlattice with $d = 200\,\text{Å}$, $f = 4.8 \times 10^{12}\,\text{s}^{-1}$ for $F_z = 10^4\,\text{V/cm}$. The amplitude of the periodic motion [see Fig. 9.2 and Eq. (9.3)] in terms of the bandwidth W_n is expressed by

$$l = \int_0^{T/2} \frac{1}{\hbar} \frac{dE_n}{dk_z} dt = \int_0^{w} \frac{1}{eF_z} dE_n = \frac{W_n}{eF_z} \, . \tag{9.15}$$

In a usual solid, $W_n \sim 10\,\text{eV}$ and $d \sim 5\,\text{Å}$. Therefore, this "Stark localization length" for $F_z = 10^4\,\text{V/cm}$ is about $10\,\mu\text{m}$, a huge distance compared with the lattice constant or the mean free path; similarly the spacing of the Stark ladder is $\sim 0.5\,\text{meV}$. These figures account for the difficulty in observing the predicted effect in practice. On the other hand, in superlattices with $d \sim 100\,\text{Å}$ and $W_n \sim 10\,\text{meV}$, we find, for the same F_z, $l \sim 100\,\text{Å}$, corresponding to localization in one cell, and a ladder energy spacing for the same order of magnitude as the band width. Recently, optical and Raman experiments [9.56, 57] revealed energy features with separation linear in the applied electric field, in agreement with the Wannier-Stark ladder prediction. Furthermore, recent theoretical work [9.59] shows that the semiclassical predictions agree extremely well with full quantum mechanical calculations.

Many revealing experiments on heterostructures are performed in the presence of an external magnetic field, which has a profound influence on the energy spectrum. It produces an enrichment and a sharpening of the optical properties, and striking transport phenomena, like the quantum Hall effect. To include the magnetic-field in the many-band envelope-function formalism [9.50], we follow the lines of the classic work by *Luttinger* [9.22] on the cyclotron resonance of holes in semiconductors. The field $\boldsymbol{B} = (0, 0, B)$ is described by the vector potential $\boldsymbol{A}$. (It is convenient to choose a gauge with $A_z = 0$.) In the $k \cdot p$ bulk Hamiltonian (9.5), $\boldsymbol{k}$ is to be replaced by $\boldsymbol{k}' = \boldsymbol{k} + (e/c)\boldsymbol{A}$. Then it is easy to see that the x and y components of this new operator do not commute, but instead

$$[k_x', k_y'] = -\mathrm{i}\left(\frac{e}{c}\right) B \ . \tag{9.16}$$

It is easy to see that as a consequence of (9.16) we can define operators a, $a^\dagger$

$$a = \sqrt{\frac{c}{2eB}}\left(k_x' - \mathrm{i}k_y'\right) \ , \quad a^\dagger = \sqrt{\frac{c}{2eB}}\left(k_x' + \mathrm{i}k_y'\right) \tag{9.17}$$

with commutator

$$[a, a^\dagger] = 1 \ , \tag{9.18}$$

so that all terms in k_x or k_y in the Hamiltonian can be expressed in terms of these harmonic oscillator raising and lowering operators. Furthermore, new diagonal terms in the $k \cdot p$ matrix appear, representing the direct coupling of the electron and hole spin to the magnetic field [9.50].

The resulting energy level scheme (the Landau levels) is strongly dependent on the orientation of the field relative to the layers. We will briefly discuss the cases in which the field is either perpendicular or parallel to the layers.

9.2.3 Landau Levels: Perpendicular Fields

Theoretical calculations in the case of perpendicular fields are available for the GaAs–$Al_xGa_{1-x}As$, InAs–GaSb and CdTe–HgTe systems. The Landau levels obtained are directly comparable to the results of intraband magneto-optical experiments in GaAs–$Al_xGa_{1-x}As$, but not to interband ones, which are strongly affected by exciton effects, to be discussed later. For InAs–GaSb and CdTe–HgTe, the distinction between the two types of experiments is unclear because of the strong mixing induced by the band line-up.

The structure of the valence band Landau levels, in particular, can be directly probed with intraband experiments. Cyclotron resonance experiments [9.60] on p-type GaAs–$Al_xGa_{1-x}As$ single heterojunctions were interpreted by several groups [9.61–64] in terms of Landau level calculations in the $B = 0$ self-consistent potential of the heterojunction. The agreement can be good, although this must involve some accident, because of the use of the Hartree approximation (inadequate for hole densities $\sim 5 \times 10^{11}\ \mathrm{cm}^{-2}$) and the axial approximation, etc. Comparison with cyclotron resonance results for p-type quantum wells are less favorable [9.65], whereas satisfactory semiquantitative agreement is found for intraband Raman scattering experiments [9.66] on p-type quantum wells. In general, while theory seems to correctly predict the main features of the valence band Landau levels, more work is needed to establish quantitative agreement with the results of interband magneto-optics experiments.

In the InAs–GaSb superlattices, magneto-optical experiments [9.36] were crucial in establishing the basic properties of the electronic structure. The calculated [9.50] transition energies are in good agreement with experiment. More recently [9.67], the theory was shown to give a reasonable account of the results of magneto-optical experiments performed under hydrostatic pressure, which modifies the band offsets; due to the peculiar nature of this system, the reduction of the overlap between the GaSb valence band and the InAs conduction band with hydrostatic pressure strongly affects the electronic states, and produces a change in the character of prominent transitions from interband-like to intraband-like.

9.2.4 Landau Levels: Parallel Fields

Interesting information on the electronic properties of quantum wells and superlattices can be obtained by performing magneto-optical experiments with a field parallel to the layers of the heterostructure. In an intuitive picture, the field forces the particles to perform cyclotron orbits in a plane intersecting both the well and the barrier materials. If the radius of such orbits [i.e., the magnetic length $a_{\mathrm{M}} = (c/eB)^{1/2}$] is large compared to the thickness of the layers, then the completion of the cyclotron orbit requires traversal of the barriers. This condition is easily met experimentally since, for example, a 10 T field corresponds to $a_{\mathrm{M}} \sim 8$ nm. Parallel field experiments probe therefore intriguing properties such as tunneling, perpendicular transport, etc., and are particularly fruitful in the investigation of superlattice band structure in the growth direction [9.68], which is more elusive than the in-plane dispersion discussed in the previous sections.

To preserve the notation used in the previous section, we now denote the growth direction by x, and choose the gauge $\boldsymbol{A} = (0, Bx, 0)$ for the field B parallel to z. Consider first an electron in a bulk semiconductor in a simple parabolic band with mass m^*. The Schrödinger equation can as usual be cast in the form

$$\left(\frac{p_z^2}{2m^*} + \frac{p_x^2}{2m^*} + \frac{1}{2} m^* \omega_{\mathrm{c}}^2 (x - x_0)^2\right) \psi = E\psi \,, \tag{9.19}$$

where $x_0 = -a_{\mathrm{M}}^2 k_y$, $\omega_{\mathrm{c}} = eB/m^* c$.

If we now consider a heterostructure, e.g., a superlattice, we must add in (9.19) a periodic potential $V_{\mathrm{s}}(x)$. It is then apparent that the degeneracy of the free-electron Landau levels with respect to x_0 is now lifted, because it makes a difference if x_0 is, for example, in a barrier or in a well.

The general problem of a particle in a field plus a periodic potential cannot be solved exactly. However, if the magnetic length a_{M} is large compared to the superlattice period d, one can use the semiclassical quantization scheme [9.69].

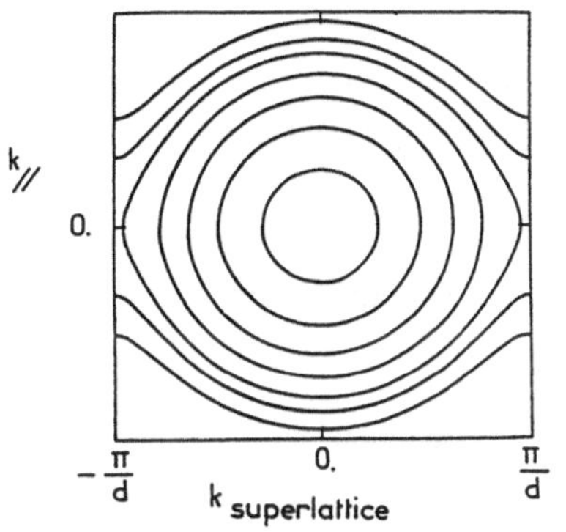

Fig. 9.3. Constant energy surfaces for a subband in a superlattice of period d. The Brillouin zone in the $k_{\|}$ direction extends to much larger values (of order π/a, where a is the lattice constant of the constituent semiconductors)

The approach is discussed in great detail by *Zilberman* [9.70]. One must first consider the subband structure problem for the superlattice in the absence of the field, and construct sections of the constant energy-surfaces in k-space with the k_z = const planes. For the $k_z = 0$ case, these are schematically shown in Fig. 9.3. This figure reflects the fact that the bandwidth in the superlattice direction, x, is much smaller than in the in-plane directions. The closed curves correspond to the energies between the subband bottom and the bandwidth in the x direction. For energies exceeding this value, the constant energy surface touches the edge of the superlattice Brillouin zone, corresponding to the "open orbit" case (one must think of a periodic repetition of the zone in the superlattice direction x).

When the field in the z direction is turned on, both open and closed orbits correspond to k-space trajectories of the representative point in the usual semi-classical Bloch dynamics [9.69]. To get a more accurate picture of the quantized spectrum, one can regard these k-space trajectories as phase-space trajectories of some kind. Indeed, (9.17) shows that k'_x and k'_y are like canonical variables (i.e., like "y" and "k_y"). We then construct the potential in the "y" direction that would produce the classical phase space trajectories of Fig. 9.3. Clearly, it is a periodic potential, with potential wells in which closed phase space orbits are observed at low energies, and unbound orbits for energies higher than the bandwidth. The corresponding quantized spectrum is obtained by solving the equation of motion in this potential, in a WKB-like approach.

Energies within the bandwidth (closed orbit region) in the x direction correspond to essentially discrete levels, or to very narrow bands, as the probability of tunneling through k-space to the next equivalent well is small [9.70]. The gap between these discrete levels is $\approx \omega_c$. The energies are given by the Onsager prescription:

$$A_k(E) = \frac{2\pi eB}{c}\left(n + \frac{1}{2}\right) , \tag{9.20}$$

where $A_k(E)$ is the k-space area of the closed orbit. For energies close to the bandwidth, the orbits get larger, the tunneling probability increases and discrete levels begin to broaden appreciably. At larger energies, the nearly flat open orbits correspond to nearly free motion in phase space, i.e., to a weak periodic potential, in which wide energy bands separated by narrow gaps occur. *Zilberman* [9.70] showed that a "band" extends approximately over the energy region:

$$\frac{2\pi eB}{c} n < A_k(E) < \frac{2\pi eB}{c}(n+1) , \tag{9.21}$$

where $A_k(E)$ is the area in the first Brillouin zone between two open orbits with energy E. Gaps of order $\sim \omega_c (d/a_M)^2$ separate each band from the neighboring bands.

The results of this analysis are in excellent agreement with the magneto-optical interband experiments of *Belle* et al. [9.71] in GaAs–AlGaAs superlattices with 5 nm period (as well as with their numerical results for the Kronig-Penney model). A series of sharp absorption peaks is observed in the region corresponding to transitions between the quasi-discrete energy levels of the valence and conduction subband. Above this energy range, no sharp structure is observed, corresponding to the broad "bands" discussed above. The intraband results of *Duffield* et al. [9.72, 73] in n-type samples are also in agreement with this general picture. Again, semiclassical Bloch dynamics confirms that bands in superlattices have the same significance and importance as in ordinary solids.

It is however important to point out that the semiclassical analysis cannot be invoked for the valence subbands, because they originate from a degenerate bulk band. However, one would suspect that the identification of the k_x-band width with the discrete portion of the spectrum should hold, since it just says that a hole of that energy does not "see" the barriers. This was confirmed by a full quantum mechanical numerical calculation [9.74].

Another interesting situation in which the semiclassical approach is not applicable is when the magnetic length is much smaller than the period, or, equivalently, when the spacing of the magnetic levels (cyclotron frequency) exceeds the bandwidth. This limit was recently studied experimentally [9.75, 76] by subjecting the two-dimensional electron gas at a GaAs–AlGaAs heterojunction to a one-dimensional potential modulation and to a perpendicular magnetic field (0–1 T range). The modulation was induced by a holographic grating, with typical periods $d \approx 0.3$–$0.5\,\mu$m. In this case, the discrete Landau levels are broadened into bands (removal of the x_0 degeneracy) with a bandwidth which is an oscillatory function of the Landau-level index and is revealed by characteristic oscillations in the magnetoresistance.

9.3 Excitons in Quantum Wells

Excitons are bound electron–hole pairs and are the lowest electronic excited states of non-metallic crystals. They are easily detected in optical spectra, because they give rise to sharp line structures, in contrast to broad continuum transitions. In a semiconductor like GaAs, bulk excitons have a binding energy of about 4 meV. The theoretical determination of this binding energy may at first seem an even more challenging problem than the acceptor one. In the bulk, however, an important simplification is possible [9.77], which makes a very good approximate calculation relatively easy. The bulk effective mass Hamiltonian for excitons can be written

$$H_{exc} = H_e(\boldsymbol{p}_e) - H_h(\boldsymbol{p}_h) - \frac{e^2}{\varepsilon|\boldsymbol{r}_e - \boldsymbol{r}_h|} . \tag{9.22}$$

Here the first term is simply $p_e^2/2m_e^*$, the second is the Luttinger Hamiltonian for the hole [i.e. Eq. (9.10) with $p_h = -i\partial/\partial r_h$ replacing $\boldsymbol{k}$]. We can define a relative coordinate $r = r_e - r_h$, its conjugate momentum p, and a coordinate $R = 1/2(r_e + r_h)$, with conjugate momentum $P = p_e + p_h$. P is actually a constant of the motion, corresponding to the overall momentum of the pair (we avoid mentioning the "center of mass" because, due to the structure of the valence band edge, there is no center of mass for the electron–hole pair [9.78]). Optical transitions can only create or destroy $P = 0$ excitons, for which the Hamiltonian becomes

$$H_{exc} = H_e(\boldsymbol{p}) - H_h(\boldsymbol{p}) - \frac{e^2}{\varepsilon r} . \tag{9.23}$$

Notice that this Hamiltonian is a 4×4 matrix. The electron term and the Coulomb term are assumed multiplied by the unit matrix, i.e., they are added to the diagonal terms of the Luttinger Hamiltonian. Now, apart from the Coulomb term, there are two kinds of terms: the terms proportional to $(\gamma_1 + 1/m_e^*)$, and other terms proportional to γ_2 or γ_3. For GaAs, $m_e^* = 0.067$, $\gamma_1 = 6.85$, $\gamma_2 = 2.10$, $\gamma_3 = 2.90$. Therefore, $\gamma_1 + 1/m_e^* \gg \gamma_2, \gamma_3$. Neglecting all γ_2, γ_3 terms in the first approximation, one finds four degenerate exciton states with binding energy

$$R^* = \frac{e^4}{2(\gamma_1 + 1/m_e^*)\hbar^2\varepsilon^2} . \tag{9.24}$$

The off-diagonal terms can be included as a second-order perturbation [9.77], which does not remove the fourfold degeneracy, but provides a correction to the binding energy.

This procedure cannot be applied to acceptor state calculations, because the large value of $1/m_e^*(\approx 15)$ is essential to the argument. Unfortunately it is also not easy to carry out this procedure for a quantum well. First of all, the translation invariance in the z direction is lost: we can set the component $P_\parallel$ of P along the interfaces equal to zero, but P_z is no longer a good quantum number. Also, the boundary conditions depend separately on z_e and z_h, not only on $z_e - z_h$. The two-particle nature of the problem actually gives rise to a situation opposite to the bulk case: the exciton problem is more complicated than the acceptor problem.

A correct treatment of the degeneracy of the valence band is necessary for a quantitative understanding of the spectra observed, especially in presence of external electric [9.79] or magnetic [9.80–83] fields. A qualitative general picture, on the other hand, is obtainable by simple considerations on the physics of a hydrogenic atom between two plane barriers which enhance the binding energy (relative to the subband edge) by keeping the electron and the hole closer together. After some early attempts to quantify this picture by more refined variational treatments of the hydrogenic problem, full-scale calculations have recently been undertaken, which aim at a complete quantitative description of the experiments, with essentially no approximations except those intrinsic to the envelope-function method. Among the most recent and complete are those of *Zhu* and *Huang* [9.84, 85], *Andreani* and *Pasquarello* [9.86], and *Bauer* and *Ando* [9.87], who went quite far in the inclusion of external fields. The results [9.88] for the transition energies in a magnetic field are in good agreement with experiments of *Ossau* et al. [9.89].

The problem of the exciton in quantum wells seems therefore well understood, at least at the level of extensive numerical diagonalizations. One problem raised by recent experiments is the doublet splitting of the lowest-lying exciton state [9.90], reportedly as large as 6 meV for 10 Å GaAs–AlGaAs quantum wells. Possible explanations in terms of the electron-hole exchange interaction [9.91], which in bulk GaAs produces an effect of the order of 0.02 meV, are not convincing; confinement in one direction can hardly produce such a large enhancement. More recent experimental [9.92] and theoretical [9.93] work in thin-layer GaAs–AlGaAs multiple quantum well systems seems to suggest a much smaller splitting due to exchange effects, although more direct evidence would be desirable.

References

9.1 T. Ando, A.B. Fowler, F. Stern: Rev. Mod. Phys. **54**, 437 (1982)
9.2 J.N. Schulman, Y.C. Chang: Phys. Rev. B **24**, 4445 (1981)
9.3 J.N. Schulman, Y.C. Chang: Phys. Rev. B **27**, 2346 (1983)
9.4 J.C. Chang: Phys. Rev. B **37**, 8215 (1988)
9.5 M. Jaros, K.B. Wong, M.A. Gell: Phys. Rev. B **31**, 1205 (1985)
9.6 M. Jaroș, A. Zoryk, D. Ninno: Phys. Rev. B **35**, 8277 (1987)
9.7 C.G. Van de Walle, R.M. Martin: Phys. Rev. B **24**, 5621 (1986)
9.8 L.J. Sham, M. Schlüter: Phys. Rev. B **32**, 3883 (1985)
9.9 M. Altarelli: In *Heterojunctions and Semiconductor Superlattices*, ed. by G. Allan, G. Bastard, N. Boccara, M. Lannoo, M. Voos (Springer, Berlin, Heidelberg 1986) p. 12
9.10 G. Bastard, J.A. Brum: IEEE J. QE-**22**, 1625 (1986)
9.11 D.J. Ben Daniel, C.B. Duke: Phys. Rev. **152**, 683 (1966)
9.12 See, e.g., E.O. Kane: In *Semiconductors and Semimetals*, ed. by. R.K. Willardson, A.C. Beer (Academic, New York 1966) p. 75
9.13 C. Mailhiot, D.L. Smith: Phys. Rev. B **35**, 1242 (1987)
9.14 H. Akera, S. Wakahara, T. Ando: Surf. Sci. **196**, 694 (1988)
9.15 P.F. Fewster, J. Phillips: Phys. Rev. B **34**, 268 (1986)
9.16 M. Sauvàge, C. Delalande, P. Voisin, P. Etienne, P. Delescluse: Surf. Sci. **174**, 573 (1986)
9.17 T.W. Ryan, P.D. Hatton, S. Bakes, M. Watt, C. Sokomayor-Torres, P.A. Claxton, J.S. Roberts: Semicond. Sci. Technol. **2**, 241 (1987)
9.18 M.F.H. Schuurmans, G.W. 't Hooft: Phys. Rev. B **31**, 8041 (1985)
9.19 Y.C. Chang, J.N. Schulman, G. Bastard, Y. Guldner, M. Voos: Phys. Rev. B **31**, 2557 (1985)
9.20 M.G. Burt: Semicond. Sci. Technol. **2**, 401 and 706 (Erratum) (1987)
9.21 M.G. Burt: Semicond. Sci. Technol. **3**, 739 (1988)

9.22 J.M. Luttinger: Phys. Rev. **102**, 1030 (1956)
9.23 E.O. Kane: In *Handbook of Semiconductors*, Vol. 1, ed. by W. Paul (Academic, New York 1982) p. 193
9.24 D.A. Broido, L.J. Sham: In *Proc. of the 17th Int. Conf. on the Physics of Semiconductors*, San Francisco, 1984, ed. by J.D. Chadi, W.A. Harrison (Springer, New York 1985) p. 337
9.25 M. Altarelli, U. Ekenberg, A. Fasolino: Phys. Rev. B **32**, 5138 (1985)
9.26 LC Andreani, A. Pasquarello, F. Bassani: Phys. Rev. B **36**, 5887 (1987)
9.27 S.S. Nedorezov: Sov. Phys. – Solid State **12**, 1814 (1971)
9.28 J.B. Xia: Phys. Rev. B **38**, 8365 (1988)
9.29 R. Wessel, M. Altarelli: Phys. Rev. B **39**, 12802 (1989)
9.30 M. Altarelli: In *Festkörperprobleme*, Vol. 25, ed. by P. Grosse (Vieweg, Braunschweig 1985) p. 381
9.31 T. Ando: J. Phys. Soc. Jpn. **54**, 1528 (1985)
9.32 Y.C. Chang, G.D. Sanders: Phys. Rev. B **32**, 5521 (1985)
9.33 A. Pinczuk, D. Heiman, R. Sooryakumar, A.C. Gossard, W. Wiegmann: Surf. Sci. **170**, 573 (1986)
9.34 R. Sooryakumar, D.S. Chemla, A. Pinczuk, A. Gossard, W. Wiegmann, L.J. Sham: J. Vac. Sci. Technol. B **2**, 349 (1984)
9.35 See, e.g., L.L. Chang: In *Heterojunctions and Semiconductor Superlattices*, ed. by G. Allan, G. Bastard, N. Boccara, M. Lannoo, M. Voos (Springer, Berlin, Heidelberg 1986) p. 152
9.36 J.C. Maan, Y. Guldner, J.P. Vieren, P. Voisin, M. Voos, L.L. Chang, L. Esaki: Solid State Commun. **39**, 683 (1981)
9.37 M. Altarelli: Phys. Rev. B **28**, 842 (1983)
9.38 E.E. Mendez, L. Esaki, L.L. Chang: Phys. Rev. Lett. **55**, 2216 (1985)
9.39 M. Altarelli, J.C. Maan, L.L. Chang, L. Esaki: Phys. Rev. B **36**, 9867 (1987)
9.40 Y. Guldner, G. Bastard, J.P. Vieren, M. Voos, J.P. Faurie, A. Million: Phys. Rev. Lett. **51**, 907 (1983)
9.41 G.S. Boebinger, J.P. Berroir, Y. Guldner, J.P. Vieren, M. Voos, J.P. Faurie: In Proc. of the MSS III Conf., Montpellier, 1987; J. de Phys. **C5**, 301 (1987)
9.42 D.J. Olego, J.P. Faurie, P.M. Raccah: Phys. Rev. Lett. **55**, 328 (1985)
9.43 S.P. Kowalczyk, J.T. Cheung, E.A. Kraut, R.W. Grant: Phys. Rev. Lett. **56**, 2755 (1986)
9.44 Tran Minh Duc, C. Hsu, J.P. Faurie: Phys. Rev. Lett. **58**, 1127 (1987); see also [9.41, p. 307]
9.45 Y.R. Lin-Liu, L.J. Sham: Phys. Rev. B **32**, 5561 (1985)
9.46 M. Jaros: Phys. Rev. Lett. **60**, 2560 (1988)
9.47 G. Bastard: Phys. Rev. Lett. **60**, 2561 (1988)
9.48 N.F. Johnson, P.M. Hui, H. Ehrenreich: Phys. Rev. Lett. **61**, 1993 (1988)
9.49 R. Ferreira, G. Bastard: Phys. Rev. B **38**, 8406 (1988)
9.50 A. Fasolino, M. Altarelli: Surf. Sci. **142**, 322 (1984); and in *Two-Dimensional Systems, Heterostructures, and Superlattices*, ed. by G. Bauer, F. Kuchar, H. Heinrich, Springer Ser. Solid-State Sci., Vol. 53 (Springer, Berlin, Heidelberg 1984) p. 176
9.51 G.D. Sanders, Y.C. Chiang: Phys. Rev. B **32**, 4282 (1985)
9.52 G. Platero, M. Altarelli: Phys. Rev. B **36**, 6591 (1987)
9.53 C. Mailhiot, Y.C. Chang, T.C. McGill: Phys. Rev. B **26**, 4449 (1982)
9.54 W.T. Masselink, Y.C. Chang, H. Morkoç: Phys. Rev. B **28**, 7373 (1984); J. Vac. Sci. Technol. B **2**, 376 (1984)
9.55 E.E. Mendez, F. Agulló-Rueda, J.M. Hong: Phys. Rev. Lett. **60**, 2426 (1988)
9.56 P. Voisin, J. Bleuse, C. Bouche, S. Gaillard, C. Alibert, A. Regrenyi: Phys. Rev. Lett. **61**, 1639 (1988)
9.57 F. Agulló-Rueda, E.E. Mendez, J.M. Hong: Phys. Rev. B **38**, 12720 (1988)
9.58 See, e.g., N.W. Ashcroft, N.D. Mermin: *Solid State Physics* (Holt, Rinehart and Winston, New York 1976)
9.59 G. Bastard, R. Ferreira, J. Bleuse, P. Voisin: In *Physics of Superlattices and Quantum Wells*, ed. by S.H. Tsai, Wang Xun, X.-C. Shen, X.-L. Lei (World Scientific, Singapore 1989)

9.60 H.L. Störmer, Z. Schlesinger, A. Chang, D.C. Tsui, A.C. Gossard, W. Wiegmann: Phys. Rev. Lett. **51**, 126 (1983)
9.61 A. Broido, L.J. Sham: Phys. Rev. B **31**, 888 (1985)
9.62 U. Ekenberg, M. Altarelli: Phys. Rev. B **32**, 3712 (1985)
9.63 E. Bangert, G. Landwehr: Superlattices Microstruct. **1**, 363 (1985); Surf. Sci. **170**, 593 (1986)
9.64 U. Ekenberg: Surf. Sci. **170**, 601 (1986)
9.65 Y. Iwasa, N. Miura, S. Tarucha, H. Okamoto, T. Ando: Surf. Sci. **170**, 587 (1986)
9.66 D. Heiman, A. Pinczuk, A.C. Gossard, A. Fasolino, M. Altarelli: In Proc. of the 18th Int. Conf. on the Physics of Semiconductors, Stockholm, 1986, ed. by E. Engström (World Scientific, Singapore 1987)
9.67 M.L. Claessen, J.C. Maan, M. Altarelli, P. Wyder, L.L. Chang, L. Esaki: Phys. Rev. Lett. **57**, 2556 (1986)
9.68 J.C. Maan: In *Festkörperprobleme*, Vol. 27, ed. by P. Grosse (Vieweg, Braunschweig 1987)
9.69 See, e.g., J.C. Slater: *Insulators, Semiconductors and Metals* (McGraw-Hill, New York 1967)
9.70 G.E. Zilberman: Sov. Phys. – JETP **5**, 208 (1957); ibid. **6**, 299 (1958)
9.71 G. Belle, J.C. Maan, G. Weimann: Solid State Commun. **56**, 65 (1985)
9.72 T. Duffield, R. Bhat, M. Koza, F. De Rosa, D.M. Hwang, P. Grabble, S.J. Allen, Jr.: Phys. Rev. Lett. **56**, 2724 (1986)
9.73 T. Duffield, R. Bhat, M. Koza, F. De Rosa, D.M. Rush, S.J. Allen, Jr.: Phys. Rev. Lett. **59**, 2693 (1987)
9.74 A. Fasolino, M. Altarelli: In Proc. of the 19th Int. Conf. on the Physics of Semiconductors, Warsaw, 1988, ed. by W. Zawadzki (Inst. of Phys., Polish Acad. of Sciences, Warsaw 1988) p. 361
9.75 R.R. Gerhardts, D. Weiss, K. v. Klitzing: Phys. Rev. Lett. **62**, 1173 (1989)
9.76 R.W. Winkler, J.P. Kotthaus, K. Ploog: Phys. Rev. Lett. **62**, 1177 (1989)
9.77 A. Baldereschi, N.O. Lipari: Phys. Rev. B **3**, 439 (1971)
9.78 M. Altarelli, N.O. Lipari: Phys. Rev. B **15**, 4898 (1977)
9.79 See, e.g., Line Viña, R.T. Collins, E.E. Mendez, W.I. Wang, L.L. Chang, L. Esaki: In *Excitons in Confined Systems*, ed. by R. Del Sole, A. D'Andrea, A. Lapiccirella, Springer Proc. Phys., Vol. 25 (Springer, Berlin, Heidelberg 1988) p. 230
9.80 J.C. Maan, G. Belle, A. Fasolino, M. Altarelli: Phys. Rev. B **30**, 2253 (1984)
9.81 N. Miura, Y. Iwasa, S. Tarucha, H. Okamoto: In Proc. of the 17th Int. Conf. on the Physics of Semiconductors, San Francisco, 1984, ed. by J.D. Chadi, W.A. Harrison (Springer, New York 1985) p. 359
9.82 W. Ossau, B. Jäkel, E. Bangert, G. Landwehr, G. Weimann: Surf. Sci. **174**, 188 (1986)
9.83 D.C. Rogers, J. Singelton, R.J. Nicholas, C.T. Foxon, K. Woodbridge: Phys. Rev. B **34**, 4002 (1987)
9.84 B. Zhu, K. Huang: Phys. Rev. B **36**, 8102 (1987)
9.85 B. Zhu: Phys. Rev. B **37**, 4689 (1988)
9.86 L.C. Andreani, A. Pasquarello: Europhys. Lett. **6**, 259 (1988)
9.87 G.E.W. Bauer, T. Ando: Phys. Rev. B **37**, 3130 (1988)
9.88 G.E.W. Bauer, T. Ando: Phys. Rev. B **38**, 6015 (1988)
9.89 W. Ossau, B. Jäkel, E. Bangert, G. Weimann: In *The Basic Properties of Impurity States in Superlattice Semiconductors*, ed. by C.Y. Fong (Plenum, New York 1988)
9.90 R. Buaer, D. Bimberg, J. Christen, D. Oertel, D. Mars, J.N. Miller, T. Fukunaga, H. Nagashima: In Proc of the 18th Int. Conf. on the Physics of Semiconductors, ed. by O. Engström (World Scientific, Singapore 1987) p. 525
9.91 Y. Chen, B. Gil, P. Lefebvre, H. Mathieu, T. Fukunaga, H. Nagashima: In [9.79, p. 200]
9.92 H.W. van Kesteren, E.C. Cosman, F.J.A.M. Greidanus, P. Dawson, K.J. Moore, C.T. Foxon: Phys. Rev. Lett. **61**, 129 (1988)
9.93 B.R. Salmassi, G.E.W. Bauer: Phys. Rev. B **39**, 1970 (1989)

10. Photonic and Electronic Devices Based on Artificially Structured Semiconductors

Fabio Beltram, Federico Capasso, and *Susanta Sen*

With 41 Figures

In the last decade solid state electronics has continued its remarkable progress towards faster and smaller devices and circuits. Its ongoing development consistently defies the predictions of the many people who elaborate on the "ultimate" device performance. All this has been achieved through the continued refinement of silicon technology [10.1].

In parallel with this work, an increasing number of studies have focused on alternative technologies such as superconducting Josephson junctions, optics, organic materials, and the exploitation of modern epitaxial growth techniques. Here we shall focus on the latter and in particular on new devices that depend on quantum effects (e.g., resonant tunneling) as well as on more "traditional" devices in which growth capabilities are exploited to optimize the performance.

The first four sections of this chapter will be dedicated to some of the most promising applications of resonant tunneling (RT). The basic physical phenomena arising when the layer thicknesses become comparable to the electron de Broglie wavelength (quantum size effect) have been qualitatively understood since the early 1970s with the pioneering work of *Chang* et al. [10.2]. With the development of modern growth techniques, the experimental study of these phenomena has proceeded in parallel with their exploitation in novel devices such as quantum well lasers.

One of the most exciting applications of RT is in transistor structures. RT transistors allow the implementation of a large class of circuits (e.g., analog-to-digital converters, parity checkers, frequency multipliers) with greatly reduced complexity (i.e., fewer transistors per function as compared to circuits using conventional transistors) [10.3]. The inherent functionality of these and other quantum electron devices has led to a potentially intriguing scenario for the future of electronics [10.4]. The progress of integrated circuits has so far been characterized by increased levels of miniaturization, to the point that nowadays certain VLSI chips contain some ten million components. Due to interconnect limitations, this scaling strategy will probably approach practical limits for patterned geometries of lateral dimensions of $\sim 0.10\,\mu$m [10.4] some time in the next fifty years. Electronics will then have to find new evolution paths. RT transistors and quantum coupled devices may provide a solution, in light of their functionality and the possibility of direct device interconnections via tunneling [10.4]. It has also been pointed out that the inherent multistate nature of RT transistors could lead to new computer architectures using multiple valued logic [10.3].

In RT structures, the ability of modern epitaxial growth techniques to produce almost atomically flat interfaces is exploited. Particularly with molecular beam epitaxy (MBE) [10.5], another interesting feature is available: materials can be grown with controlled compositional variations over distances of $\lesssim 100\,\text{Å}$ and different functional forms of grading (linear, parabolic, etc.) can be obtained by accurately controlling the growth. Band-gap grading [10.6] is a powerful tool for engineering the energy-band diagram of a device and thus modifying its electrical transport properties. One of the most interestig properties, which has far-reaching consequences for devices made of such materials, is that electrons and holes experience different electric forces so that their transport properties are effectively independently tuned.

The resulting device applications include ultrahigh-speed phototransistors, transistors with a graded-gap base, and heterojunction bipolar transistors (HBTs) with a graded emitter–base interface. Unipolar single and multiple sawtooth graded-gap structures have shown interesting physical properties and device applications. For example, because of the lack of reflection symmetry, sawtooth superlattices can be electrically polarized or used as rectifying elements. Grading of the high field region in valanche photodiodes has been used to enhance the ionization rate ratio. One of the most exciting applications of graded materials is the "staircase" potential profile which can be used in solid-state photomultipliers and in repeated velocity overshoot devices. The last four sections of this chapter will be dedicated to a review of the electronic transport properties of compositionally graded materials and their applications.

10.1 Resonant Tunneling Bipolar Transistors with a Double Barrier in the Base

Resonant tunneling through heterojunction double barriers (DBs) was first observed by *Chang* et al. in 1974 [10.2], however, the observed negative differential resistance (NDR) effects were too small to be useful in device applications. The impressive RT experiments at terahertz frequencies by *Sollner* et al. in 1983 [10.7] stimulated renewed interest in NDR. Here we shall focus in particular on RT transistor structures. The reader is referred to the many recent reviews covering the physics as well as the dc and high frequency performance of RT diodes [10.8, 9].

The concept of a resonant tunneling bipolar transistor (RTBT) originated with the general idea, conceived by *Capasso* and *Kiehl* in 1984, of associating with each state of quantum system, for example the energy levels of a quantum well, a corresponding logic level [10.3]. This general scheme leads naturally to the idea of multiple valued logic. Although such a logic has been the subject of considerable investigation [10.10], all circuits employing two-state devices require complex and cumbersome architectures to implement it. The above correspondence (energy level/logic state) led to the conception of a class of bipolar devices with inherent multistate operation [10.3].

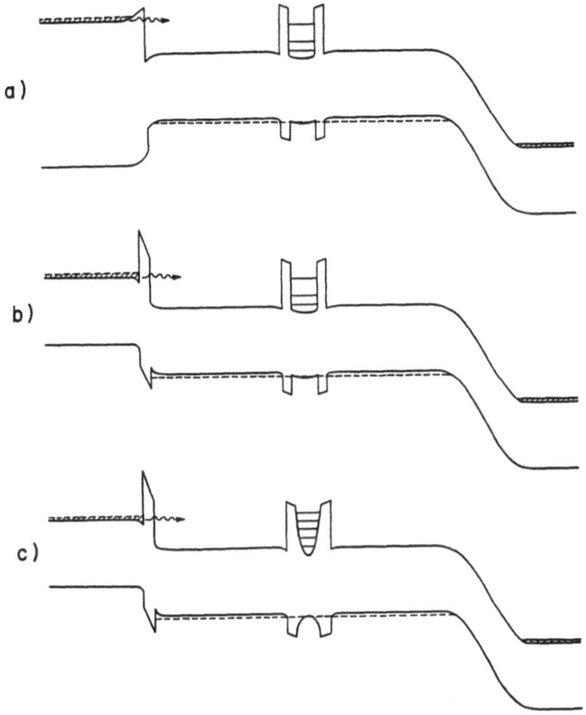

Fig. 10.1. Band diagram of resonant tunneling transistors with abrupt (a), and tunneling (b, c) emitters

Figure 10.1 illustrates this type of device. It should be noted that although the transistors in this figure utilize nonequilibrium injection, the underlying operating principle is at the basis of the operation of all other RTBTs with the quantum well in the base, irrespective of the details of the base contact. As the base-emitter voltage is increased, RT through each subband first reaches a maximum and is then quenched as the bottom of each quantum well (QW) subband is lowered below the conduction band edge in the emitter. This produces multiple peaks in the collector current, i.e., multiple negative transconductance. The tunnel emitter device with a parabolic well in the base can be used to generate equally spaced peaks. For example, using a well of width 200 Å with $Al_{0.45}Ga_{0.55}As$ barriers, one finds an energy level separation ≈ 64 meV. this gives a total of five states in the well. In a recent experiment [10.11] as many as sixteen resonances were observed in resonant tunneling diodes with a parabolic well.

10.1.1 Design Considerations for RTBTs with Ballistic Injection

The design of the transistors in Fig. 10.1 is critical. This is due to the various requirements that must simultaneously be satisfied in order to achieve acceptable current densities ($\geq 10^4$ A/cm^2), current gains ($\gtrsim 10$), and peak-to-valley ratios ($> 2:1$). We assume first that the RT through the DB is coherent; this can be achieved by designing the DB so that $\hbar/\Gamma \ll \tau_\Phi$, where Γ is the resonance full width at half maximum and τ_Φ is the phase relaxation time [10.12]. An estimate of τ_Φ can be obtained from the reciprocal of the *total* scattering rate $1/\tau_T$ (inelastic + elastic) at the energy of the incident particle [10.12]. As an example, in GaAs at a concentration $p = 5 \times 10^{18}\,cm^{-3}$, $1/\tau_T$ is in the range (2–2.5$\times 10^{13}\,s^{-1}$ for injection energies in the range 0.1–0.3 eV [10.13]. thus to ensure that the above condition for coherent transport is satisfied, Γ should be larger than $\sim$ 10 meV. This can be achieved with the ground state resonance (E_1 = 133 meV) of an $Al_{0.40}Ga_{0.60}As$ (15 Å)/GaAs (30 Å) DB, for which tunneling resonance calculations show $\Gamma \approx 64$ meV. The coherence of the RT process and the lateral momentum conservation during tunneling ensure that in symmetric DBs incident electrons with a perpendicular energy $E_\perp$ equal to the bottom of one of the subbands of the well traverse the DB with unity transmission. However, in any experimental situation the incident perpendicular energy distribution $n_\perp(E_\perp)$ has a finite width ΔE; in order to exploit coherent RT and achieve a high base transport factor, ΔE must be smaller than Γ. If $\Delta E \gg \Gamma$ (a situation commonly encountered in DB diodes) only a small fraction of the incident electrons $\sim \Gamma/\Delta E$ contributes to the RT current J_R. J_R is then approximately given by

$$J_R = e v_R n_\perp(E_R) \Gamma T_R \,, \tag{10.1}$$

where $T_R \approx 1$, E_R is the transmission resonance energy, and v_R is the perpendicular component of the velocity corresponding to $E_\perp = E_R$. Since $\Gamma \approx E_R T_B$, where T_B is the transmission of the individual barriers (usually $\ll 1$), (10.1) shows that, for a broad incident distribution, it is the transmission of the individual barriers and not the overall transmission of the DB that determines the current [10.14]. Therefore, to maximize J_R, ΔE must be smaller than Γ. To achieve this, the energy distribution in the emitter should be narrower than the resonance width and electrons should traverse the distance between the DB and the emitter quasi-ballistically. The width of the emitter energy distribution perpendicular to the barrier is approximately $k_B T + \Phi_n$, where $k_B T$ is the thermal energy and Φ_n is the quasi-Fermi energy in the emitter, which is comparable to the equilibrium Fermi level. Consider first the structure of Fig. 10.1a. The emitter composition should be chosen in order to have a conduction band discontinuity $\Delta E_c \approx E_1$ (E_1 being the energy of the first resonance) so that under resonance conditions the conduction band in the emitter is nearly flat, in order to maximize the peak collector current. If ΔE_c is significantly smaller than E_R, the base–emitter junction must be biased beyond flat band to achieve resonance. In this case the electric field in the emitter will heat the injected distribution and broaden it; an unwanted effect in light of the above discussion. On the other hand, if ΔE_c is significantly larger than E_R, at resonance the emitter current will not be large neough due to the residual base emitter barrier. Consider, for example, the $Al_{0.4}Ga_{0.6}As$/GaAs DB previously discussed. The emitter composition should be chosen to be approximately $Al_{0.20}Ga_{0.60}As$, which corresponds to $\Delta E_c \approx E_1$ = 133 meV. For an emitter doping density of $5 \times 10^{17}\,cm^{-3}$ the width of the distribution ballistically launched with an energy $E_\perp = \Delta E_c$ by the abrupt emitter is ≈ 50 meV, which is close to the resonance width in the DB.

One must, however, consider also the effects of scattering in the region between the DB and the emitter, which broadens the distribution. In order to achieve a high peak-to-valley ratio, scattering in this region must be minimized. Electrons are launched by the emitter with a forward velocity $(2\Delta E_c/m^*)^{1/2} \approx 8\times10^7$ cm/s limited by the band structure. Since the scattering rate at the injection energy is $\approx 2\times10^{13}\,s^{-1}$ (for $p = 5\times10^{18}\,cm^{-3}$ in the base), the mean free path for these electrons is $\lambda \approx 400$ Å [10.13]. If the distance between the DB and the emitter (L) is kept ≈ 300 Å this implies that only half of the carriers ($\approx e^{-L/\lambda}$) traverse this distance without collisions. Electrons that have lost a portion [$\gtrsim$ optical phonon energy, ≈ 35 meV, for the $Al_{0.4}Ga_{0.6}As$ (15 Å)/GaAs (30 Å) DB case under consideration] of their energy normal to the DB as a result of these collisions will see a significantly reduced transmission through the DB or "miss" the first resonance altogether if the width of the latter is too small.

One way to considerably reduce the scattering rate is to dope the region between the DB and the emitter very heavily ($> 10^{20}\,cm^{-3}$). *Levi* [10.13] has shown theoretically that the inelastic scattering rate of minority carrier electrons in p-type GaAs first increases with increasing doping and rapidly decreases for doping levels well above $10^{19}\,cm^{-3}$, due to the decreased phase space available for scattering. These levels can be achieved by carbon doping [10.15], which also has the advantage of a small diffusion coefficient. At the same time, elastic scattering, which increases rapidly with increasing doping, can be strongly suppressed by placing impurities in a periodic sublattice by delta doping techniques [10.16]. Scattering rate calculations show that at injection energies of ≈ 15 eV and for $p = 2\times10^{20}\,cm^{-3}$ mean free paths as long as 1500 Å can be achieved in GaAs [10.13]. These considerations of course also apply to the structures of Fig. 10.1b and c.

The above discussion clearly demonstrates that the design of a RTBT with quasi-ballistic injection is an extremely difficult task. The task may be somewhat simpler for the AlInAs/GaInAs and InP/GaInAs systems in light of the larger electron mean free path in p-type $Ga_{0.45}In_{0.53}As$ (approximately twice that of GaAs at the same injection energy) and the smaller effective masses in the barrier and well layers. The preferred structure in this case would have a tunnel emitter of the type shown in Fig. 10.1b, c, consisting of a $Ga_{0.45}In_{0.53}As$ layer followed by an InP or an $Al_{0.48}In_{0.52}As$ tunnel barrier. Tunnel emitters allow one to bias the base–emitter junction well beyond flat-band conditions while still maintaining a narrow incident energy distribution.

10.1.2 Quasi-Ballistic Resonant Tunneling in a Tunneling Emitter RTBT

In this section we present results on the RTBT of Fig. 10.1b, fabricated in the AlInAs/GaInAs system. The collector layer is 3000 Å thick undoped $Ga_{0.47}$-$In_{0.53}As$. The base layer comprises a 600 Å region on the emitter side (doped to $p = 3\times10^{18}\,cm^{-3}$) and a 2000 Å region, doped to $p = 5\times10^{18}\,cm^{-3}$, on the collector side, separated ba an undoped $Al_{0.48}In_{0.53}As$ (50 Å)/$Ga_{0.47}In_{0.53}As$ (100 Å) DB. The emitter consists of an $Al_{0.48}In_{0.52}As$ 30 Å tunnel barrier separated from

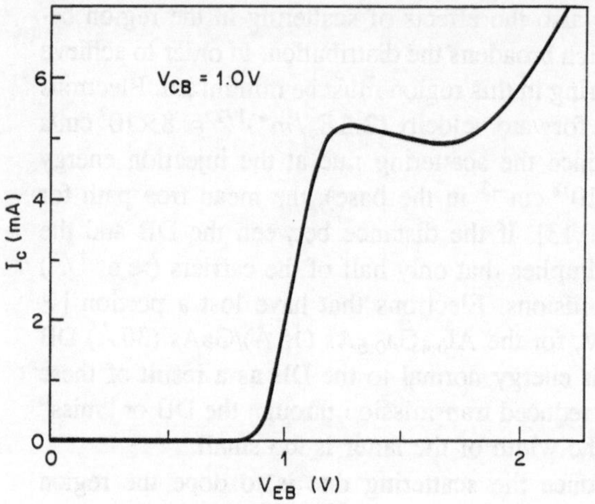

Fig. 10.2. Transfer characteristics of the resonant tunneling transistor of Fig. 10.1b at 10 K

the base by a 50 Å undoped space layer and followed by an $n = 1 \times 10^{18}\,\mathrm{cm}^{-3}$ 3000 Å-thick $Ga_{0.47}In_{0.53}As$ layer. The device transfer characteristics in the common base configuration are shown in Fig. 10.2 for $V_{CB} = 1.0V$, at cryogenic temperature. The collector current rises rapidly above the built-in voltage ($V_{bi} = 0.8V$) and peaks at ≈ 1.25 V. This value equals the calculated voltage required to line up the bottom of the first quantized subband in the accumulation layer (on the emitter side of the tunnel barrier) with the second resonance of the well ($E_2 = 193$ meV). Thus the peak corresponds to ballistic RT of electrons injected from the emitter in the first excited state of the quantum well (QW). Note that the negative transconductance region after the peak is broad (≈ 0.2 eV) and the peak-to-valley ratio is small.

The features in Fig. 10.2 can be understood as follows. The mean free path for electrons with kinetic energies of the order of 100 meV in p^+-InGaAs can be estimated to be about 500 Å [10.17]. A large fraction of the electron distribution incident on the DB is therefore nonballistic due to scattering in the region between the emitter and the DB. The dominant scattering mechanisms for electrons in p^+-InGaAs doped to densities $> 10^{18}\,\mathrm{cm}^{-3}$ are inelastic collisions with holes and coupled phonon-plasmon modes [10.17]. Assume now that the emitter base junction is biased beyond the second resonance of the QW so that the electron injection energies exceed E_2. The injected electrons which ballistically approach the DB have their energy mostly associated with perpendicular motion so that $E_\perp > E_2$. These electrons therefore cannot resonantly tunnel into the QW. On the other hand, the part of the electron distribution which evolves following scattering includes electrons with reduced $E_\perp$. Electrons with $E_\perp$ equal to E_1 or E_2 are still able to resonantly tunnel into the QW and yield an increasing background to the collector current as V_{EB} is increased. We therefore observe a rather broad peak region and a small peak-to-valley ratio in Fig. 10.2. Since the DB region in this device has no intentional dopants, possible effects of elastic scattering centers in the QW on the peak-to-valley ratio [10.18] are relatively unimportant in the present case. Previous work [10.19] on RT spectroscopy of electrons injected via band discontinuities in p^+-GaAs wells showed the formation of a broad, scattering-induced hot electron distribution, but failed to detect ballistic effects, probably due to the short electron mean free path in GaAs.

It is interesting to note that the peak corresponding to the first resonance of the well ($E_1 = 48\,\mathrm{meV}$) is not observed in Fig. 10.2. This is primarily due to the fact that the peak current associated with the first resonance is reduced relative to the current through the second resonance by the ratio (~ 10) of the transmission coefficients of the indivudal barriers at the two resonant energies [see (10.1)], so that the first peak is masked by the rapidly rising emitter current for $V_{\mathrm{EB}} > V_{\mathrm{bi}}$; in addition, the effect of scattering on the incident distribution will reduce to the peak-to-valley and broaden the peak, as previously discussed.

10.1.3 Thermionic Injection RTBTs Operating at Room Temperature

The first operating RTBT was designed to have minority electrons thermally injected into the DB [10.20]. This made the design parameters of the device much less critical, and the structure, implemented in the AlGaAs system, operated at room temperature. The band diagram of this transistor is shown in Fig. 10.3a, b under operating conditions. The alloy composition of the region adjacent to the emitter was adjusted in such a way that the conduction band in this region lines up with or is slightly below the bottom of the ground-state subband of the QW. For a 74 Å-thick well with 21.5 Å AlAs barriers the first quantized energy level is $E_1 = 65\,\mathrm{meV}$. Thus the Al mole fraction was chosen to be $x = 0.07$ (corresponding to $E_{\mathrm{g}} = 1.521\,\mathrm{eV}$) so that $\Delta E_{\mathrm{c}} \approx E_1$. The QW was undoped; nevertheless it is easy to show that there is a high concentration ($\approx 7\times10^{11}\,\mathrm{cm}^{-2}$) two-dimensional hole gas in the well. These holes have transferred by tunneling from the nearby $Al_{0.07}Ga_{0.93}As$ region. This reduces scttering in the well by essentially eliminating elastic scattering with doping impurities [10.18]. Electrical contact was made to both the well region and the GaAs portion of the base adjacent to the DB, but not to the $Al_{0.07}Ga_{0.93}As$ region. The wide gap emitter ($Al_{0.25}Ga_{0.75}As$) provides the well-known advantages of HBTs. Details of the structure and processing are given in [10.20].

In order to understand the operation of the device, consider a common emitter bias configuration (Fig. 10.3a, b). Initially the collector–emitter voltage V_{CE} and the base current I_{B}, are chosen in such a way that the base–emitter and the base–collector junctions are respectively forward and reversed biased. If V_{CE} is kept constant and the base current I_{B} is increased, the base–emitter potential also increases until the flat conduction band condition in the emitter–base *pn* junction region is reached (Fig. 10.3a). The device behaves in this regime like a conventional transistor with the collector current linearly increasing with the base current (Fig. 10.3c). The slope of this curve is, of course, the current gain β of the device. In this region of operation, electrons in the emitter overcome, by thermionic injection, the barrier of the base–emitter junction and undergo RT through the DB. If the base current is further increased above the value I_{Bth}

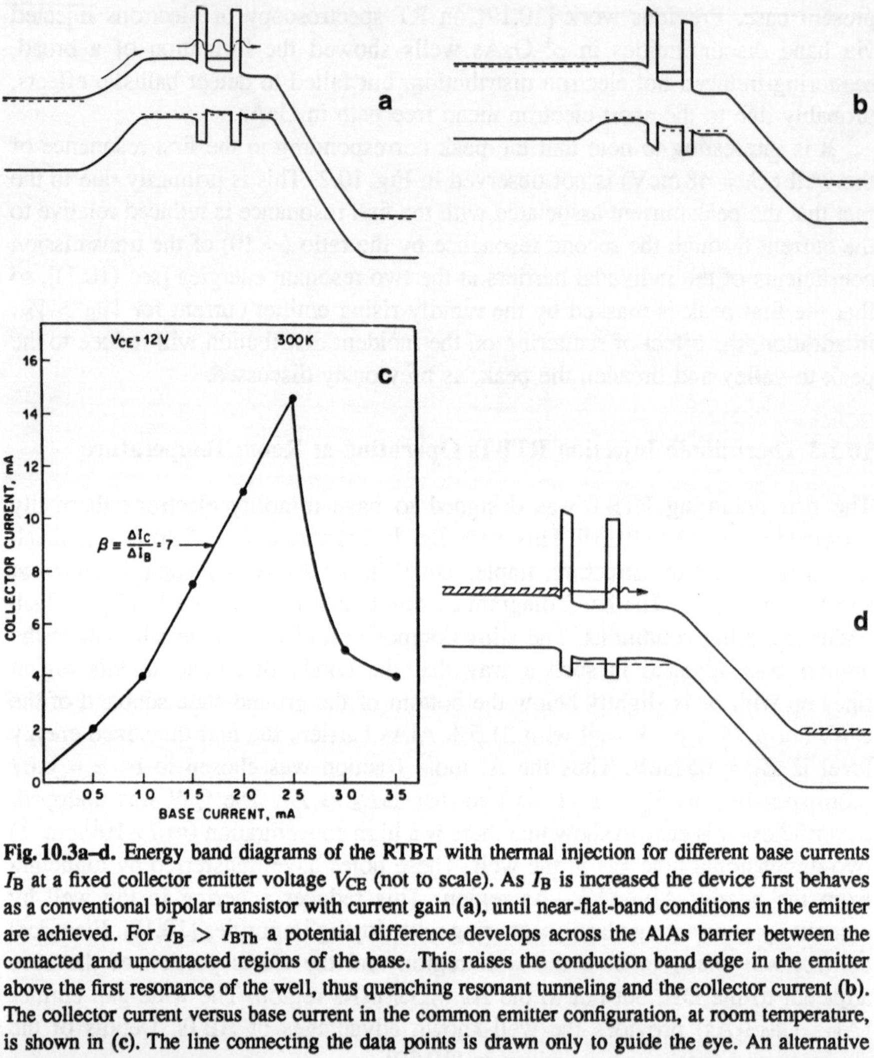

Fig. 10.3a–d. Energy band diagrams of the RTBT with thermal injection for different base currents I_B at a fixed collector emitter voltage V_{CE} (not to scale). As I_B is increased the device first behaves as a conventional bipolar transistor with current gain (a), until near-flat-band conditions in the emitter are achieved. For $I_B > I_{BTh}$ a potential difference develops across the AlAs barrier between the contacted and uncontacted regions of the base. This raises the conduction band edge in the emitter above the first resonance of the well, thus quenching resonant tunneling and the collector current (b). The collector current versus base current in the common emitter configuration, at room temperature, is shown in (c). The line connecting the data points is drawn only to guide the eye. An alternative RTBT design is shown in (d)

corresponding to the flat band condition, the additional potential difference drops primarily across the first semi-insulating AlAs barrier (Fig. 10.3b), between the contacted and uncontacted portions of the base, since the highly doped emitter is now fully conducting. This pushes the conduction band edge in the $Al_{0.07}Ga_{0.93}As$ above the first energy level of the well, thus quenching the RT. The net effect is that the base transport factor and the current gain are greatly reduced. This causes an abrupt drop of the collector current as the base current exceeds the threshold value I_{Bth} (Fig. 10.3c). Thus the device has negative transconductance.

It should be clear that this device is not equivalent to a series combination of a RT diode and a bipolar transistor since electrical contact is made to the QW.

Thus the base–emitter voltage directly modulates the energy difference between the states of the well and the emitter quasi Fermi level (Fig. 10.3). A RTBT based on this operating principle but with the base layer restricted to the GaAs quantum well has been reported [10.21], as originally proposed by *Ricco* and *Solomon* [10.22]. *Futatsugi* and coworkers [10.23] reported a RTBT with a DB between the base and the emitter, exhibiting negative transconductance at liquid nitrogen temperature. Since the quantum well is not contacted and is placed out of the base, this device, unlike those in [10.20, 21], can be thought of as a monolithic series integration of a DB and a bipolar transistor.

Several alternative RTBT designs are possible; one is shown in Fig. 10.3d. Here the p region between the DB and the emitter is eliminated. The well is heavily doped with low diffusivity acceptors (e.g. C). The operating principle is the same as for the device of Fig. 10.3a, b.

10.1.4 Speed and Threshold Uniformity Considerations in RTBTs

The insertion of a DB in a HBT structure offers new interesting circuit opportunities, but also raises questions concerning the resulting effects on speed and threshold uniformity.

The introduction of a DB in the base or in the emitter will increase the emitter–collector delay time τ_{EC} and therefore reduce the cutoff frequency f_T. This is due to the tunneling delay time, which, in general, is a complicated function of the shape of the incident perpendicular energy distribution. If the latter is much broader than the resonance width and nearly centered on one of the resonances, it can be shown that τ_T is approximately given by [10.8]

$$\tau_T = \frac{d}{v_G} + \frac{2\hbar}{\Gamma} . \qquad (10.2)$$

The first term represents the semiclassical transit time across the RT structure of width d (v_G is the drift velocity) and is $\lesssim 0.1$ ps for the RTDBs of interest here. The second term is the so-called phase time (Γ is the width of the resonance). In the RT transistor structures with potential practical impact (e.g., Fig. 10.3) the first resonance width is much smaller than the quasi Fermi energy in the emitter, thus satisfying the first assumption underlying (10.2). The condition that the tunneling wavefunction be nearly centered on the resonance is only partially valid, so that (10.2) can only be used for a rough estimate of the delay time associated with RT. It is clear from this expression that to minimize τ_T the resonant width Γ_1, which depends exponentially on the barrier thickness, must be maximized. Consider a RTBT structure of the type previously discussed (Fig. 10.3). For a 17 Å AlAs barrier thickness and a 45 Å GaAs well, tunneling resonance calculations give $E_1 = 0.136$ eV for the first energy level and $2\hbar/\Gamma_1 = 0.45$ ps [10.8]. The first term in (10.2) is 0.08 ps (assuming a drift velocity $\geq 10^7$ cm/s since overshoot effects following unjection in the DB are possible [10.8]). Thus $\tau_T \approx 0.5$ ps. It is well known that AlGaAs/GaAs HBTs without a RTDB and uniform composition in the base can achieve $f_T > 50$ GHz. The introduction of the above DB in a

HBT with $f_T = 50\,\text{GHz}$ will increase τ_{EC} by 0.5 ps, thus yielding $f_T \approx 43\,\text{GHz}$. This example shows that RTBTs with suitably designed DBs should have cutoff frequencies and overall response speed comparable to those of state-of-the-art HBTs. A $Ga_{0.47}In_{0.53}As$ HBT with an $Al_{0.48}In_{0.52}As$ (44 Å)/$Ga_{0.47}In_{0.53}As$ (38 Å) DB in the emitter having an f_T of 12.5 GHz has recently been reported (see the review article cited in [10.23]). The microwave performance of multistate RTBTs will be discussed in the next section.

Concerning the threshold (V_{Bth}) uniformity issue, let us recall that a conventional HBT has excellent uniformity (a few millivolts) both on the same wafer and from wafer to wafer since V_{Bth} is given by the base–emitter built-in voltage. The latter is proportional to the band gap and weakly (logarithmically) dependent on doping. The introduction of a DB in an HBT will induce greater fluctuations in V_{Bth}. To estimate the latter consider the case of an RTBT with a DB in either the emitter or the base. The voltage position of the collector current peak (transistor fully on) is approximately given by $V_{BE} + 2E_1/e$, where E_1 is the energy of the first resonance of the well. E_1 can fluctuate across a wafer primarily due to in-plane thickness fluctuations, Δl. Using the formula for an infinite potential well (this assumption provides an upper limit for the fluctuations in E_1) one finds that $\Delta E_1 = 2E_1 \Delta l/l$. Thus the corresponding fluctuation in the peak position is

$$\Delta V_p = \Delta V_{BE} + \frac{4E_1}{e}\frac{\Delta l}{l} \,. \tag{10.3}$$

Thickness fluctuations in state-of-the-art MBE material are of the order of one monlayer, i.e., $\Delta l \approx 2.5$ Å. For a RTBT with the double barrier considered in this section, (10.3) gives therefore $\Delta V_p = 30\,\text{mV}$. One obtains $\Delta V_p = 21.5\,\text{mV}$ for a RTBT containing an AlInAs (25 Å)/GaInAs (50 Å) DB. Values $\lesssim 100\,\text{mV}$ are adequate for the circuits envisioned.

10.2 Devices with Multiple Peak I-V Characteristcs and Multiple-State RTBTs

A simple approach to the realization of multiple-peak I-V characteristics is the integration of a number of RT diodes. In this method, a single resonance of different quantum wells is used to generate multiple peaks. Hence the peaks occur at almost the same current level and exhibit similar peak-to-valley ratios as requird by the circuit applications that will be discussed in what follows. However, these devices do not have the gain and input–output isolation of three-terminal devices.

There are two different ways to integrate RT diodes to achieve multiple-peak I-V characteristics. One is horizontal integration, so that the diodes are in parallel in the equivalent circuit [10.24, 25]. The other is to vertically integrate them so that they are in series [10.26, 27]. In this section we shall discuss only the latter approach, since it is important for the design of the multistate RTBT. The rest of the section is devoted to RTBTs with multiple peaks in the transfer characteristics and their digital and analog circuit applications.

10.2.1 Vertical Integration of RT Diodes

Vertical integration of RT structures is achieved by stacking a number of DBs in series, separated by heavily doped cladding layers to quantum mechanically decouple the adjacent DBs from each other [10.26, 27]. The DBs are designed so that the ground state in the QW is substantially above the Fermi level in the adjacent cladding layers. The band diagram of the structure under bias is shown in Fig. 10.4. When bias is applied, the electric field is higher at the anode end of the device (Fig. 10.4a) because of charge accumulated in the QWs under bias. Quenching of RT is thus initiated across the DB adjacent to the anode and then sequentially propagates to the other end, as the high-field region widens with increasing applied voltage, as shown in Fig. 10.4a and b. Once RT has been suppressed across a DB, the voltage drop across it quickly increases with bias because of the increased resistance. The non-RT component through this DB provides continuity for the RT current through the other DBs on the cathode side. A NDR region is obtained in the I-V characteristic, corresponding to the quenching of RT through each DB. Thus with n diodes, n peaks are present in the I-V characteristic.

Generating multiple-peak I-V characteristics by combining tunnel diodes in series is a well-known method [10.28]. However, the mechanism in that ar-

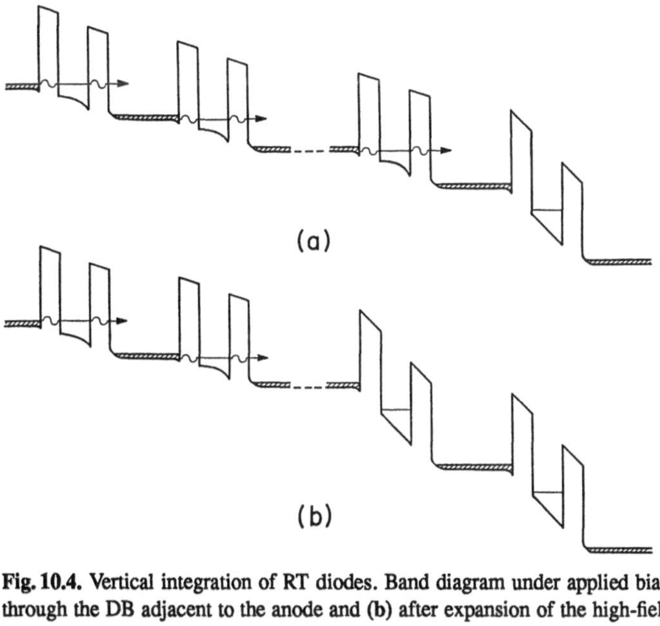

Fig. 10.4. Vertical integration of RT diodes. Band diagram under applied bias (a) with RT quenched through the DB adjacent to the anode and (b) after expansion of the high-field region to the adjacent DB with increasing bias. The arrows indicate the RT component of the current

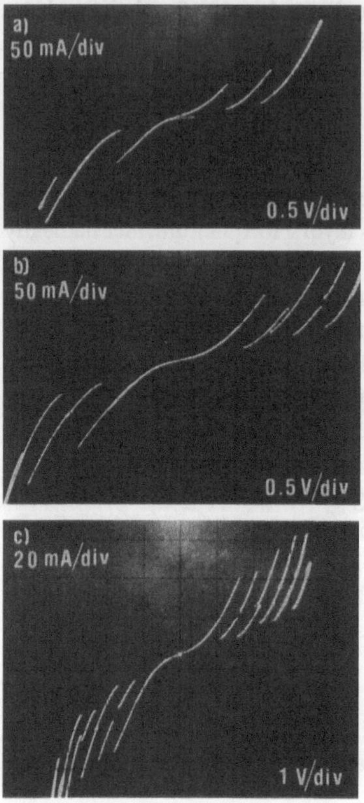

Fig. 10.5. Current–voltage characteristics of devices with two, three, and five vertically integrated RT double barriers, taken for both bias polarities at 300 K

rangment is different. The tunnel diodes used in such a combination must have different characteristics with successively increasing peak currents, so that each of them can go into the NDR region only when a specific current level is reached [10.28]. Structures using RT diodes have a number of significant advantages over those which use tunnel diodes [10.24].

We have tested devices consisting of two, three, and five RT $Al_{0.48}In_{0.52}As$ (50 Å)/$Ga_{0.47}In_{0.53}As$ (50 Å) DBs in series, separated by 1000 Å thick n^+-$Ga_{0.47}$-$In_{0.53}As$ regions. The resulting I-V characteristics taken for both polarities of the applied voltage at room temperature are shown in Figs. 10.5a–c for two, three, and five RTDBs in series, respectively. Positive polarity here refers to the top of the mesa being biased positively with respect to the bottom. Note that, in this polarity, the devices show two, three, and five peaks in the I-V characteristic as expected. For negative polarity, the third peak is not observed in the device with three DBs because of rapidly increasing background current. This is likely to be due to structural asymmetries unintentionally introduced during growth.

10.2.2 Multiple-State RTBTs

The stacked RT structure discussed above was used to design a RTBT exhibiting multiple NDR and negative transconductance characteristics [10.29, 30]. A schematic of this transistor is shown in Fig. 10.6. The device essentially consists of a $Ga_{0.47}In_{0.53}As/Al_{0.48}In_{0.52}As$ *n-p-n* transistor with a stack of two $Ga_{0.47}In_{0.53}As$ (50 Å)/$Al_{0.48}In_{0.52}As$ (50 Å) RT DBs embedded in the emitter. Details of the structure (doping and layer thicknesses) are described in [10.29]. The operation of the transistor can be understood from the band diagrams in the common-emitter configuration shown in Fig. 10.7. The collector–emitter bias (V_{CE}) is kept fixed and the base-emitter voltage (V_{BE}) is increased. For V_{BE} smaller than the built-in voltage ($V_{bi} \approx 0.7$ eV at 300 K) of the $Ga_{0.47}In_{0.53}As$ *p-n* junction, most of the bias voltage falls across this junction (Fig. 10.7a), since its impedance is much greater than that of the two DBs in series, both of which are conducting via RT. The device in this region behaves as a conventional bipolar transistor with the emitter and hence the collector current increasing with V_{BE} (Fig. 10.8) until the base–emitter junction reaches the flat-band condition. Beyond flat band, most of the additional increase in V_{BE} will fall across the DBs (Fig. 10.7b), and as RT through these is sequentially quenched, abrupt drops are observed in the emitter and collector current (Fig. 10.8). The highest peak-to-valley ratio in the transfer characteristics at room temperature is 4 : 1, it increases to about 20 : 1 at 77 K.

Figure 10.9 shows the common-emitter output characteristcs of the transistor (I_C vs V_{CE} at different I_B) at room temperature (top) and 77 K (bottom). At low

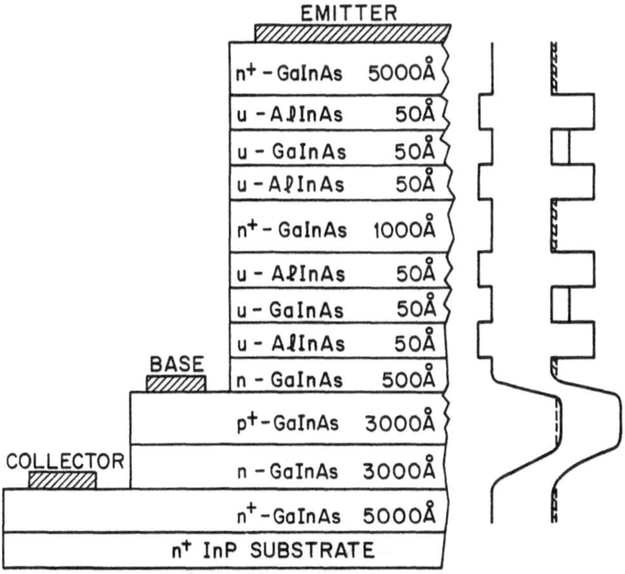

Fig. 10.6. Schematic structure of the multiple-state RTBT and equilibrium conduction-band diagram

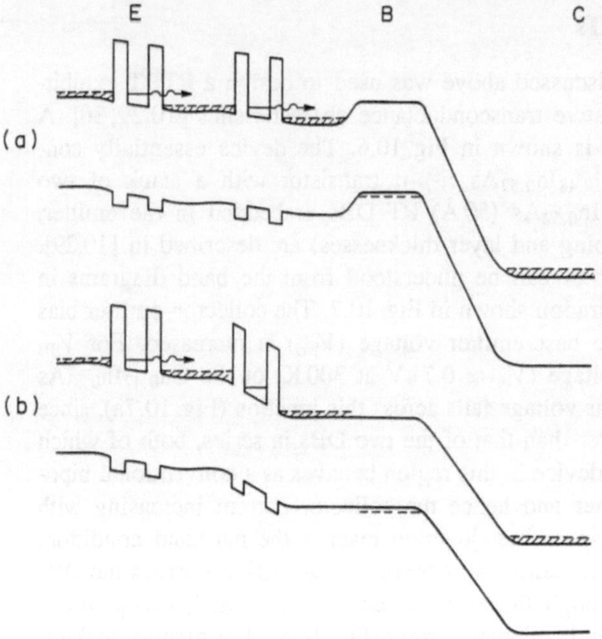

Fig. 10.7a, b. Band diagram of the multiple-state RTBT in the common emitter configuration for different base–emitter bias conditions. (a) Electrons resonantly tunnel through both DBs; in this regime the device operates as a conventional bipolar transistor. (b) Quenching of RT through the DB adjacent to the *pn* junction gives rise to a negative differential resistance region in the collector current. Quenching of RT through the other DB produces a second peak in the I-V characteristic

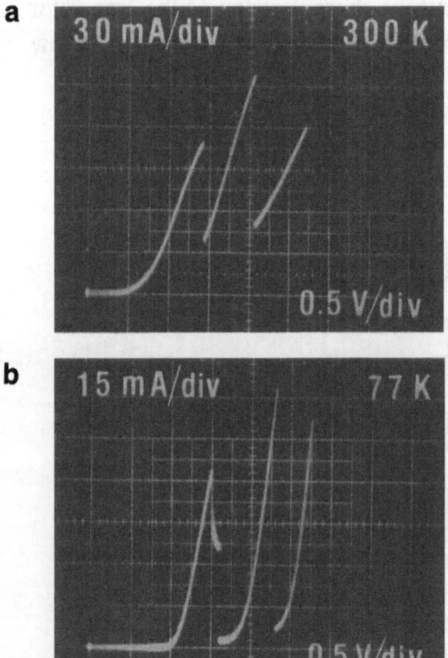

Fig. 10.8. Collector current vs base–emitter voltage in the common-emitter configuration for $V_{CB} = -0.1$ V at 300 K (a) and 77 K (b)

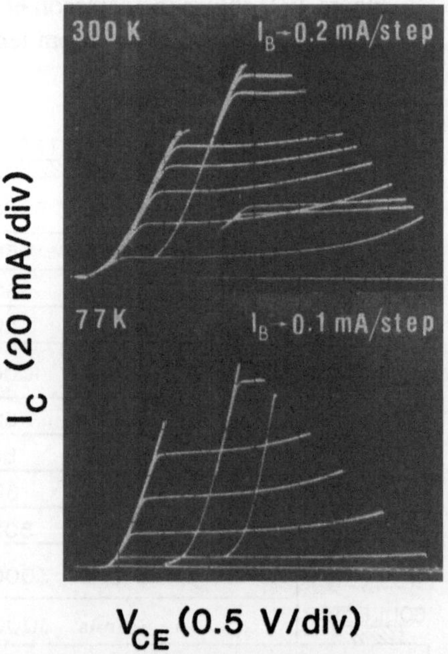

Fig. 10.9. Common emitter output characteristics of the multiple-state RTBT. Collector current vs collector–emitter voltage for different base currents, at 300 K (*top*) and 77 K (*bottom*)

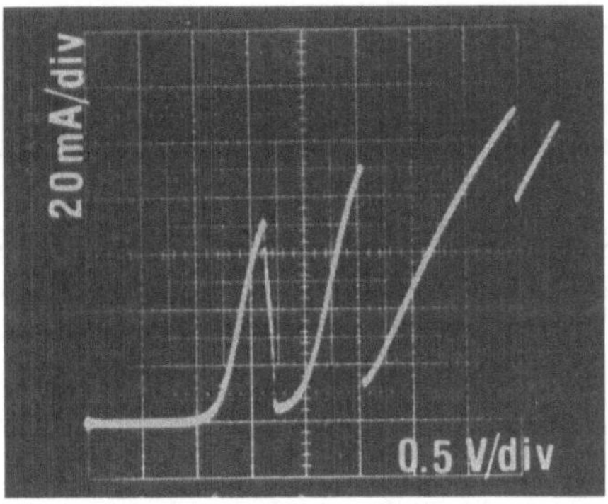

Fig. 10.10. Common-emitter transfer characteristics of a multiple-state RTBT with three DBs in the emitter at 77 K. I_C vs V_{BE} is shown for $V_{CE} = 4.75$ V

base currents I_B (and hence low base–emitter voltages V_{BE}), the device behaves as a conventional bipolar transistor with large current gain (200 at 77 K and 70 at 300 K). On increasing I_B (V_{BE}) beyond the flat-band condition, the excess applied voltage V_{BE} starts appearing across the series of DBs in the emitter. As RT through the DB is sequentially quenched (at the threshold base currents $I_{Bth\,1}$ and $I_{Bth\,2}$), the electron current acros the base–emitter junction drops abruptly, while the hole current, flowing by thermionic emission, continues to increase. This results in sudden quenching of the current gain and hence of the collector current I_C, giving rise to two NDR regions (Fig. 10.9). The highest peak-to-valley ratios observe are 6 : 1 at room temperature and 22 : 1 at 77 K. It should be noted that the small-signal current gain of the transistor at room temperature in its second ($1.2\,\mathrm{mA} < I_B < 1.6\,\mathrm{mA}$) and third ($I_B > 1.6\,\mathrm{mA}$) operation regions is reduced to 40 and 20 respectively. This is expected since the hole current flowing from the base towards the emitter increases with increasing V_{BE}, thus reducing the injection efficiency. This reduction of the current gain is less pronounced at 77 K, since the thermionic flow of holes is much lower.

Figure 10.10 shows the common-emitter transfer characteristics at 77 K of a similar transistor with three DBs in the emitter. The third peak is shifted to a significantly higher voltage relative to the other two. Systematic studies also indicate large hysteresis associated with the same structure. Such a behavior is not uncommon in RT devices whenever there is a large parasitic resistance [10.31]. When three DBs are in series, the parasitics also add up and enhance the effect. The structure with three peaks has to be optimized to minimize these effects.

To minimize the flow of holes from the base to the emitter an n^+-$Al_{0.48}In_{0.52}As$ layer can be inserted between the stack of DBs and the base, somewhat in the fashion of a wide gap emitter in HBTs. Perliminary results with an $n = 1 \times 10^{18}\,\mathrm{cm}^{-3}$ 500 Å AlInAs layer, followed by the growth of an equally thick and doped AlInGaAs grading layer before the DBs, give a β of 4000 at 77 K.

10.2.3 Microwave Performance of Multiple-State RTBTs

In the high-frequency operation of RTBTs [10.32], the typical device structure, grown lattice-matched to an InP substrate by MBE, is very similar to the one previously discussed in Sect. 10.2.2. For microwave evaluation, the structure was grown in a semi-insulating InP substrate instead of an n^+ substrate. Furthermore, the base layer thickness was reduced by a factor of two (down to 1500 Å) and the doping was doubled (4×10^{18} cm^{-3}) to reduce the base transit time without increasing the base resistance.

The emitter-up transistor in a mesa configuration was obtained by successive steps of photolithography and wet chemical etching to expose base and collector layers. After planarization of the structuress with dielectric deposition, a nonalloyed metallization (Ti/Au) was deposited for the contact fabrication. The emitter area was 42 μm^2. The I-V characteristics of the devices are virtually identical to those of Figs. 10.8 and 10.9. The only difference is that the collector currents are considerably smaller due to the scaled-down area.

Scattering (S) parameter measurements were performed in the frequency range 0.5–26.5 GHz using a wafer prober in conjunction with an automatic network analyzer (HP8510B). Figure 10.11 displays the current gain h_{21} as a function of frequency for different bias conditions. The pad structure without a device was also measured. Pad corrections were made by subtracting the admittance parameters Y of the pad structure from those of the device-plus-pad combination. These corrections had negligible effects on the measured gain ($\lesssim$ 1 dB) and on the frequency roll-off (−6 dB/octave in Fig. 10.11). Curve a in Fig. 10.11 refers to an operating point after the second peak in the common emitter characteristics. The f_T obtained by extrapolation using a −20 dB/decade straight line is 24 GHz, which is the highest ever achieved in a RTBT. Previously a comparable cutoff frequency was obtained in a Resonant tunneling Hot Electron Transistor (RHET) with a single collelctor current peak [10.33]. Curve b was obtained for a bias point between the two peaks. For curve c, the base current (40 μA) is such that no NDR appears in the common-emitter characteristics.

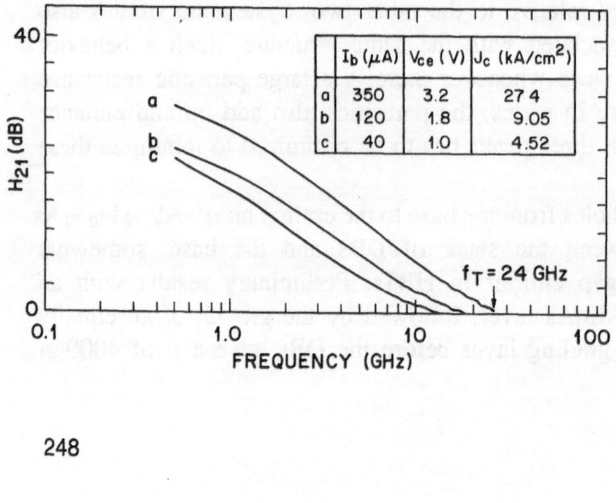

	Ib (μA)	Vce (V)	Jc (kA/cm²)
a	350	3.2	27.4
b	120	1.8	9.05
c	40	1.0	4.52

Fig. 10.11. Current gain (h_{21}) as a function of frequency for different bias points in tshe common-emitter configuration. The corresponding collector current density J_C is also indicated

10.3 Circuit Applications of Multiple-State RTBTs

10.3.1 Frequency Multiplier

The transfer characteristic of Fig. 10.8 was exploited in the frequency multiplier circuit shown in Fig. 10.12a. As the input voltage is increased, the collector current increases, resulting in a decrease in the collector voltage until the device reaches the negative transconductance regions, where sudden drops in the collector current and hence increases in the output voltage are observed. Thus under a suitable bias V_{BB} (base–emitter junction biased between the two peaks of the common-emitter transfer characteristic), triangular input waves will be multiplied by a factor of three, and sine waves by a factor of five [10.30]. Unlike two-terminal multipliers, the output signal in this case is referenced to ground and isolated from the input. These advantages are obtained because the multiple peaks are present in the transfer characteristics of a transistor rather than in the I-V curve of a two terminal device as in [10.24]. It should be noted

a)

b)

c)

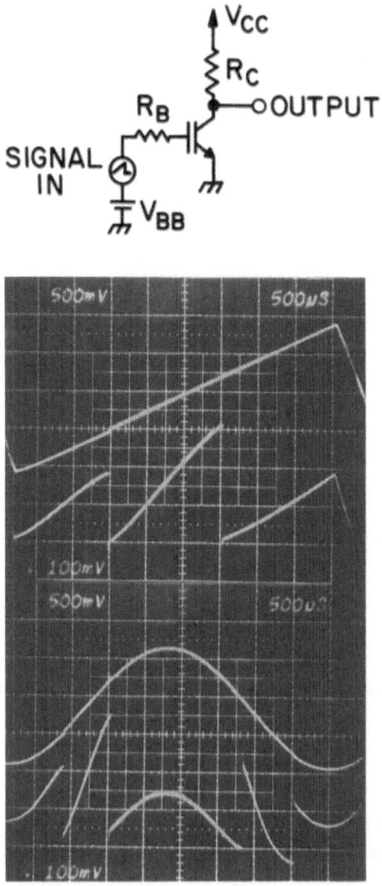

Fig. 10.12. (a) Frequency multiplier using the multiple-state RTBT, and the experimental results of multiplying (b) sawtooth and (c) sine-wave input signals, at room temperature

that in applications where two-terminal devices cannot be used, conventional frequency-independent multipliers require the use of a phase-lock loop and a digital frequency divider.

The gain of the circuit is determined by the transconductance of the transistor and the collector resistance R_C. However, too large a value for R_C will lead to saturation of the device at large input voltages. The situation can be avoided by choosing a larger supply voltage V_{CC}, but the maximum usable V_{CC} is presently limited by the collector–base breakdown voltage of the device. Figure 10.12 shows triangular (b) and sine-wave (c) input signals together with the experimental output after frequency multiplication (V_{CC} = 3.0 V, V_{BB} = 1.8 V, R_C = 5 Ω and R_B = 50 Ω). The polarity of the output signals (bottom traces) have been inverted in the display for clarity of presentation.

For frequency multiplication at high frequency, the devices described in Sect. 10.2.3 were biased in the common-emitter configuration with V_{CE} = 3.2 V and the characteristic impedance of the 50 Ω line as the load. The base-emitter junction was dc biased at 2.0 V via a bias tee. A 350 MHz sine wave was applied to the base and its amplitude was adjusted to achieve a base–emitter voltage swing large enough to bring the device into the negative transconductance regions of the transfer characteristic. The output power vs frequency was displayed on a spectrum analyzer (Fig. 10.13). Note that the amplitude of the fifth harmonic is much larger than that of the fourth and the sixth. The efficiency of the multiplier (power ratio of the fifth harmonic to the fundamental) is $\approx$ 15 %, which is

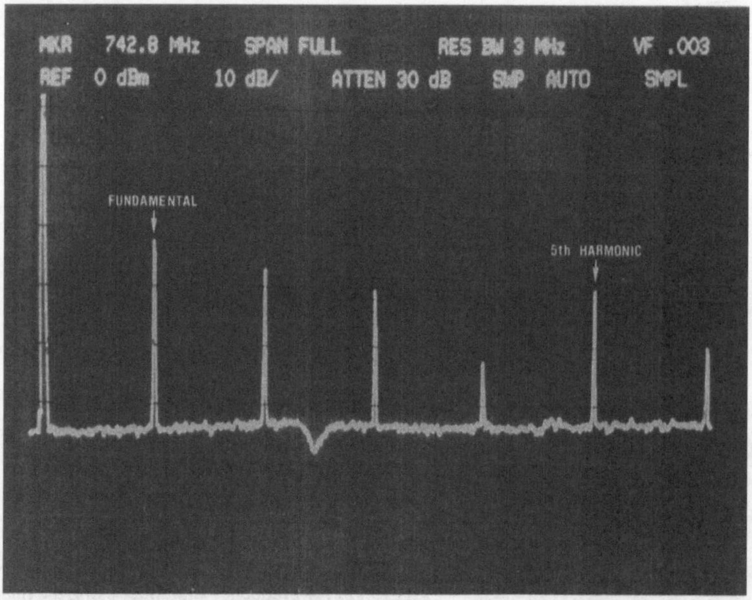

Fig. 10.13. Power output spectral response of the RTBT used as a frequency multiplier. Vertical scale is −10 dB/div. measured from the top horizontal line (0 dBm reference). The frequency span is 1.8 GHz (180 MHz/div.)

close to the maximum achievable (20 %) in a resistive multplier ($1/n$, where n is the harmonic under consideration).

10.3.2 Parity Generator

Figure 10.14a shows a four-bit parity generator circuit employing the multiple-state RTBT [10.34]. The voltages of the four input bits of the digital word are added up at the base node of the transistor by the resistive network, to generate a step-like waveform. The quiescent bias of the transistor, adjusted by the resistance R_{B1}, and the values of the resistances R_0 are chosen to select the operating points of the transistor alternately at low and high collector current levels (i.e., valleys and peaks of the transfer characteristics) at the successive steps of the summed up voltage. In the circuit, $R_0 = 15\,\mathrm{k}\Omega$, $R_{B1} = 6.9\,\mathrm{k}\Omega$, $R_{B2} = 2.4\,\mathrm{k}\Omega$, $R_C = 15\,\Omega$ and $V_{CC} = 4.5\,\mathrm{V}$. The output voltage at the collector would thus be high or low depending on the number of input bits set high being even or odd,

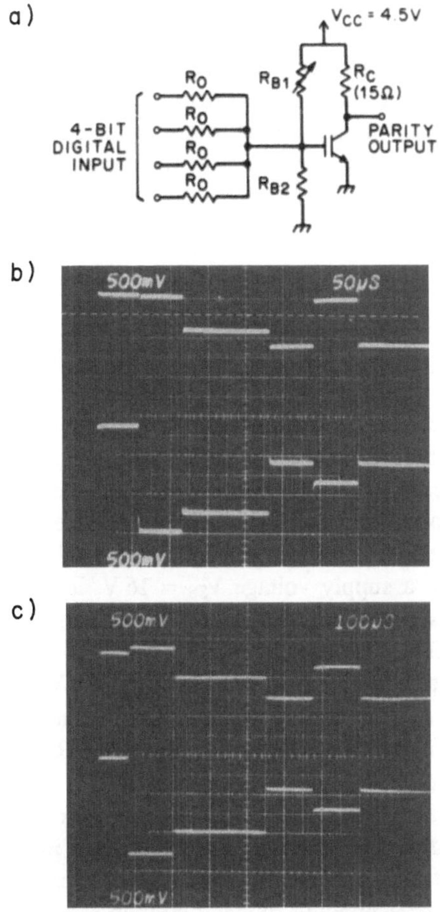

Fig. 10.14. (a) Four-bit parity generator circuit using a RTBT ($R_0 = 15\,\mathrm{k}\Omega$, $R_{B1} = 6.9\,\mathrm{k}\Omega$, $R_{B2} = 2.4\,\mathrm{k}\Omega$ and $R_C = 15\,\Omega$). (b) Collector (*top trace*) and base (*bottom trace*) waveforms in parity generator circuit at 77 K. (c) Collector (*top trace*) and base (*bottom trace*) waveforms at 300 K

respectively. Thus we obtain a 4-bit parity generator using only one transistor as compared to 24 needed in an optimized conventional circuit using three exclusive ORs. Also note that the 4-bit binary data are first converted to a multistate signal that is then processed by the device. This is equivalent to processing all four bits in parallel, which results in improved speed compared to conventional sequential processing of binary logic. Such multistate processing elements thus show potential in replacing clusters of circuits in existing binary logic systems. Parity generators using horizontally [10.24] and vertically integrated [10.35] RT diodes were demonstrated before. The advantage of the present circuit is that a separate summing amplifier is not required, resulting in further reduction in complexity.

To test the circuit, a pseudo-random sequence of 4-bit binary words was used rather than a monotonically increasing staircase waveform [10.35] since the latter does not take into account the effect of any hysteresis in the I-V characteristics. The train of input data produced both positive and negative steps at the base of the transistor. Experimental results at 77 K and 300 K are shown in Fig. 10.14b and c respectively, where the top traces show the output waveforms and the bottom traces the base waveforms of the transistor. Considering the dotted line in the upper trace as a logic threshold level, we find that the output is low for the second and the fourth voltage levels at the base while it is high for the others. It may also be noted that, at room temperature, the differential transconductance of the device decreases appreciably at higher voltages, making the design of the circuit more critical.

10.3.3 Multistate Memory

A suitable load line drawn on an I-V characteristic with n peaks will intersect the latter at $n+1$ ponts in the positive slope part, as illustrated in Fig. 10.15 in the case of two peaks [10.24]. Thus the circuit shown in the inset of Fig. 10.15 will have $n+1$ stable operating points and hence can be used as a memory element in an $n+1$ state logic system. Even in a binary computer, the storage system could be built around an $n+1$ logic to increase the packing density, and the data converted to and from binary at the input/output interface. This scheme has been demonstrated using the horizontally integrated RT structures exhibiting two peaks in the I-V curve described in [10.24]. With a supply voltage V_{SS} = 16 V, load resistance R_L = 215 Ω and the device biased to V_{BA} = 0.7 V, the three stable states were found at 3.0 V, 3.6 V, and 4.3 V. The corresponding load line at V_{BA} = 0.7 V intersects the measured characteristics of the device at 2.8 V, 3.4 V, and 4.1 V, respectively, which are in close agreement with the measured values of the three stable operating points. Similar memory cells utilizing triple-well RT diodes have been demonstrated by *Tanoue* et al. [10.36].

The three-state memory cell discussed above is also suitable for integration in memory ICs with read/write and decoding networks laid out as shown in Fig. 10.16. The memory cells are placed in a matrix array and a particular element in the array is addressed by activating the corresponding row and column select

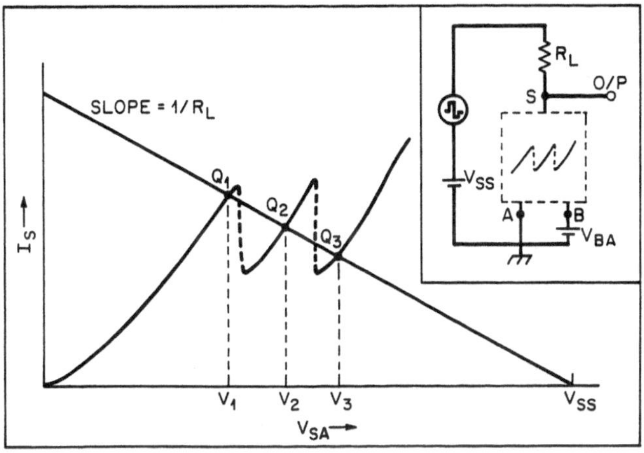

Fig. 10.15. Schematic of a 3-state memory cell. The load on the I-V characteristic shows three stable operating points Q_1, Q_2, and Q_3

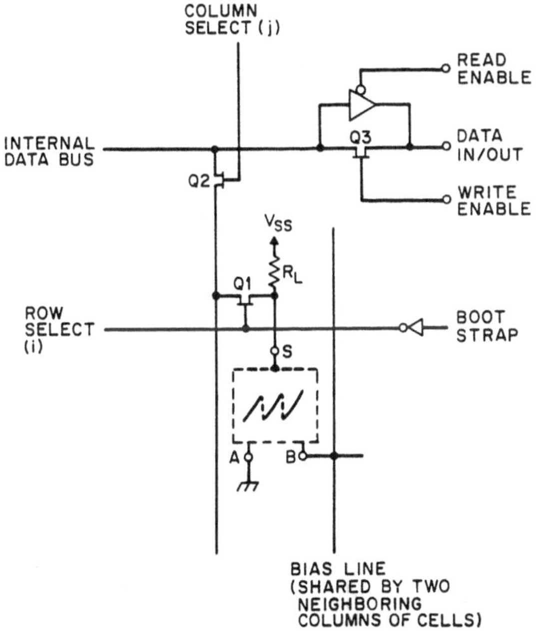

Fig. 10.16. Typical layout of an IC using the 3-state memory cells

lines. A row select connects each device in that row to the corresponding column lines. The column select finally connects the selected column to the data bus. Consider the element (ij) of the memory matrix shown in Fig. 10.16. When the row select line is activated, it turns the driving switch $Q1$ on. It also turns on the switches for every element in the ith row. The column select logic now connects the jth column only to the data bus. The ternary identity cell T acts as the buffer between the memory element and the external circuit for reading data. For reading data from the memory, the identity cell is activated with the read enable line, and the datum from the (i, j)th element in the matrix goes, via the data bus, to the in/out pin of the IC. When the write enable line is activated, the datum from the external circuit is connected ot the data bus and is subsequently forced on the (i, j)th element in the array and is written there.

10.3.4 Analog-to-Digital Converter

Among the circuit applications of multiple-state RTBTs, the analog-to-digital converter, briefly mentioned in [10.4] and shown in Fig. 10.17, is potentially the most significant. The analog input is simultaneously applied to an array of RTBT circuits having different voltage scaling networks. To understand the operation of the circuit, consider the simplest system, comprising only the two transistors Q_1 and Q_2. The voltages at different points of this circuit are shown in Fig. 10.18a, for various input voltages V_i. Consider that the resistances R_0, R_1 and R_2 are so chosen that the base voltages V_{B1} and V_{B2} of the transistors Q_1 and Q_2 vary with V_i according to the curves V_{B1} and V_{B2}, respectively. With the input voltage at V_1, the output of both transistors will be at the operating point P_1 (hi-state). With the input changing to V_2, the output of Q_1 will become low (P_2), while that of Q_2 will still remain high (closer to P_1). Applying this logic to the input voltages V_3 and V_4, it can be easily shown that this circuit indeed follows the truth table of Fig. 10.18b. The outputs of the RTBT array thus constitute a binary code representing the quantized analog input level. The system can be extended to more bits with a larger number of peaks in the I-V characteristic. Note that this is a flash converter requiring only n transistors for n-bit conversion as compared to 2^n analog comparators in conventional flash

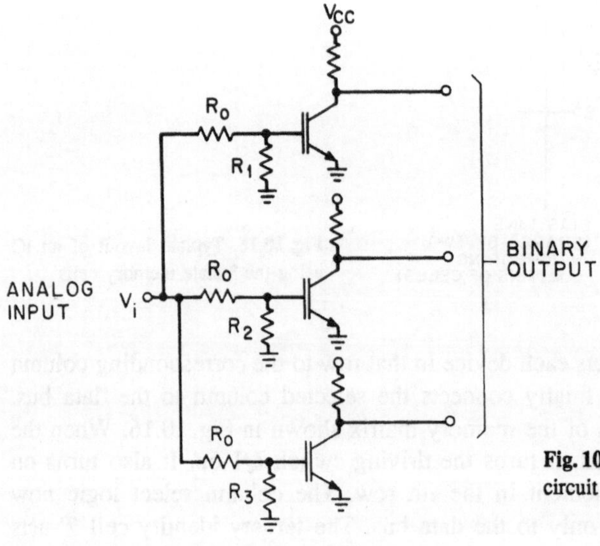

Fig. 10.17. Analog-to-digital converter circuit using multiple-state RTBTs

(a)

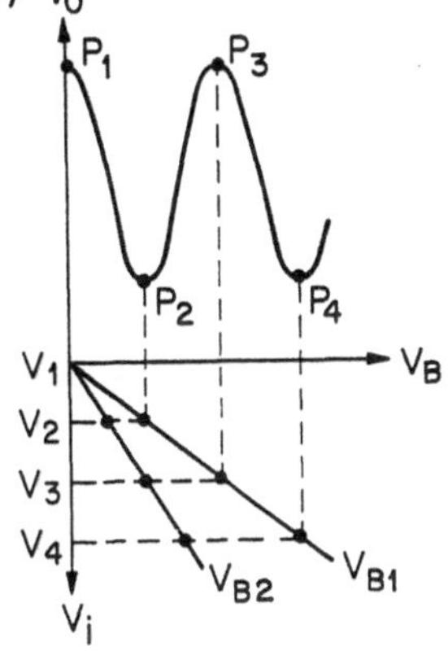

(b) TRUTH TABLE

INPUT	OUTPUT	
V_i	Q_2	Q_1
V_1	1	1
V_2	1	0
V_3	0	1
V_4	0	0

Fig. 10.18. Schematic operation of the analog-to-digital converter circuit of Fig. 10.17, involving only 2 bits: **(a)** the voltages at different points of the circuit at various input voltages; **(b)** the truth table

converters. Furthermore, the RTBTs in the array not only work as comparators, but also give the digital output directly, eliminating the 2^n-to-n bit decoder needed in conventional circuits. This further reduces the circuit complexity and enhances the speed of operation. However, it is very difficult to implement this circuit with present RTBTs. Successful operation of the circuit relies on a transfer characteristic where the current remains at a high or low level for a significant span of the base–emitter voltage, while in the multiple-state RTBTs implemented so far, the current increases gradually with the input voltage and then suddenly drops. The drop is followed by another gradual rise.

It should be mentioned that in all these circuits the minimum allowable collector voltage is determined by the maximum input signal voltage applied at the base terminal. This is actually higher than in normal bipolar transistors. In fact, in the multiple-state RTBTs demonstrated so far, the QWs are positioned between the base and the emitter contact so that the applied base-to-emitter voltage is used to bias the emitter pase p-n junction *and* the QWs. The base potential is then elevated to a relatively high value under operating conditions (in the common-emitter mode). As a result, the quiescent collector bias must be large in order to allow for sufficient output signal swing without forward biasing the collector–base junction. This requires proper care on the part of the circuit designer and careful device design so as to achieve a sufficiently high breakdown voltage.

10.4 Gated Quantum Well and Superlattice-Base Transistors

Other than the RHET, several unipolar three-terminal devices have been proposed and implemented utilizing RT structures as electron injectors to generate voltage-tumbel NDR and negative transconductance characteristics. These include the quantum wire transistor, a device in which the resonant tunneling involves two-dimensional electrons injected into a one-dimensional quantum well [10.37], and the Resonant Tunneling gate Field-Effect Transistor (RT-FET) [10.38–40]. The integration of RT diodes and FETs [10.41–43] and their circuit applications [10.44] have also been demonstrated.

10.4.1 Gated Quantum Well Transistor

Band-gap engineering allows the utilization of RT also in other transistor structures. One such structure is the gated QW transistor [10.45]. This is the first transistor in which negative transconductance was achieved by directly controlling the potential of the QW. The structure was proposed by *Bonnefoi* et al. under the name Stark effect transistor [10.46]. The key ideas of the transistor were the use of a QW collector and the inverted sequence of layers in which the controlling electrode (here referred to as the *gate* [10.45]) was placed "behind" the collector layer. It was predicted that the gate field would modify the position of the collector subbands with respect to the emitter Fermi level, and thus modulate the tunneling current. As demonstrated in [10.45], the structure offers additional advantages, namely NDR and negative transconductance. Moreover the operation of the device is only partly goverened by the Stark effect. In fact another mechanism, the *quantum capacitance* [10.47], is essential for its operation.

The device grown by MBE in the AlGaAs material system consisted of an undoped quantum well collector 120 Å thick to which contact was made. This layer was separated from the n^+-doped emitter by a 40-Å-thick undoped AlAs tunneling barrier. On the other side of the collector a 1200-Å-thick undoped AlAs barrier was followed by the n^+ gate. The nominal doping of the 5000-Å-thick n^+ layers was $2 \times 10^{18}\,\mathrm{cm}^{-3}$. The energy diagram of the device is sketched in Fig. 10.19 in the common-collector configuration.

The emitter-collector I-V characteristics of the device are expected to peak at biases which maximize RT of the emitter electrons into the 2-D collector

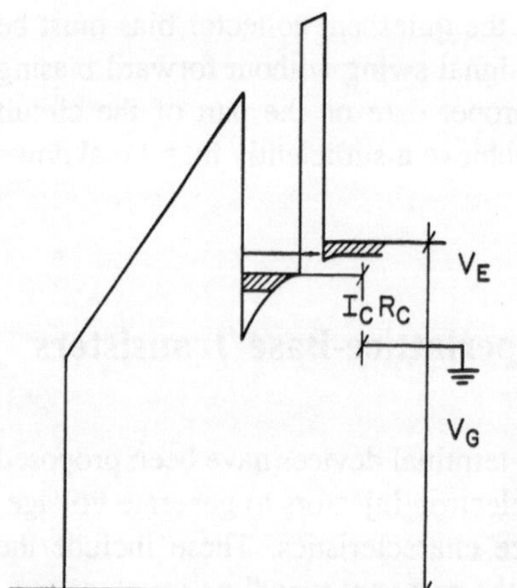

Fig. 10.19. Band diagram of the gated quantum well resonant tunneling transistor with the collector at reference and the biases $V_G > 0$ and $V_E < 0$ corresponding to peak resonant tunneling of emitter electrons into the second subband of the well

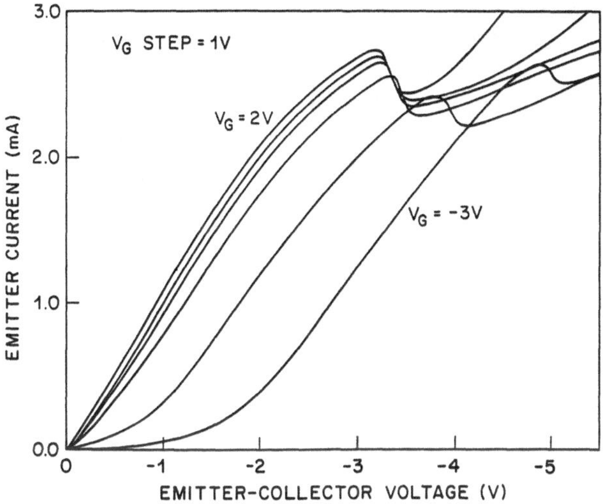

Fig. 10.20. Common-collector characteristics of the resonant tunneling transistor of Fig. 10.19 at various V_G $(2, 1, 0, -1, -2, -3V)$. The measurements were performed at 7 K

subbands. Transistor action in the structure is obtained via the influence of the gate field on the alignment of the 2-D electron gas energy levels relative to the emitter Fermi level. This occurs, as anticipated above, for the combined action of the generalized Stark effect and the quantum capacitance effect. This contribution to the capacitance, not present in a classical metal, arises from the energy that has to be spent in order to raise the Fermi energy in the well, as the carrier concentration is increased by the increasing gate field. This causes the gate field to penetrate beyond the 2-D metal in the quantum well and induce charges on the emitter electrode [10.47].

In Fig. 10.19 the band-diagram of the device is shown in the common-collector configuration with applied biases $V_B > 0$ and $V_E < 0$ such that the bottom of the conduction band in the emitter is in resonance with the second collector subband; this corresponds to a peak in the current. The RT current can be subsequently reduced by increasing either V_E (in modulus) or V_G. The former leads to the observation of NDR, the latter of negative transconductance.

Experimental data of cryogenic temperature are shown in Fig. 10.20. The expected features are indeed present, and were observed up to liquid nitrogen temperature, although less pronounced. In particular the data correspond to a transconductance value of the order of ≈ 1 mS. This device has the advantage of a negligible gate current (it is always several orders of magnitude smaller than the emitter current), which gives a large current transfer ratio, but suffers from the drawback of a relatively small transconductance.

10.4.2 Superlattice-Base HBT

Negative transconductance can also be obtained using suitably designed minibands in the superlattice base of a transistor. The emitter is degenerately doped so that electrons can be injected by tunneling into the miniband (Fig. 10.21a). When the base–emitter voltage exceeds the bias required to line up the bottom of the conduction band in the emitter with the top of the miniband (Fig. 10.21b) the collector current is expected to drop. This negative transconductance arises in a straightforward manner from the conservation of lateral momentum and energy during tunneling into the miniband, similarly to what happens in RTDBs. This effect has been observed in an InP/GaInAs superlattice HBT [10.48].

The structure, grown by chemical beam epitaxy, was grown on an n^+ InP substrate. A 5000 Å $Ga_{0.47}In_{0.53}As$ buffer layer ($n = 5 \times 10^{17}\,cm^{-3}$) was followed by an undoped n-type $Ga_{0.47}In_{0.53}As$ 1.8 μm thick collector. The base consists of a p^+ ($2 \times 10^{18}\,cm^{-3}$) $Ga_{0.47}In_{0.53}As$ 500 Å thick region, adjacent to the collector layer, followed by a 20-period $Ga_{0.47}In_{0.53}As$ (70 Å)/InP (20 Å) superlattice. The barrier layers are undoped while all the GaInAs wells are heavily doped ($2 \times 10^{18}\,cm^{-3}$) p-type. A 20 Å undoped InP doping set back layer separates the superlattice from the 5000 Å thick n^+ ($\approx 2 \times 10^{18}\,cm^{-3}$) InP emitter. This superlattice design ensures the formation of relatively wide minibands, in order to guarantee Bloch conduction of injected electrons through the base. The calculations show that the ground state electron miniband extends from 41 to 96 meV,

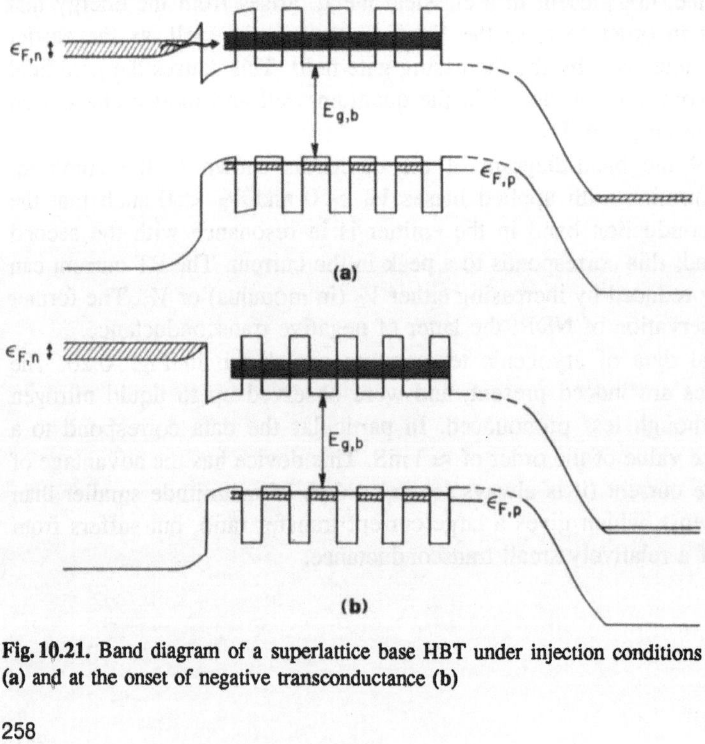

Fig. 10.21. Band diagram of a superlattice base HBT under injection conditions into the miniband (a) and at the onset of negative transconductance (b)

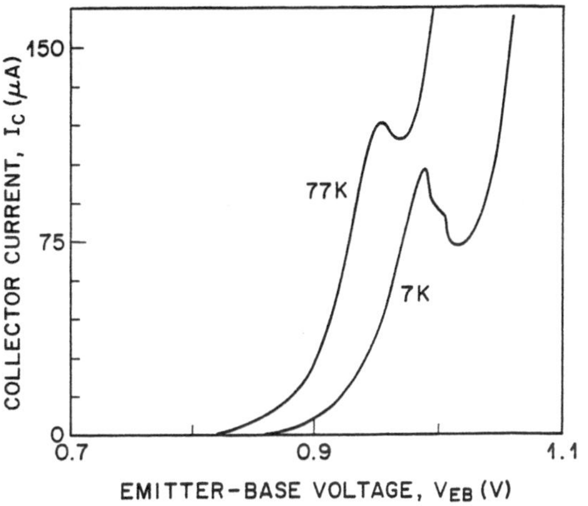

Fig. 10.22. Transfer characteristic of the superlattice base HBT in the common-base configuration at two different temperatures

while the heavy hole miniband lextends from 11.9 meV to 12 meV. Energies are measured from the classical bottom of the conduction and valence band wells, respectively. Conduction band nonparabolicity was included in these envelope-function-type calculations; $\Delta E_c = 0.23$ eV and $\Delta E_v = 0.39$ eV were used for the band discontinuities.

Figure 10.22 shows the common base transfer characteristics at 7 K and 77 K. Consider first the collector current at 7 K. At such low temperatures thermionic emission from the emitter is completely negligible, and injection from the emitter is dominated by tunneling. From the band diagram (Fig. 10.21) it is clear then that no electron current can flow from the emitter into the base until the quasi Fermi energy in the emitter is lined up with the bottom of the miniband. This requires a bias V_{EB} equal to $(E_{F,p} + E_{1,hh} + E_{g,b} + E_{1,e})/e = 0.88$ V, where $E_{F,p}$ is the Fermi energy in the base (= 15 meV), $E_{1,hh}$ the bottom of the heavy hole miniband (= 12 meV), $E_{g,b}$ the GaInAs bulk bandgap (= 0.812 eV at 7 K) and $E_{1,e}$ the bottom of the first electron miniband (= 41 meV). The data of Fig. 10.22 indeed show that conduction starts at $V_{BE} \approx 0.87$ V, in excellent agreement with the calculated onset voltage. Below the onset the measured current is $I_C \approx 5$ pA, which is very close to the detection limit of the parameter analyzer. The suppression of injection into the miniband requires increasing the emitter base voltage by $> \Delta E_1 + E_{F,n} = 135$ meV, where $\Delta E_1 = 55$ meV is the width of the first electron miniband and $E_{F,n} = 80$ meV the quasi Fermi energy in the emitter. Thus the onset of negative transconductance (voltage position of the peak) is expected to be at 1.015 V, in excellent agreement with the experimental value (approximately 1.00 V). Following the drop in the collector current, the latter rises rapidly for $V \geq 1.02$ V. This is expected, since at a bias $\approx (E_{g,b} + \Delta E_c)/e \approx 1.04$ V, the conduction band edge in the emitter becomes flat, leading to a steep increase in the injection efficiency. Note that at 77 K the peak shifts to a lower voltage. The shift ($\approx$ 25 mV) is close, as expected, to the GaInAs bandgap lowering ($\approx$ 30 meV) as the temperature is varied from 7 to 77 K.

The common-base characteristics I_C, I_E vs V_{EB} show a maximum base transport factor $\alpha = I_C/I_E = 0.76$ at $V_{BE} \approx 0.98$ V. This value is consistent with the maximum gain $\beta = I_C/I_B \approx 3.2$ measured in the common-emitter configuration. These values of α and β, although far from optimal, indicate that transport in the base is via miniband conduction rather than hopping. Previously *Palmer* et al. had reported a HBT with an AlGaAs/GaAs superlattice base [10.49]. Although miniband conduction in the base was demonstrated, no negative transconductance was shown since the structure did not use a tunneling emitter for injection.

10.4.3 Unipolar Superlattice-Base Transistor

Negative transconductance can also be achieved by controlling injection into minibands above the top of the barriers. Recently *Lent* proposed a tunneling emitter transistor in which hot electrons transfer through the base by miniband conduction in a continuum state [10.50]. Here we present the operation of a superlattice-base unipolar transistor in which electrons are injected into a miniband in the classical continuum [10.51].

The structure, whose equilibrium conduction band energy diagram is sketched in Fig. 10.23a, was grown by MBE. It consisted of a 8000 Å n^+ collector followed by an undoped $Al_xGa_{1-x}As$ layer 5000 Å thick with x varying from 0 to 0.25. The SL base comprised 5.5 periods of 40 Å n^+ GaAs/200 Å undoped $Al_{0.31}Ga_{0.69}As$. An undoped $Al_xGa_{1-x}As$ injector layer 500 Å thick was then grown with x varying from 0.33 to 0 (corresponding to a band discontinuity ΔE_c = 273 meV, roughly at the bottom of the chosen miniband). Finally an n^+ emitter layer 3000 Å topped the structure. In all of the doped layers $n = 2 \times 10^{18}\,\mathrm{cm}^{-3}$.

The operation of the device is easily understood with the help of Fig. 10.23, where the common-base operation mode is illustrated. At a fixed positive collector–base bias, the negative emitter–base bias (V_{EB}) is increased and the

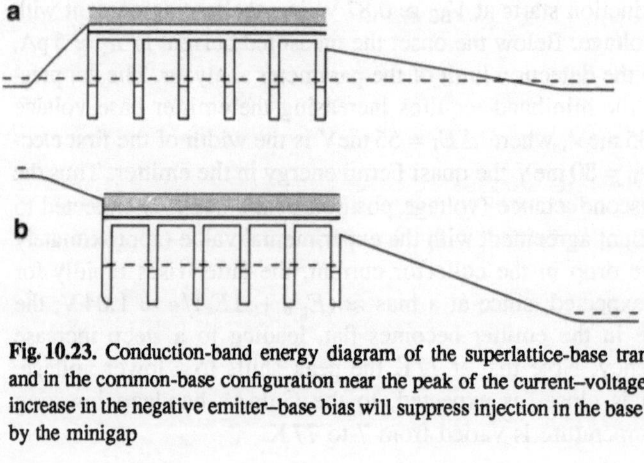

Fig. 10.23. Conduction-band energy diagram of the superlattice-base transistor: at equilibrium (a), and in the common-base configuration near the peak of the current–voltage characteristic (b). Further increase in the negative emitter–base bias will suppress injection in the base due to quantum reflections by the minigap

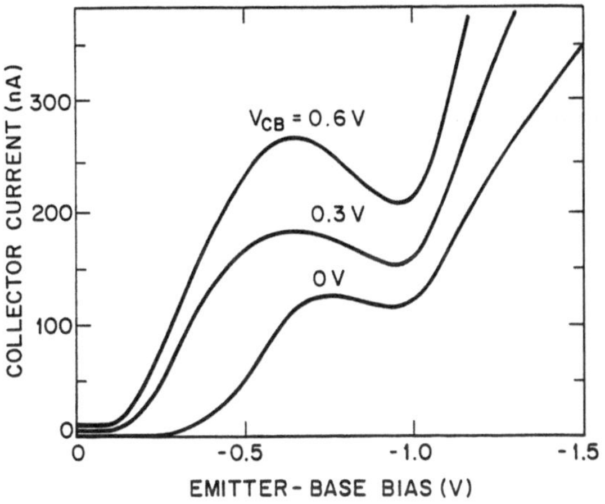

Fig. 10.24. Collector current as a function of emitter–base bias at fixed collector–base (V_{CB}) bias at 30 K. The curves shown are for $V_{CB} = 0$, 0.3, 0.6 V

collector current is measured. Because of the appropriately tailored compositionally graded emitter barrier, electrons are injected into the third miniband. The energy dispersion of this miniband was calculated in the envelope-function approximation, taking into account band nonparabolicities, and was found large enough (≈ 23 meV) to guarantee miniband conduction. Increasing V_{EB} will further flatten the triangular injector and increase the injection current. However, part of the bias will appear in a depletion region in the base, thus shifting the top of the band discontinuity with respect to the miniband. When this shift is larger than the miniband width, the injected electrons will not be able to satisfy the energy and lateral momentum conservation conditions across the interface and will experience strong quantum mechanical reflections. Consequently, injection efficiency will drop together with the collector current and the I-V characteristic will exhibit negative transconductance. This effect is shown in the experimental curves shown in Fig. 10.24. In fact a 23 meV shift (of the order of the miniband width) of the top of the injector band discontinuity is required to suppress electron injection. This corresponds to an almost total depletion of the first well of the base, and gives (by a simply electrostatic computation) an emitter–base bias of ≈ 0.6 V, which is in agreement with the peak position observed for $V_{CB} > 0$ in Fig. 10.24.

In the structure examined the base transport factor is $\ll 1$, leading to a small I_C. Optimization of the design and elimination of stray leakage paths must be achieved in order to enhance α.

In general, the structures discussed in this section are primarily of interest as tools to investigate transport in two-dimensional systems and superlattices. Their operation has only been demonstrated at cryogenic temperatures. In terms of sheer performance, these devices have several shortcomings in comparison with the advanced RTBTs described in the previous section.

10.5 Quasi-Electric Fields in Graded-Gap Materials

Kroemer [10.52] first considered the problem of transport in a graded-gap semiconductor. As a result of compositional grading, electrons and holes experence "quasi-electric" fields, F, of different intensities,

$$F_e = -\frac{dE_c}{dz}, \qquad F_h = +\frac{dE_v}{dz}, \tag{10.4}$$

where $E_c(z)$ and $E_v(z)$ are the conduction and valence band edges, respectively. The forces resulting from these fields push electrons and holes in the same direction. This is illustrated in Fig. 10.25a for the case of an intrinsic material. Such a graded material can be thought of as a stack of many isotype heterojunctions of progressively varying band gap. If the conduction and valence band edge discontinuities ΔE_c and ΔE_v of such heterojunctions are known and relatively independent of the alloy compositions (as in the case of $Al_xGa_{1-x}As$ heterojunctions), then one can expect that for the structur in Fig. 10.25a the ratio of the quasi-electric fields F_e/F_h will be equal to $\Delta E_c/\Delta E_v$.

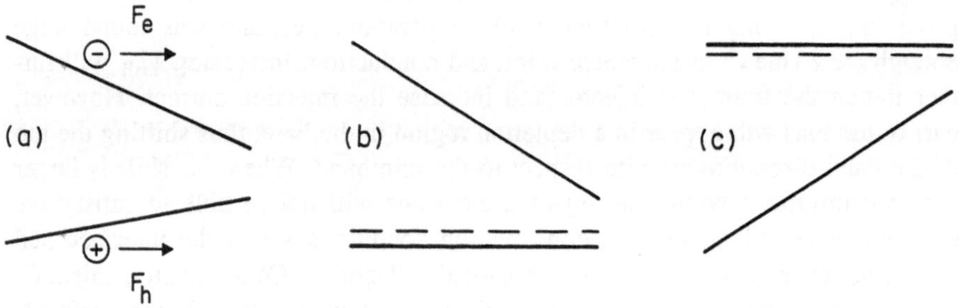

Fig. 10.25. Energy band diagram of compositionally graded materials: (a) intrinsic, (b) *p*-type, and (c) *n*-type

For a *p*-type graded-gap material the situation is different; the energy band diagram is shown in Fig. 10.25b. The valence band edge is now horizontal so no effective field acts on the holes, while the effective field for the electrons is $F_e = -dE_g/dz$, which can be significantly greater than in the intrinsic case. In other words, all the band gap grading is transferred to the conduction band. This can be interpreted physically using the following heuristic argument. Consider the effect of *p*-type doping on an initially intrinsic material of the type in Fig. 10.25a. The acceptor atoms will introduce holes, which under the action of the valence-band quasi-electric field will be spatially separated from their negatively ionized parent acceptor atoms. This separation produces an electrostatic (space-charge) field. Holes accumulate (on the right-hand side of Fig. 10.25b) until this space-charge field equals the magnitude of, and cancels, the hole quasi-electric field, F_h, thus achieving the thermodynamic equilibrium configuration (flat valence band) of Fig. 10.25b. Note, however, that as a result of this process the equilibrium hole density is spatially nonuniform. The electrostatic field of magnitude $|dE_v/dz|$ produced by the separation of holes and acceptors adds instead to the conduction band quasi-electric field to give a total effective field acting on an electron of

$$F_c = -\left(\frac{dE_c}{dz} + \frac{dE_v}{dz}\right) = -\frac{dE_g}{dz} . \tag{10.5}$$

Thus in a p-type material the conduction band field is made up of a nonelectrostatic (quasi-electric field) and an electrostatic (space-charge) contribution.

For an n-type material the same kind of argument can be applied to electrons. This yields the band diagram of Fig. 10.25c and to the same effective field acting on the hole as given by (10.5). Consider, for example, the case of an $Al_xGa_{1-x}As$ graded-gap p-type semiconductor. If we assume that in an $Al_xGa_{1-x}As$ heterojunction 62 % of the band gap difference is in the conduction band, it follows that 62 % of the effective conduction band field $F = -dE_g/dz$ will be quasi-electric in nature and the rest (38 %) electrostatic. The opposite occurs in the case of an n-type $Al_xGa_{1-x}As$ graded material, where 62 % of the valence band effective field is electrostatic in nature.

So far we have only considered quasi-electric fields arising from band gap grading. When the composition of the alloy is changed, however, the carrier effective masses m_c^* and m_h^* also change, giving rise to additional quasi-electric fields for electrons and holes [10.53].

The quasi-electric fields in direct-band-gap graded-composition $Al_xGa_{1-x}As$ are primarily due to band gap grading; the quasi-electric fields due to the effective mass gradients being negligible in this case [10.53]. However, effective mass gradients can make a substantial contribution to the quasi-electric field for $Al_xGa_{1-x}As$ graded materials in which the composition x is varied through the direct–indirect transition at $x = 0.45$. This is because the effective mass of the electron varies by about one order of magnitude in the direct–indirect transition region. Similar considerations also apply to other III-V alloys.

10.5.1 Electron Velocity Measurements

Quasi-electric fields are particularly important because they can be used to enhance the velocity of minority carriers that would otherwise move by diffusion (a relatively slow process) rather than by drift. *Kroemer* [10.52] first proposed the use of a graded-gap p-type layer (Fig. 10.25b) for the base of a bipolar transistor to reduce the minority carrier (electron) transit time in the base.

Levine et al. [10.54, 55], using an all-optical method, measured for the first time the electron velocity in a heavily p^+-doped compositionally graded $Al_xGa_{1-x}As$ layer grown by MBE. The energy band diagram of the sample and the principle of the experimental method are sketched in Fig. 10.26. The measurement technique was a "pump and probe" scheme. The pump laser beam, transmitted through one of the AlGaAs window layers, is absorbed in the first few thousand angstroms of the graded layer. Optically generated electrons un-

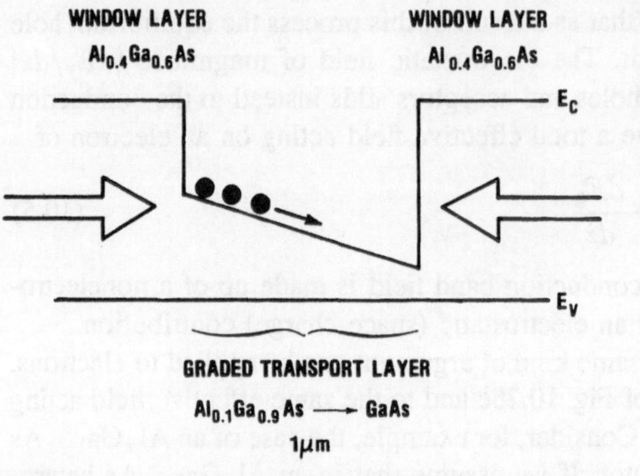

Fig. 10.26. Band diagram of the sample used for electron velocity measurements and schematic illustration of the pump-and-probe measurement technique

der the influence of the quasi-electric field drift towards the right in Fig. 10.26 and accumulate at the end of the graded layer. This produces a change in the refractive index at the interface with the second window layer. This refractive index variation produces a reflectivity change that can be probed with a counter-propagating laser beam. This reflectivity change is measured as a function of the delay between pump and probe beams using phase-sensitive detection techniques. The reflectivity data obtained in [10.54] for a sample with a 1-μm-thick transport layer graded from $Al_{0.1}Ga_{0.9}As$ to GaAs and doped to $p \approx 2 \times 10^{18}\,cm^{-3}$ are shown in Fig. 10.27. The grading corresponds to a quasi-electric field of 1.2 kV/cm. The laser pulse width was 15 ps, and the time zero in Fig. 10.27 represents the center of the pump pulse as determined by two-photon absorption in a GaP crystal near the sample. The approximate transit time is given by the shift of the half height of the reflectivity curve from zero: $\tau = 33\,ps$. The drift

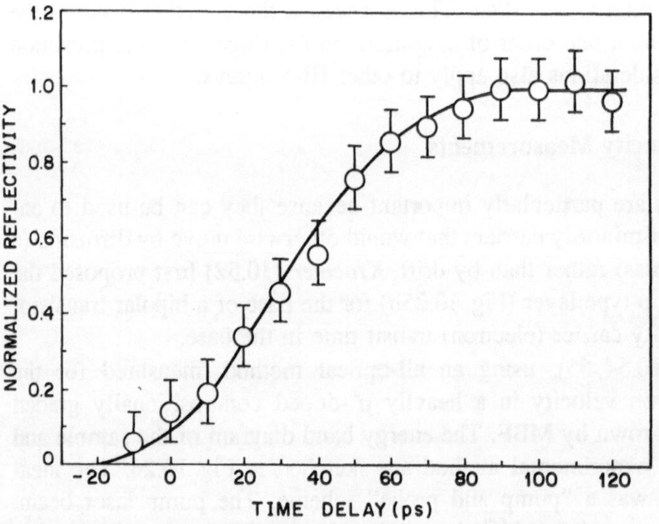

Fig. 10.27. Normalized experimental results for pump-induced reflectivity change vs time delay obtained in 1 μm thick, graded-gap, p^+ AlGaAs at a quasi-electric field $F = 1.2\,kV/cm$

length was taken as the thickness of the graded layer minus the absorption length of the pump beam ($1/\alpha \approx 2500$ Å). The resulting minority carrier velocity was $v \approx (L - \alpha^{-1})/\tau \approx 2.3 \times 10^6$ cm/s. In this relatively thick sample, carrier diffusion was important and caused a spread in the electron arrival time at the end of the sample, which is roughly the risetime of the reflectivity curve from 10 % to 90 %, i.e., 63 ps. It is interesting to note that the drift mobility obtained from the measurement is $\mu_d = v_e/F = 1900\,cm^2/V\,s$, which is comparable with the usual mobility of $2200\,cm^2/V\,s$ at the doping level of the graded layer in GaAs.

Electron velocity measurements were also made in a 0.42-μm-thick, strongly graded ($F_e = 8.8$ kV/cm), highly doped ($p = 4 \times 10^{18}\,cm^{-3}$) $Al_xGa_{1-x}As$ layer graded from $Al_{0.3}Ga_{0.7}As$ to GaAs. A transit time of only 1.7 ps was measured, more than an order of magnitude shorter than that for $F = 1.2$ kV/cm, and corresponding to a velocity $v_e \approx 2.5 \times 10^7$ cm/s [10.55]. The velocity can be obtained rigorously and accurately (within ± 10 %) from the reflectivity data by solving the drift-diffusion equation and taking into account the effects of the pump absorption length (especially important in the thin sample) and the partial penetration of the probe beam into the graded material. If one includes all of these effects, one finds that the reflectivity data can be fitted using only one adjustable parameter, the electron drift velocity [10.55]. The velocity was found to be $v_e = 2.8 \times 10^6$ cm/s for $F = 1.2$ kV/cm and $p = 2 \times 10^{18}\,cm^{-3}$, and $v_e = 1.8 \times 10^7$ cm/s for $F = 8.8$ kV/cm and $p = 4 \times 10^{18}\,cm^{-3}$.

We see that when we increase the quasi-electric field from 1.2 to 8.8 kV/cm (a factor of 7.3) the velocity increases from 2.8×10^6 to 1.8×10^7 cm/s (a factor of 6.4). That is, we observe that the relation $v = \mu F$ is approximately valid. Using $\mu = 1700\,cm^2/V\,s$ (for $p = 4 \times 10^{18}\,cm^{-3}$) we calculate $v = 1.5 \times 10^7$ cm/s for $F = 8.8$ kV/cm, which is in reasonable agreement with the experimental result. The measured velocity of 1.8×10^7 cm/s (in the quasi-electric field) is significantly larger than that obtainable in undoped GaAs ($v = 1.2 \times 10^7$ cm/s for an ordinary electric field of $F = 8.8$ kV/cm). Our measured velocity is comparable to the peak velocity reached in GaAs for $F = 3.5$ kV/cm before the transfer from the Γ to the L valley occurs and to the maximum possible phonon-limited velocity in the Γ minimum of GaAs. The latter is given by $v_{max} = [(E_p/m^*)\tanh(E_p/2k_BT)]^{1/2} = 2.3 \times 10^7$ cm/s, where $E_p = 35$ meV is the optical phonon energy and the effective mass $m^* = 0.067 m_0$.

This high velocity can be understood without reference to transient effects because the transit time is much larger than the momentum relaxation time of 0.3 ps. The high velocity results from the fact that the electrons spend most of their time in the high velocity central Γ valley rather than in the low velocity L valley. This may result from the injected electron density being so far below the hole doping density that strong hole scattering can rapidly cool the electrons without excessively heating the holes. Furthermore, the electrons remain in the Γ valley throughout their transit across the graded layer since the total conduction band edge drop ($\Delta E_g = 0.37$ eV) is comparable to the GaAs Γ–L separation ($\Delta E_{\Gamma L} = 0.33$ eV), so that electrons do not have sufficient excess energy for significant transfer to the L valley.

10.6 Heterojunction Bipolar Transistors with Graded-Gap Layers

10.6.1 High-Speed Graded-Base Transistors

The first device to utilize the high electron velocity found in p-type graded materials was a photo-transistor [10.56] with an AlGaAs graded-gap base with a quasi-electric field of about 10^4 kV/cm. The device was grown by MBE on a Si-doped ($\approx 4 \times 10^{18}\,\mathrm{cm}^{-3}$), n^+-type GaAs substrate. A buffer layer of n^+-type GaAs was grown first, followed by a Sn-doped, n-type ($\approx 10^{15}\,\mathrm{cm}^{-3}$), 1.5-$\mu$m-thick GaAs collector layer. The 0.45-μm-thick base layer was compositionally graded from GaAs (on the collector layer side) to $Al_{0.20}Ga_{0.80}As$ (E_g = 1.8 eV) and was heavily doped with Be ($p^+ \approx 5 \times 10^{18}\,\mathrm{cm}^{-3}$). The abrupt wide gap emitter consisted of an $Al_{0.45}Ga_{0.55}As$ (E_g = 2.0 eV), 1.5-μm-thick, window layer with Sn in the range $n = 2\text{–}5 \times 10^{15}\,\mathrm{cm}^{-3}$. Figure 10.28b shows the energy band diagram of the phototransistor.

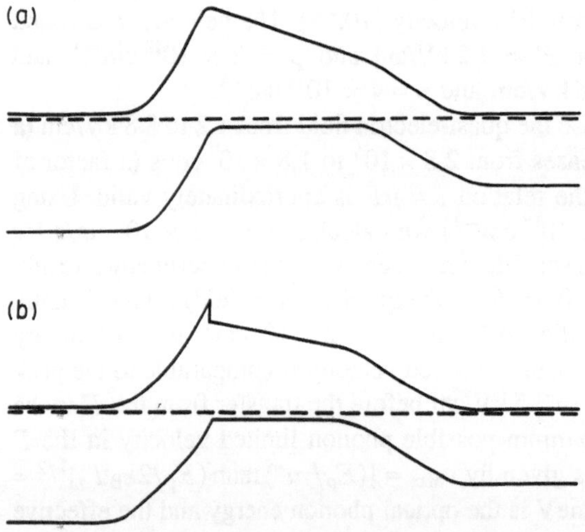

Fig. 10.28. Band diagram of a graded-gap base bipolar transistor: (a) with graded emitter–base interface, and (b) with ballistic launching ramp for even higher velocity in the base

To study the effect of grading in the base on the speed of the device, 4-ps laser pulses were used. The wavelength (λ = 6400 Å) was chosen so that the light could only be absorbed in the base layer. The incident power was kept relatively high (100 mW) to minimize the effective emitter charging time. Under these conditions the speed-limiting factors are the RC time constant and the base transit time. Figure 10.29 shows the pulse response of the device as monitored by a fast sampling scope (the response was signal averaged; note the symmetrical rise and fall time and the absence of long tails, which are normally very difficult to achieve in picosecond photodetectors). From the observed 10 %–90 % response time of 30 ps, a sum-of-squares approximation was used to estimate an intrinsic

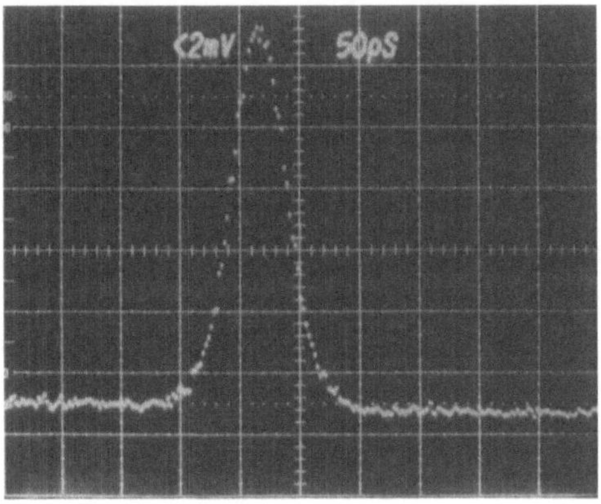

Fig. 10.29. Pulse response of graded-gap-base AlGaAs/GaAs phototransistor to a 4-ps laser pulse displayed after signal averaging the sampling scope signal

detector response time of about 20 ps. In the absence of a quasi-electric field in the heavily doped p^+ base, a broadened response followed by a tail with a square root of time dependence (due to slow diffusion) is expected. The diffusion time t_D is given by $W^2/2D$, where W is the base thickness and D is the diffusion coefficient. For a GaAs phototransistor with a base doped at $p^+ = 10^{18}\,cm^{-3}$, D is approximately $16\,cm^2/s$. In this structure D is likely to be smaller both because AlGaAs has a lower mobility than GaAs, and because of the higher doping; one estimates $t_D \gtrsim 50$ ps. The fact that the expected broadening is not observed indicates that the quasi-electric field in the base sweeps out the electrons in a time much shorter than the diffusion time. From the velocity measurements previously discussed we know that the base transit time is about 2 ps, which is indeed much less than t_D. Thus the pulse response of this device is the first experimental verification [10.56] of Kroemer's prediction [10.52].

Finally, the combination of the graded-gap base and the abrupt wide gap emitter (Fig. 10.28b) suggests a new high-speed ballistic transistor [10.56, 57]. The conduction band discontinuity can be used to ballistically launch electrons into the base with an initial velocity substantially higher than 10^7 cm/s. If no electric field were introduced in the base, ballistic launching alone, using the abrupt base emitter heterojunction, may not be sufficient to achieve a very high velocity in the base because collisions with plasmons or coupled plasmon-phonon modes in the heavily doped base would rapidly relax the initial forward momentum. For an initial high velocity it is sufficient that the conduction band discontinuity used for the launching be a few $k_B T$ (typically 50 mV at 300 K).

The first bipolar transistor with a compositionally graded base was reported by *Hayes* et al. [10.58] and *Miller* et al. [10.59]. Incorporation of a graded-gap base yields much shorter base transit times because of the induced quasi-electric field for electrons, which allows a valuable tradeoff against the base resistance. To understand this last point consider a base of width W linearly graded from one alloy with a band gap of E_{g1} to another with a band gap of E_{g2}. The quasi-electric field for electrons $(E_{g1} - E_{g2})/eW$ results in a base transit time (neglecting diffusion effects) of

$$\tau_b' \approx \frac{eW^2}{\mu(E_{g1} - E_{g2})} . \tag{10.6}$$

We have made use of the experimental fact that the velocity in the graded base nearly equals μF_e, where F_e is the quasi-electric field. This time must be compared with the diffusion-limited base transit time of a transistor with an ungraded GaAs base of the same thickness and doping level

$$\tau_b = \frac{W^2}{2D} , \tag{10.7}$$

where D is the ambipolar diffusion coefficient. If we compare (10.6) and (10.7) and use Einstein's relationship $D = \mu k_B T/e$, we find that the base transit time is shortened by the factor

$$\frac{\tau_b}{\tau_b'} = \frac{E_{g1} - E_{g2}}{2k_B T} , \tag{10.8}$$

using a graded-gap base. Although (10.8) is rigorously valid only in the limit $E_{g1} - E_{g2} \gg k_B T$, it can be employed as a useful "rule of thumb" in cases where $E_{g1} - E_{g2}$ is several times $k_B T$. Thus the band gap difference must be made as large as possible without exceeding the intervalley energy separation $\Delta E_{\Gamma L}$, which would greatly reduce the electron velocity. Using $E_{g1} - E_{g2} = 0.2\,\mathrm{eV}$, the transit time is reduced by a factor of about four at 300 K relative to an ungraded base of the same thickness. This allows a valuable tradeoff against the base resistance (R_b), since the base thickness can be increased to reduce R_b, while still keeping a reasonable base transit time. Finally, an added advantage of the quasi-electric field is the increased base transport factor that comes about because the short transit time reduces minority carrier recombination in the base.

Devices grown by MBE on an n^+ substrate [10.58] had a 1.5 μm GaAs buffer layer followed by a 5000-Å-thick collector doped to $n \approx 5 \times 10^{16}\,\mathrm{cm^{-3}}$. The p-type ($2 \times 10^{18}\,\mathrm{cm^{-3}}$) base was graded from $Ga_{0.98}Al_{0.02}As$ to $Ga_{0.8}Al_{0.2}As$ over 4000 Å. The grading corresponds to a field of about 5.6 kV/cm. The lightly doped ($n \approx 2 \times 10^{16}\,\mathrm{cm^{-3}}$), wide gap emitter consisted of an $Al_{0.35}Ga_{0.65}As$ layer 3000 Å thick and a region adjacent to the base graded from $Ga_{0.8}Al_{0.2}As$ to $Ga_{0.65}Al_{0.35}As$ over 500 Å. This corresponds to a base/emitter energy gap difference of approximately 0.18 eV. This grading removes a large part of the conduction band spike, allowing most of the band gap difference to fall across the valence band and blocking the unwanted injection of holes from the base [10.60]. Figure 10.28a shows the energy band diagram of the structure in the equilibrium (unbiased) configuration.

These devices had a current gain of 35 at a base current of 1.6 mA, and the collector characteristics were nearly flat with minimum collector–emitter offset voltage. More recently, high current gain, graded-base bipolars with good high-frequency performance have been reported by *Malik* et al. [10.60]. The base layer was linearly graded over 1800 Å, from $x = 0$ to 0.1, yielding a quasi-electric field of 5.6 kV/cm, and was doped with Be to $p = 5 \times 10^{18}\,\mathrm{cm^{-3}}$. The emitter–base junction was graded over 500 Å from $x = 0.1$ to 0.25 to enhance hole confinement in the base. The 0.2-μm-thick $Al_{0.25}Ga_{0.75}As$ emitter and the 0.5-μm-thick collector were doped n-type at $2 \times 10^{17}\,cm^{-3}$ and $2 \times 10^{16}\,cm^{-3}$, respectively. The $Al_xGa_{1-x}As$ layers were grown at a substrate temperature of 700° C. It was determined empirically that the insertion of an undoped setback layer of 200–500 Å between the base and emitter compensated for the Be diffusion and resulted in significantly increased current gains. Zn diffusion was used to contact the base and provided a low base contact resistance.

With a dopant setback layer in the base of 300 Å the maximum differential dc current gain was 1150, obtained at a collector current density of $J_C = 1.1 \times 10^3\,A/cm^2$, a higher gain than previously reported for graded-gap base HBTs. Such a gain can be compared with those found in previous work, which were consistently < 100 in HBTs without the setback layer [10.58, 59]. Several wafers were processed with undoped setback layers of 200–500 Å in the base, and all transistors exhibited greater current gains.

Graded-gap base HBTs were fabricated for high-frequency evaluation using the Zn diffusion process. A single 5-μm-wide emitter strip contact with dual adjacent base contacts was used. The areas of the emitter and collector junctions were approximately $2.3 \times 10^{-6}\,cm^2$ and $1.8 \times 10^{-5}\,cm^2$, respectively. The transistors were wire bonded in a microwave package, and automated s-parameter measurements were made with an HP 8409 network analyzer. The transistor had a current gain cutoff frequency of $f_T \approx 5\,GHz$ and a maximum oscillation frequency of $f_{max} \approx 2.5\,GHz$. Large-signal pulse measurements indicated risetimes $\tau_r \approx 150\,ps$ and pulsed collector currents $I_C > 100\,mA$, suitable for high-current laser driver applications.

10.6.2 Emitter Grading in Heterojunction Bipolar Transistors

The essential feature of the heterojunction bipolar transistor is the use of part of the energy band gap difference between the wide band gap emitter and the base to suppress hole injection. This allows the base to be more heavily doped than the emitter, which leads to the low base resistance and low emitter–base capacitance necessary for high-frequency operation while still maintaining a high emitter injection efficiency [10.57]. In this section we discuss the performance of $Al_{0.48}In_{0.52}As/Ga_{0.47}In_{0.53}As$ bipolar transistors with graded versus ungraded emitters [10.61] and consider the optimum grading of the emitter.

Most of the work on MBE-grown heterojunction bipolar transistors has concentrated on the AlGaAs/GaAs system. The first vertical npn $Al_{0.48}In_{0.52}As/Ga_{0.47}In_{0.53}As$ heterojunction bipolar transistors grown by MBE with high current gain were reported by *Malik* et al. [10.61]. The (Al,In)As/(Ga,In)As layers were grown by MBE lattice-matched to an Fe-doped semi-insulating InP substrate. Two HBT structures were grown: the first with an abrupt emitter of $Al_{0.48}In_{0.53}As$ on a $Ga_{0.47}In_{0.53}As$ base, and a second with a graded emitter comprising a quaternary layer of AlGaInAs 600 Å wide linearly graded between the two ternary layers. Grading from $Ga_{0.47}In_{0.53}As$ to $Al_{0.48}In_{0.52}As$ was achieved

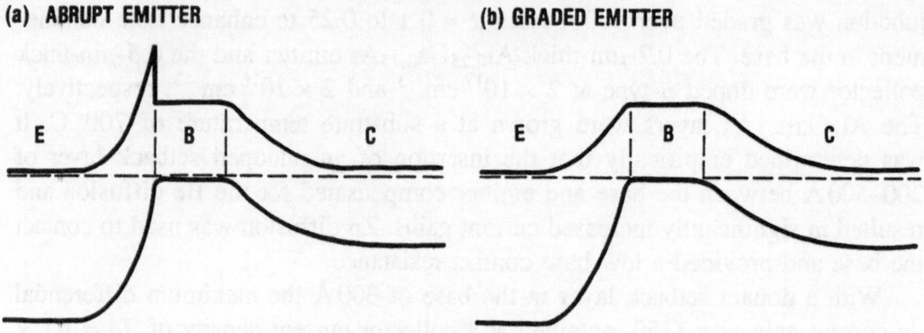

Fig. 10.30. Band diagrams under equilibrium conditions of heterojunction bipolar transistors with (a) an abrupt emitter and (b) a graded emitter. Note that the conduction band notch is eliminated through the use of a graded emitter and the increase of the emitter–base valence band barrier

by simultaneously lowering the Ga and raising the Al oven temperatures in such a manner as to keep the total Group III flux constant during the transition. It should be noted that this was the first use of a graded quaternary alloy in a device structure.

The energy band diagrams for the abrupt and graded emitter transistors are shown in Fig. 10.30a and b, respectively. The effect of grading is to eliminate the conduction band notch in the emitter junction. This in turn leads to a larger emitter–base valence band difference under forward bias injection. The following material parameters were used in both types of transistors. The $Al_{0.48}In_{0.52}As$ emitter and $Ga_{0.47}In_{0.53}As$ collector were doped n-type with Sn at levels of $5 \times 10^{17}\,cm^{-3}$ and $5 \times 10^{16}\,cm^{-3}$, respectively. The $Ga_{0.47}In_{0.53}As$ base was doped p-type with Be to a level of $5 \times 10^{18}\,cm^{-3}$. Experimental determination of the band edge discontinuities in the $Al_{0.48}In_{0.52}As/Ga_{0.47}In_{0.53}As$ heterojunction indicates $\Delta E_c \approx 0.50\,eV$ and $\Delta E_v \approx 0.20\,eV$ [10.62]. This value of ΔE_v is large enough to allow the use of an abrupt $Al_{0.48}In_{0.52}As/Ga_{0.47}In_{0.53}As$ emitter at 300 K. Nevertheless, a current gain increase by a factor of two (from $\beta = 200$ to $\beta = 400$) was achieved through the use of the graded-gap emitter, attributed to a larger valence band difference between the emitter and base under forward bias injection.

The common emitter characteristic of HBTs exhibits a relatively large collector–emitter offset voltage. This voltage is equal to the difference between the built-in potential for the emitter–base p-n junction and that of the base–collector p-n junction. Therefore no such offset is present in homojunction Si bipolars. By appropriately grading the emitter near the interface with the base such an offset can be reduced and even totally eliminated [10.63]. The other advantage of grading the emitter is that the potential spike in the conduction band can be reduced, thus increasing the injection efficiency. The conduction band potential has two components: the electrostatic potential ϕ_{es} equal to V_{bi} (the built-in potential) minus V_{BE} (the base–emitter voltage), which varies parabolically, and the grading potential ϕ_g. If linear grading is used there is always unwanted structure in the conduction band (spikes or notches, see Fig. 10.31). The "notches"

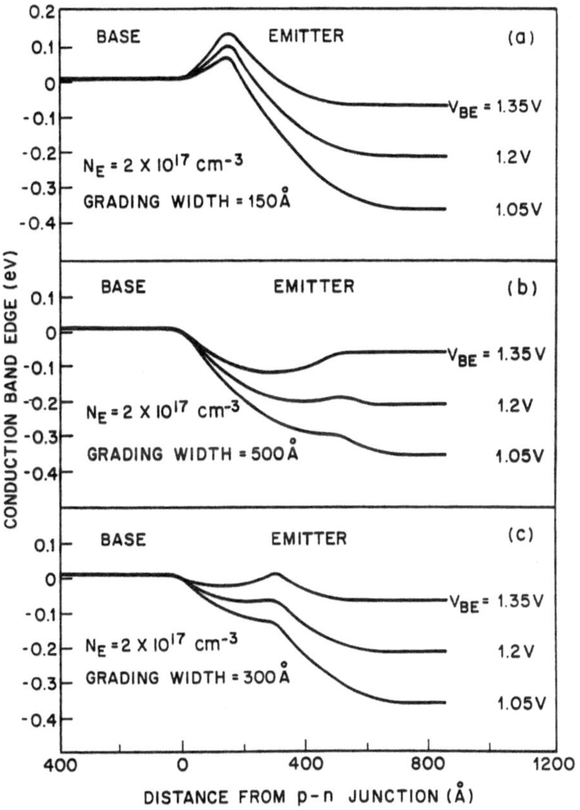

Fig. 10.31a–c. Conduction band edge vs distance from the p^+-n base–emitter junction for three different linear grading widths at different base emitter forward bias voltages

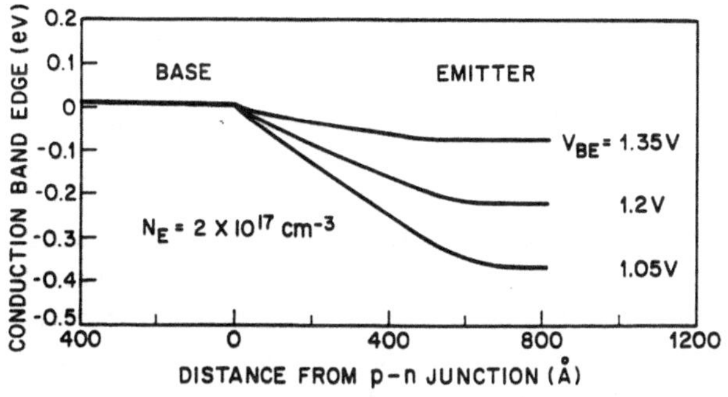

Fig. 10.32. Conduction band edge vs distance from the p^+-n junction, using a parabolically graded layer 500 Å wide at different forward bias voltages

can reduce the injection efficiency by promoting carrier recombination. It has become obvious that such structures can be eliminated by grading with the complementary function of the electrostatic potential in the emitter region $(1 - \phi_{es})$ over the depletion layer width at a forward bias equivalent to the base band gap (Fig. 10.32). If the base emitter junction is then forward biased at 1.42 eV, the two potentials (grading and electrostatic) give rise to a smooth conduction band edge and the flat band condition is attained with a built-in voltage for the base emitter equivalent to the band gap in the base (1.42 eV).

A HBT with such a parabolic grading was fabricated, using MBE, with a $Ga_{0.7}Al_{0.3}As$ emitter and GaAs base and collector [10.63]. The emitter–base junction was graded from $x = 0$ to 0.3 on the emitter side over a distance of 600 Å; the parabolic grading function was approximated by linear grading over nine regions. It was found that the collector–emitter offset voltage was very small (about 0.03 V).

10.7 Multilayer Sawtooth Materials

In this section we examine the electronic transport properties of sawtooth structures obtained by periodically varying the composition of III-V semiconductor alloys in an asymmetric fashion. The key feature of such structures is the lack of reflection symmetry [10.64], which has several important consequences. For example, these devices can be used as rectifying elements or, under suitable conditions, it is possible to optically generate in these structures a macroscopic electrical polarization that gives rise to a cumulative photovoltage across the uniformly doped sawtooth material. In addition, under an appropriate bias they can give rise to a staircase potential, which has several intriguing applications.

10.7.1 Rectifiers

The basic principle of sawtooth rectifiers, demonstrated by *Allyn* and coworkers [10.65, 66] is shown in Fig. 10.33. A sawtooth-shaped potential barrier is created by growing a semiconductor layer of graded chemical composition followed by an abrupt composition discontinuity. The adjoining layers, to which contact is made, are of the same conductivity type. In the present case, the barrier material is $Al_xGa_{1-x}As$ in which the Al content is graded and the adjoining layers are n-type GaAs (Fig. 10.33a). Near zero bias (Fig. 10.33b), conducting in the direction perpendicular to the layer is inhibited by the barrier. When the device is biased in the forward direction (Fig. 10.33c) the voltage drop initially occurs across the graded layer. This reduces the slope of the potential barrier, thus allowing increased thermionic emission. when the applied voltage exceeds the barrier height, the device will conduct completely, as in the case of a Schottky barrier. In the reverse direction (Fig. 10.33d) electrons will be attracted to, but inhibited from passing through, the abrupt potential discontinuity at the sharp edge of the sawtooth. The width of the interface and potential discontinuity is known to be only 5–10 Å. Thus, the primary reverse current-carrying mechanism will be tunneling. The barrier can be either doped or undoped, although depletion of carriers from within the barrier (in the case of doped barriers) leads to band bending, which reduces the equilibrium height and width of the barrier. Multiple sawtooth barriers with five periods were also fabricated [10.65]. These showed a turn-on voltage equal to five times that of the single barrier, thus demonstrating the additivity of the technique.

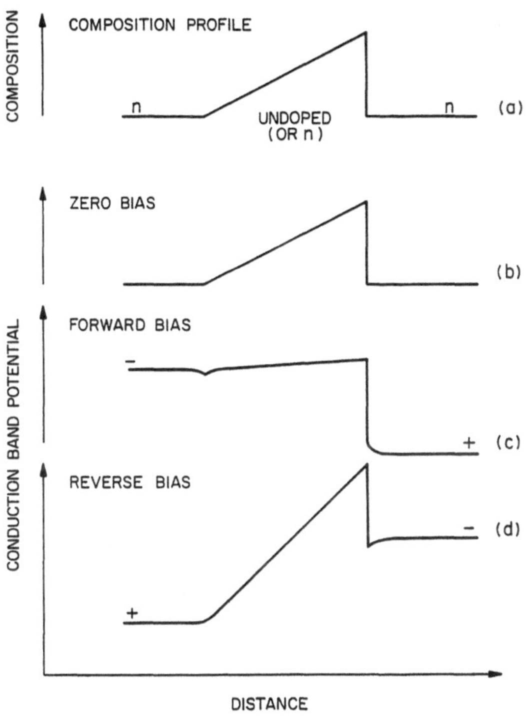

Fig. 10.33. (a) Compositional structure of a sawtooth barrier rectifying structure; **(b)** potential distribution for band-edge conduction electrons at zero bias (undoped barrier case); **(c)** potential distribution under forward bias; and **(d)** potential distribution under reverse bias

10.7.2 Electrical Polarization Effects in Sawtooth Superlattices

The lack of planes of symmetry in sawtooth superlattice material, relative to conventional superlattices with rectangular wells and barriers, can lead to electrical polarization effects. Generation of a transient macroscopic electrical polarization extending over many periods of the superlattice has been observed [10.67]. This effect is a direct consequence of the above-mentioned lack of reflection symmetry in these structures.

The energy band diagram of a sawtooth p-type superlattice is sketched in Fig. 10.34a, in which a negligible valence band offset was assumed. Layer thicknesses are typically a few hundred angstroms, and a suitable material is graded-gap $Al_xGa_{1-x}As$. The superlattice is sandwiched between two highly doped p^+ contact regions.

Let us assume that electron–hole pairs are excited by a very short light pulse, as shown in Fig. 10.34a. Due to the grading, electrons experience a higher quasi-electric field than holes do. For this reason, and because of their much higher velocity, electrons separate from holes and reach the low gap in subpicosecond times ($\approx 10^{-13}$ s). This sets up an electrical polarization in the sawtooth structure, which results in the appearance of a photovoltage across the device terminals (Fig. 10.34b). The macroscopic dipole moment and its associated voltage subsequently decay in time by a combination of (a) dielectric relaxation and (b) hole drift.

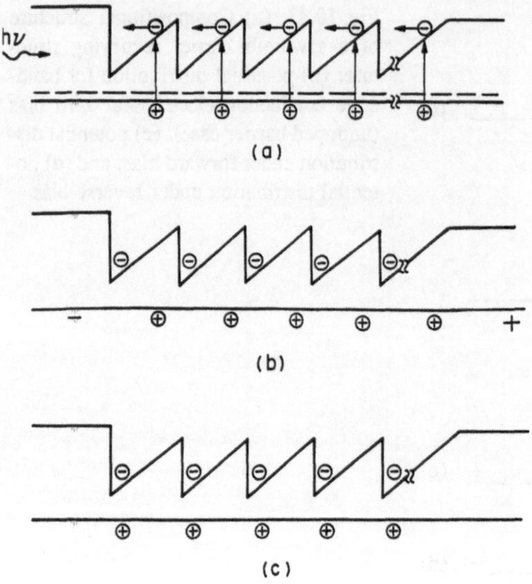

Fig. 10.34a–c. Formation and decay of the macroscopic electrical polarization in a sawtooth superlattice

The excess hole density decays by dielectric relaxation to restore a flat valence band (equipotential) condition, as illustrated in Fig. 10.34c. Note that in this final configuration holes have redistributed to neutralize the electrons at the bottom of the wells. Thus the net negative charge density on the low gap side of the wells decreases with the same time constant as the positive charge packet (the dielectric relaxation time). The other mechanism by which the polarization decays is hole drift caused by the electric field created by the initial spatial separation of electrons and holes.

The graded-gap superlattice structure shown in Fig. 10.34 and the underlying p^+-GaAs buffer layer were grown by MBE. A total of ten graded periods were grown with a period of ≈ 500 Å. The layers were graded from GaAs to $Al_{0.2}Ga_{0.8}As$. A heavily doped GaAs contact layer of ≈ 700 Å was grown on top of the 1-μm-thick $Al_{0.45}Ga_{0.55}As$ ($p \approx 5 \times 10^{18}\,cm^{-3}$) window layer. Unbiased devices were mounted in a microwave stripline and illuminated with short light pulses (4 ps) of wavelength $\lambda = 6400$ Å. The absorption length was ≈ 3500 Å. In this particular wafer the carrier concentration was $10^{16}\,cm^{-3}$. It was found that the rise-time of the pulse response was ≤ 25 ps, while the fall time (at the 1/e point) was ≈ 200 ps. Unlike conventional detectors, the current carried in this photodetector is of displacement rather than conduction nature since it is associated with a time-varying polarization. This current, by continuity, equals the conduction current in the external load.

10.7.3 Staircase Structures

The novel concept of a staircase potential [10.68–71] is finding several interesting applications. We shall concentrate on the staircase avalanche photodiode (APD) [10.1, 68–71] and on the repeated velocity overshoot device [10.72].

a) Staircase Solid-State Photomultipliers and Avalanche Photodiodes

Figure 10.35a shows the band diagram of the graded-gap multilayer structure (assumed intrinsic) at zero applied field. Each stage is linearly graded in composition from a low (E_{g1}) to a high (E_{g2}) band gap, with an abrupt step back to low band gap material. The conduction band discontinuity shown accounts for most of the band gap difference, as is typical of many III-V heterojunctions. The materials are to be chosen in order to obtain a conduction-band discontinuity comparable to or greater than the electron ionization energy E_{ie} in the low gap material following the step. The biased detector is shown in Fig. 10.35b. Consider a photoelectron generated near the p^+ contact: the electron does not cause impact ionization in the graded region before the conduction band step because the net electric field is too low. At the step, however, impact ionization will take place. This process will be repeated at every stage. Note that the steps perform the function of the dynodes of a phototube. Holes created by electron impact ionization at the steps do not impact ionize, since the valence band steps are of the wrong sign to assist ionization and the electric field in the valence band is too low to cause hole-initiated ionization. Holes, however, do multiply since at every step both an electron and a hole are created. The gain is $M = (2 - \delta)^N$, where δ is the fraction of electrons that do not ionize per stage. The noise per unit bandwidth on the output signal, neglecting dark current, is given by $\langle i^2 \rangle = 2eI_{ph}M^2F$, where I_{ph} is the primary photocurrent and F the avalanche excess noise factor. For the staircase APD, F is given by [10.71]

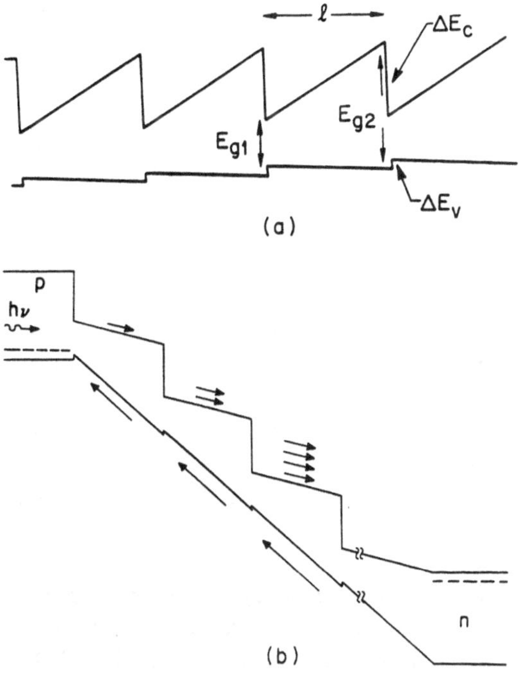

Fig. 10.35a, b. Band diagram of staircase solid-state photomultiplier. The arrows in the valence band indicate that holes do not cause impact ionization. **(a)** Zero applied field. **(b)** Biased detector

$$F = 1 + \frac{\delta[1-(2-\delta)^{-N}]}{2-\delta} . \tag{10.9}$$

Note that for small δ, $F \approx 1$ and is practically independent of the number of stages. Thus, the multiplication process is essentially noise-free. It is interesting to note that the excess noise of this structure does not follow the McIntyre theory of conventional APDs [10.73]. In a conventional APD the minimum excess noise factor at high gain (> 10) is 2 if one of the ionization rates is zero. The fact that in the staircase APD the avalanche noise is lower than in the best conventional APDs ($\alpha/\beta = \infty$) can be understood since in a conventional APD the avalanche is more random, since carriers can cause ionization everywhere in the avalanche region. In the staircase APD electrons cause ionization only at well-defined positions in space so that the multiplication process is more deterministic. Similarly, in a photomultiplier tube the avalanche is essentially noise-free ($F \approx 1$).

Finally, the low voltage operation of this device with respect to conventional APDs should be mentioned. For a five-stage detector and $\Delta E_c \approx E_{g1} \approx 1$ eV, the applied voltage required to achieve a gain of about 32 is slightly greater than 5 V. Possible material systems for the implementation of the device in the 1.3–1.6 μm region are AlGaAs/GaSb and HgCdTe. In a practical structure one should always leave an ungraded layer immediately after the step having a thickness of the order of a few ionization mean free paths ($\lambda_i \approx$ 50–100 Å) to ensure that most electrons ionize near the step.

The staircase APD has not yet been implemented, but *Capasso* et al. [10.74] demonstrated experimentally an enhancement of the α/β (≈ 8) in an AlGaAs/GaAs quantum well superlattice. The effect has been attributed to the difference between the conduction and valence band discontinuities ($\Delta E_c > \Delta E_v$). Thus electrons enter the well with a higher kinetic energy than holes and have a higher probability of causing ionization. Note that the staircase APD is the limiting case of this detector since the whole ionization energy is gained at the band discontinuity. The staircase devices are probably the best example of the potential of the band-gap engineering concept.

b) Repeated Velocity Overshoot Devices

Another interesting proposed application of staircase potentials is the repeated velocity overshoot device [10.72]. This structure has the potential to achieve average drift velocities well in excess of the maximum steady-state velocity over distances greater than 1 μm. Figure 10.36a shows a general type of staircase potential structure. The corresponding electric field, shown in Fig. 10.36b, consists of a series of high field regions of value E_1 and width d superimposed upon an background field E_0. To illustrate the electrical behavior and design considerations for a specific case, we consider electrons in the central valley of GaAs. The background field E_0 is chosen so that the steady-state electron energy distribution is not excessively broadened beyond its thermal equilibrium value, but at the same time the average drift velocity is still relatively high. For GaAs,

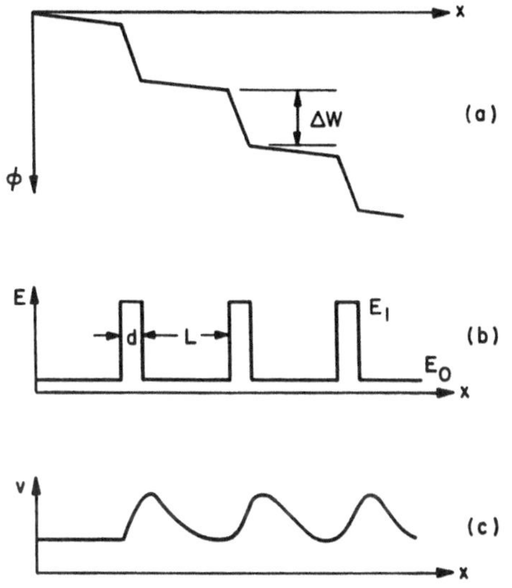

Fig. 10.36. Principle of repeated velocity overshoot staircase potential (**a**) and corresponding electric field (**b**). The ensemble velocity as a function of position is also illustrated (**c**)

an appropriate value would be around 2.5 kV/cm. At this field, the steady-state drift velocity is 1.8×10^7 cm/s and fewer than 2 % of the electrons reside in the satellite valley. Immediately downstream from the high field region, the electron distribution is shifted to higher energy by an amount $\Delta W = E_1 d$. (Note that while the distribution is shifted uniformly in energy, it is compressed in momentum in the direction of transport.) The width d is selected so that the transit time across the high field region is shorter than the mean phonon scattering time, which is about 0.13 ps in GaAs. The energy step ΔW is chosen as to maximize the average velocity of the distribution after the step while still keeping most of the distribution below the threshold energy for transfer to the satellite valley. In GaAs, the intervalley separation is about 0.3 eV, so an appropriate value of ΔW would be about 0.2 eV, resulting in an average velocity of approximately 1×10^8 cm/s immediately after the step. The momentum decays rapidly beyond the step due to scattering by polar optical phonons, so that the velocity decreases roughly linearly with distance, as shown in Fig. 10.36c. During this time the distribution is broadened considerably in momentum. After momentum and velocity have relaxed, the distribution requires additional time to relax to its original energy. Thus, the spacing L between high field regions must be large enough to allow sufficient cooling of the electron distribution before another oveshoot can be attempted. This is necessary to avoid populating the high mass satellite valleys. The effect of the resulting repeated velocity overshoot shown in Fig. 10.36c is that average drift velocities greater than the maximum stead-state velocity can be maintained over long distances. A practical way to achieve this device with graded-gap materials is shown in Fig. 10.37.

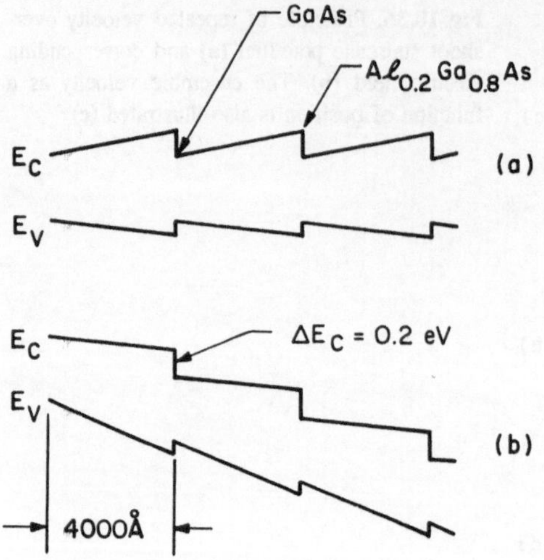

Fig. 10.37. Band diagram of a graded-gap repeated velocity overshoot device, without (a) and with (b) bias

10.8 AlGaAs Floating-Gate Memory Devices with Graded-Gap Injector

In this section we present results on memory devices in the AlGaAs/GaAs system in which an element of the type of Fig. 10.33 was used as injector [10.75, 76].

The structure was grown by MBE on a semi-insulating GaAs substrate. The zero bias Γ-edge conduction-band diagram at equilibrium (corresponding to the logical ZERO state), is shown in Fig. 10.38a. The channel layer consisted of 750 Å GaAs Si doped to $n \approx 1 \times 10^{17}\,\mathrm{cm}^{-3}$ followed by a 1000 Å undoped rectangular AlAs barrier. The undoped ($p \approx 1 \times 10^{15}\,\mathrm{cm}^{-3}$) 2000 Å thick GaAs floating-gate was grown on top of this layer and was followed by an undoped trapezoidal barrier consisting of 200 Å of AlAs and 1800 Å thick $Al_xGa_{1-x}As$ compositionally graded from $x = 1$ to $x = 0$. The GaAs n^+ contact layer was Si-doped to $2 \times 10^{18}\,\mathrm{cm}^{-3}$ and was 5000 Å thick.

With the source at reference, electrons were injected in the well by applying negative pulses of different amplitude (ΔV_{GS}) and duration (Δt_p) to the control-gate electrode. The band diagram during this WRITE ONE phase is shown in Fig. 10.38b. The width of the floating-gate is such that most of the injected electrons are collected. To monitor the charge concentration in the well, the drain current (I_D) was measured at constant positive drain-to-source bias (V_{DS}) as a function of time. After the pulse, the drain current exhibited a drop (ΔI_D) due to the depletion of the channel caused by the electron accumulation in the well (logical ONE state, Fig. 10.38c). Subsequently the current returned to a stationary value in a time linked to the discharging of the floating-gate well. One such measurement is shown in Fig. 10.39a. The device can be subsequently erased (WRITE ZERO) by applying a pulse of opposite polarity to the gate, as illustrated by the band diagram of Fig. 10.38d, or by using (visible) light. In

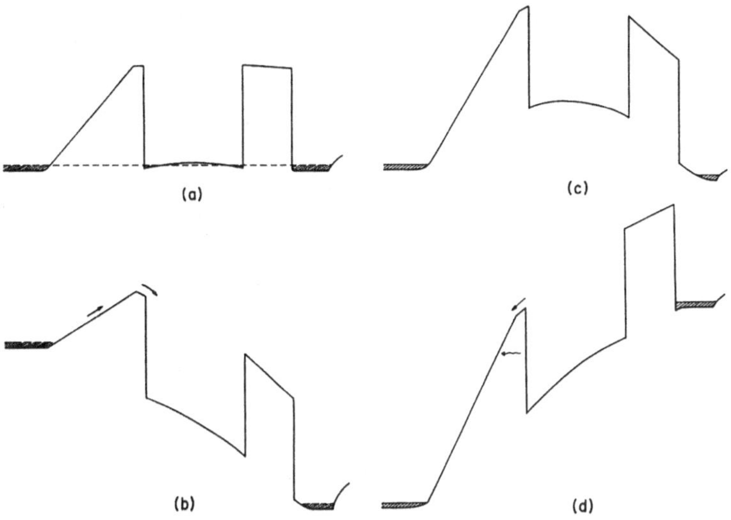

Fig. 10.38. Conduction-band diagram of the loating-gate AlGaAs/GaAs memory device under equilibrium (READ ZERO) (a), during WRITE ONE phase (b), in the READ ONE phase (c), and during the electrical ERASE (= WRITE ZERO) phase (d). In (d) the arrows indicate the two dominant transport mechanisms: tunneling and thermionic emission

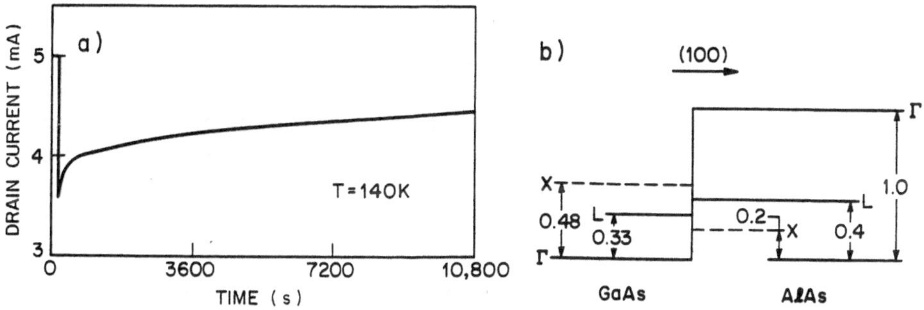

Fig. 10.39. Drain current vs time following a Δt_p = 1 ms long, −5 V pulse on the control gate at 140 K (a); band edge profiles for Γ, X and L point minima at a GaAs/AlAs interface (energies are given in eV) (b)

the latter erasure mode, while the majority of the photogenerated electrons drift away from the floating gate, the holes photogenerated throughout the structure perferentially accumulate in it and recombine with the stored electrons.

Figure 10.39a shows I_D as a function of time at 140 K, following injection with a gate pulse of $\Delta V_{GS} = -5$ V and $\Delta t_p = 1$ ms at $V_{DS} = +0.8$ V. From the values of the drain current before and immediately after the injection pulse, the doping level of the channel, and the dimensions of the structure, it is easy to estimate the charge accumulated in the GaAs well. From the above data one obtains a sheet density of $\approx 10^{12}\,\mathrm{cm}^{-2}$. Following an initial short nonexponential decay, the current decays exponentially in $\approx 4\,\mathrm{h}$, the storage or retention time of the memory. Similar measurements at liquid nitrogen temperature under identical injection and bias conditions revealed no visible decay of the drain current after several hours. In fact, as shown by the analysis of the discharge mechanism, the estimated storage time at this temperature is ≈ 700 years. These results illustrate the potential of these devices for nonvolatile memory and SRAM applications in GaAs-based low temperature electronics. At room temperature, the devices exhibited storage times of a few seconds. These time constants place these devices in the realm of interest of DRAMs even at room temperature [10.75].

It is interesting to study the discharge mechanism of the floating-gate. The Arrhenius law, attempt frequency times emission probability, is the standard starting point in modeling thermionic emission across heterojunction barriers. This gives for the emission rate $e_f = (v_{th}/L_W)\exp(-E_a/k_BT)$, where v_{th} is the electron thermal velocity perpendicular to the interface, L_W is the well thickness (or an effective mean free path, whichever is smaller), E_a is the activation energy, T is the lattice temperature and k_B is the Boltzmann constant [10.77]. By measuring the retention time (defined as the time after the pulse required to reach the quasi-stationary level) as a function of temperature, an activation energy $E_a \approx 0.2\,\mathrm{eV}$ was found [10.76]. The experimental data are in agreement with the measured GaAs-AlAs conduction band discontinuity [10.78]. However, the calculated pre-exponential factor gives a rate which is orders of magnitude larger than that experimentally observed. With the above approach (Bethe's thermionic emission theory) and in the thermionic emission–diffusion theory of Crowell and Sze [10.79], the lateral momentum is conserved in crossing the heterointerface. These models apply to the case of two materials having the conduction band minima at the same symmetry point. In this situation, the fraction of electrons with energy above the barrier [i.e., $\exp(-E_a/k_BT)$] crosses the heterointerface. The picture, however, is very different for the case of a direct–indirect heterojunction such as GaAs/AlAs, for which the experimentally determined activation energy corresponds to the Γ–X indirect discontinuity. How do Γ-valley electrons in GaAs with energy between 10.2 and 0.33 eV (the L point, see Fig. 10.39b) cross the heterointerface? Such electrons form a large fraction of the distribution, especially at low temperatures. It is not energetically possible for these electrons to scatter to the satellite valleys in GaAs and then cross the heterojunction. In [10.75] scattering events in the AlAs barrier were suggested as a possible mechanism. There is a finite probability of finding the electron in the AlAs Γ-barrier ($\approx 1\,\mathrm{eV}$) due to the penetration of the wavefunction. A scattering process in AlAs with a large momentum exchange (by either a phonon or a defect) can then transfer the electron to the X or L valleys in AlAs. If λ_s and λ_e are the intervalley scattering mean free path and the attenuation length for the electrons in the barrier, $(\lambda_e/\lambda_s)\exp(-\lambda_s/\lambda_e)$ represents an estimate of the probability of a scattering event providing the lateral momentum needed for the thermionic emission to the X valleys of AlAs, i.e., across the conduction band discontinuity. Since λ_s can be several hundred angstroms long and $\lambda_e \approx 40\,\text{Å}$, this factor can reduce the emission rate by several orders of magnitude, in accordance with the

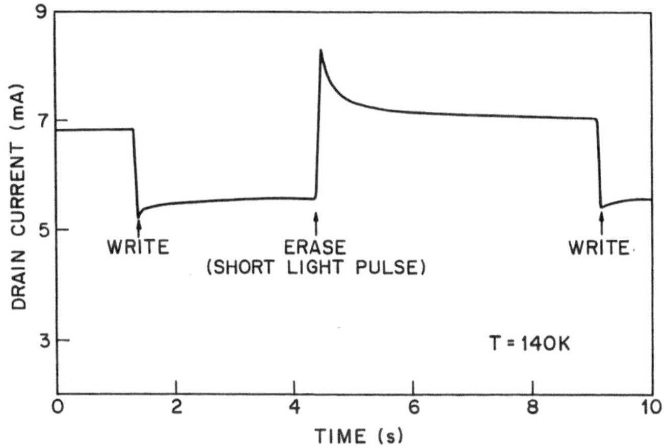

Fig. 10.40. The drain current drops following charge injection into the floating-gate with a −5 V, 1 ms long gate pulse at 140 K. The memory is then erased by applying a fraction of a second long light pulse

experimental findings. Additional mechanisms may be possible, namely mixing between Γ (GaAs) and X (AlAs) wavefunctions at the heterojunction interface as discussed by *Zohta* [10.80] and *Meynadier* et al. [10.81], and scattering due to interface roughness when electrons cross the heterojunction, as studied by *Liu* [10.82].

The above results have been confirmed in a similar structure by *Lott* et al. [10.83]. In that device a fourth contact was provided, which allowed the controlled discharge of the floating-gate either perpendicularly (as in the device above) or laterally. As predicted from the model outlined above [10.76], the prefactors were found to differ dramatically (a factor $\approx 10^7$).

To erase the memory a short light pulse from a microscope lamp was used. It was turned on some time after a negative pulse had been applied to the gate (Fig. 10.40). Electrical erasure of the devices was not possible due to excessive leakage currents. To facilitate it a triangular barrier could be used in order to allow thermionic injection of the carriers also in the WRITE ZERO phase.

10.8.1 Integration in Arrays

The memory device is suitable for integration in arrays as illustrated in Fig. 10.41. This circuit is different from that of conventional Si-based E^2PROMs: two FETs (Q_1 and Q_2) are required per cell (where Q_{FG} represents the floating gate memory device) and the control and power circuitry are significantly simplified. The operation of the circuit is as follows. In order to write a logic ONE into a given cell, the corresponding ROW SELECT voltage is set to a value that closes the Q_1's and Q_2's switches of the row (the polarity of this depends on the type of Q_1 and Q_2, i.e., positive if they are enhancement mode n-channel). Then a negative pulse is applied to the WRITE LINE in order to charge Q_{FG} through Q_1. Since

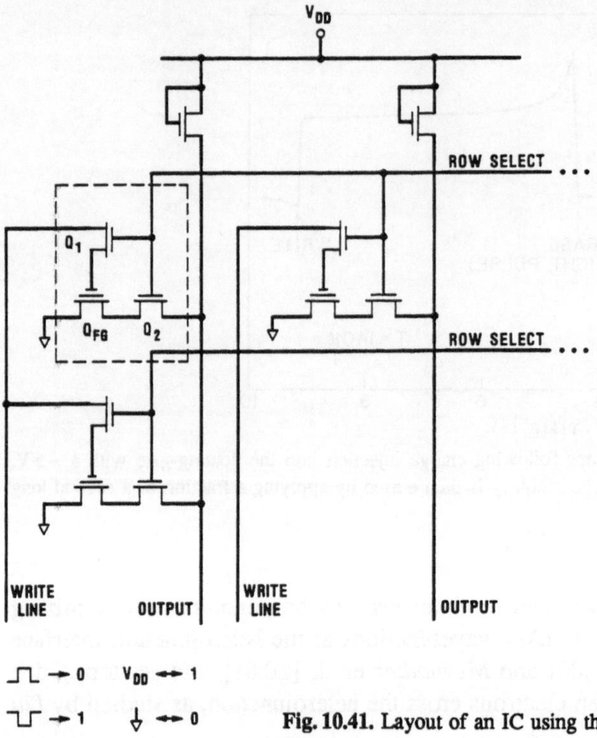

Fig. 10.41. Layout of an IC using the floating-gate memory devices

only one of the Q_1's of the column is on, all the other Q_{FG}'s will not be affected. Through the use of multipliers the memory can be written one word (i.e., ROW) at a time. Conversely, in order to write a logic ZERO (which is equivalent to the ERASE function) a positive pulse must be applied to the WRITE LINE.

To perform the READ operation, again the word is selected by proper bias of the ROW SELECT line and the desired Q_1's and Q_2's are closed. The OUTPUT LINE will be at the supply voltage V_{DD} if the selected Q_{FG} is open (logic state ONE) so that no current flows through Q_2 and Q_{FG} to reference. On the other hand, the output line is at reference if Q_{FG} is closed (logic state ZERO); i.e., Q_{FG} and Q_2 are short circuits to ground.

Compared to standard Si-based E^2PROM architecture [10.84] the circuit has the advantage of requiring less voltage sources. Moreover, with the use of band-gap engineering the WRITE and ERASE phases do not require the high fields necessary in the Si-SiO_2-based analog of these memories. Therefore wear-out problems that limit the number of cycles Si-based memories can withstand are not present. Finally, once the speed requirements are satisfied, this architecture can operate as SRAM or DRAM (in the latter case also at room temperature).

Acknowledgments. It is a pleasure to acknowledge the many colleagues who have collaborated with the authors. In particular we would like to mention A.Y. Cho, A.C. Gossard, W.T. Tsang, R.J. Malik, L.M. Lunardi, S. Luryi, A.S. Vengurlekar, J. Allam, R. Kiehl, C.G. Bethea and A.L. Hutchinson.

References

10.1 S. Sze: "Silicon VLSI Technology", in Proc. Int. Conf. Adv. Electron. Mat., Shigaku-Kaikan, Tokyo, 1988

10.2 L.L. Chang, L. Esaki, R. Tsu: Appl. Phys. Lett. **24**, 593 (1974)

10.3 F. Capasso, R.A. Kiehl: J. Appl. Phys. **58**, 1366 (1985)

10.4 R.T. Bate, G.A. Frazier, W.R. Frensley, J.K. Lee, M.A. Reed: Proc. SPIE **792**, 26 (1987)

10.5 A.Y. Cho: Thin Solid Films **100**, 291 (1983)

10.6 F. Capasso: J. Vac. Sci. Technol. **B 1**, 457 (1983)

10.7 T.C.L.G. Sollner, W.D. Goodhue,P.E. Tannenwald, C.D. Parker, D.D. Peck: Appl. Phys. Lett. **43**, 588 (1983)

10.8 E.R. Brown, T.C.L.G. Sollner, W.D. Goodhue, C.L. Chen: Proc. SPIE **943**, 2 (1988)

10.9 F. Capasso, S. Sen, F. Beltram, A.Y. Cho: "Resonant Tunneling Devices and Their Applications", in *Submicron Integrated Circuits*, ed. by R.K. Watts (Wiley, New York 1989) Chap.5

10.10 C. Rine (ed.): *Computer Science and Multiple Valued Logic* (North-Holland, Amsterdam 1977) p.101

10.11 S. Sen, F. Capasso, A.C. Gossard, R.A. Spah, A.L. Hutchinson, S.N.G. Chu: Appl. Phys. Lett. **51**, 1428 (1987)

10.12 F. Capasso, K. Mohammed, A.Y. Cho: IEEE J. QE-**22**, 1853 (1986)

10.13 A.F.J. Levi: Electron. Lett. **24**, 1273 (1988)

10.14 T. Weil, B. Vinter: Appl. Phys. Lett. **50**, 1281 (1987)

10.15 R.J. Malik, R. Nottenburg, E.F. Schubert, J.F. Walker, R.W. Ryan: Appl. Phys. Lett. **53**, 2661 (1988)

10.16 A.F.J. Levi, S.L. McCall, P.M. Platzman: Appl. Phys. Lett. **54**, 940 (1989)

10.17 A.F.J. Levi. Y. Yafet: Appl. Phys. Lett. **51**, 42 (1987)

10.18 E. Wolak, K.L. Lear, P.M. Pitner, E.S. Hellman, B.G. Park, T. Weil, J.S. Harris, Jr.: Appl. Phys. Lett. **53**, 201 (1988)

10.19 F. Capasso, S. Sen, A.Y. Cho, A.L. Hutchinson: Appl. Phys. Lett. **50**, 930 (1987)

10.20 F. Capasso, S. Sen. A.C. Gossard, A.L. Hutchinson, J.H. English: IEEE Electron. Dev. Lett. **7**, 573 (1986)

10.21 M.A. Reed, W.F. Frensley, R.J. Matyi, J.N. Randall, A.C. Seabaugh: Appl. Phys. Lett. **54**, 1034 (1989)

10.22 B. Riccò, P.M. Solomon: IBM Tech. Discl. Bull. **27**, 3053 (1984)

10.23 N. Yokoyama, K. Imamura, S. Muto, S. Hiyamizu, H. Nishi: Jpn. J. Appl. Phys. **24**, L-853 (1985)
For a comprehensive review of RHETs see: N. Yokoyama, H. Ohnishi, T. Futatsugi, S. Muto, T. Mori, K. Imamura, A. Shibatomi: Proc. SPIE **943**, 14 (1988)

10.24 S. Sen, F. Capasso, A.Y. Cho, D. Sivco: IEEE Trans. ED-**34**, 2185 (1987)

10.25 J. Söderström, T.G. Andersson: IEEE Electron. Dev. Lett. **9**, 200 (1988)

10.26 R.C. Potter, A.A. Lakhani, D. Beyea, E. Hempling, A. Fathimulla: Appl. Phys. Lett. **52**, 2163 (1988)

10.27 S. Sen, F. Capasso, D. Sivco, A.Y. Cho: IEEE Electron. Dev. Lett. **9**, 403 (1988)

10.28 *General Electric Tunnel Diode Manual*, 1st ed. (1961) p.66

10.29 F. Capasso, S. Sen, A.Y. Cho, D.L. Sivco: Appl. Phys. Lett. **53**, 1056 (1988)

10.30 S. Sen, F. Capasso, A.Y. Cho, D.L. Sivco: IEEE Electron Dev. Lett. **9**, 533 (1988)

10.31 M. Tsuchiaya, H. Sakaki, J. Yoshino: Jpn. J. Appl. Phys. **24**, L-466 (1985)

10.32 L.M. Lunardi, S. Sen, F. Capasso, P.R. Smith, D.L. Sivco, A.Y. Cho: IEEE Electron Dev. Lett. **10**, 219 (1989)

10.33 T. Mori, K. Imamura, H. Ohnishi, Y. Minami, S. Muto, N. Yokoyama: "Microwave Analysis of Resonant Tunneling Hot Electron Transistor at Room Temperature", in Extended Abstracts of the 20th Conf. on Solid State Devices and Materials, Tokyo (1988) p.507

10.34 S. Sen, F. Capasso, A.Y. Cho, D.L. Sivco: Electron. Lett. **24**, 1506 (1988)
10.35 A.A. Lakhani, R.C. Potter, H.S. Hier: Electron. Lett. **24**, 681 (1988)
10.36 T. Tanoue, H. Mizuta, S. Takahashi: IEEE Electron Dev. Lett. **9**, 365 (1988)
10.37 S. Luryi, F. Capasso: Appl. Phys. Lett. **47**, 1347 (1985); also Erratum, Appl. Phys. Lett. **48**, 1693 (1986)
10.38 F. Capasso, S. Sen, F. Beltram, A.Y. Cho: Electron. Lett. **23**, 225 (1987)
10.39 S. Sen, F. Capasso, F. Beltram, A.Y. Cho: IEEE Trans. ED-**34**, 1768 (1987)
10.40 F. Capasso, S. Sen, A.Y. Cho: Appl. Phys. Lett. **51**, 526 (1987)
10.41 A.R. Bonnefoi, T.C. McGill, R.D. Burnham: IEEE Electron Dev. Lett. **6**, 636 (1985)
10.42 T.K. Woodward, T.C. GcGill, R.D. Burnham: Appl. Phys. Lett. **50**, 451 (1987)
10.43 T.K. Woodward, T.C. McGill, H.F. Chung, R.D. Burnham: Appl. Phys. Lett. **51**, 1542 (1987)
10.44 T.K. Woodward, T.C. McGill, H.F. Chung, R.D. Burnham: IEEE Electron Dev. Lett. **9**, 122 (1988)
10.45 F. Beltram, F. Capasso, S. Luryi, S.N.G. Chu, A.Y. Cho: Appl. Phys. Lett. **53**, 219 (1988)
10.46 A.R. Bonnefoi, D.H. Chow, T.C. McGill: lAppl. Phys. Lett. **47**, 888 (1985)
10.47 S. Luryi: Appl. Phys. Lett. **52**, 501 (1988)
10.48 F. Capasso, A.S. Vengurlekar, A.L. Hutchinson, W.T. Tsang: Electron. Lett. **25**, 1117 (1989)
10.49 J.F. Palmier, C. Minot, J.L. Lievin, F. Alexandre, J.C. Harmand, J. Dangla, C. Dubon-Chevallier, D. Ankri: Appl. Phys. Lett. **49**, 1260 (1986)
10.50 C.S. Lent: Superlattices Microstruct. **3**, 387 (1987)
10.51 F. Beltram, F. Capasso, A.L. Hutchinson, R.J. Malik: Appl. Phys. Lett. **55**, 1534 (1989)
10.52 H. Kroemer: RCA Rev. **18**, 332 (1957)
10.53 J.A. Hutchby: J. Appl. Phys. **49**, 4041 (1978)
10.54 B.F. Levine, W.T.Tsang, C.G. Bethea, F. Capasso: Appl. Phys. Lett. **41**, 467 (1982)
10.55 B.F. Levine, C.G. Bethea, W.T. Tsang, F. Capasso, K.K. Thornber, R.C. Fulton: Appl. Phys. Lett. **42**, 9 (1983)
10.56 F. Capasso, W.T. Tsang, C.G. Bethea, A.L. Hutchinson, B.F. Levine: Appl. Phys. Lett. **42**, 93 (1983)
10.57 H. Kroemer: J. Vac. Sci. Technol. B **1**, 126 (1983)
10.58 J.R. Hayes, F. Capasso, A.C. Gossard, R.J. Malik, W. Wiegmann: Electron. Lett. **19**, 410 (1983)
10.59 D.L. Miller, P.M. Asbeck, R.J. Anderson, F.H. Eisen: Electron. Lett. **19**, 367 (1983)
10.60 R.J. Malik, F. Capasso, R.A. Stall, R.A. Kiehl, R. Wunder, C.G. Bethea: Appl. Phys. Lett. **46**, 600 (1985)
10.61 R.J. Malik, J.R. Hayes, F. Capasso, K. Alavi, A.Y. Cho: IEEE Electron Devices Lett. **4**, 383 (1983)
10.62 R. People, K.W. Wecht, K. Alavi, A.Y. cho: Appl. Phys. Lett. **43**, 118 (1983)
10.63 J.R. Hayes, F. Capasso, R.J. Malik, A.C. Gossard, W. Wiegmann: Appl. Phys. Lett. **43**, 949 (1983)
10.64 P.J. Price: IEEE Trans. ED-**28**, 911 (1981)
10.65 A.C. Gossard, W. Brown, C.L. Allyn, W. Wiegmann: Appl. Phys. Lett. **36**, 373 (1980)
10.66 A.C. Gossard, W. Brown, C.L. Allyn, W. Wiegmann: J. Vac. Sci. Technol. **20**, 694 (1982)
10.67 F. Capasso, S. Luryi, W.T. Tsang, C.G. Bethea, B.F. Levine: Phys. Rev. Lett. **51**, 2318 (1983)
10.68 F. Capasso, W.T. Tsang: "Superlattice, Graded Band Gap, Channeling and Staircase Avalanche Photodiodes: Towards a Solid-State Photomultiplier", in Tech. Digest Int. Electron. Devices Meeting, Washington, DC (1982) p. 334
10.69 F. Capasso: IEEE Trans. Nucl. Sci. **30**, 424 (1983)
10.70 F. Capasso: Surf. Sci. **132**, 527 (1983)
10.71 F. Capasso, W.T. Tsang, G.F. Williams: IEEE Trans. ED-**30**, 381 (1983)
10.72 J.A. Cooper, Jr., F. Capasso, K.K. Thornber: IEEE Electron Dev. Lett. **3**, 407 (1982)
10.73 R.J. McIntyre: IEEE Trans. ED-**13**, 164 (1966)
10.74 F. Capasso, W.T. Tsang, A.L. Hutchinson, G.F. Williams: Appl. Phys. Lett. **40**, 38 (1982)
10.75 F. Capasso, F. Beltram, R.J. Malik, J.F. Walker: IEEE Electron Dev. Lett. **9**, 377 (1988)

10.76 F. Beltram, F. Capasso, J.F. Walker, R.J. Malik: Appl. Phys. Lett. **53**, 376 (1988)
10.77 K. Hess: *Advanced Theory of Semiconductor Devices* (Prentice Hall, Englewood Cliffs, NJ 1988) Chap. 12
10.78 S.R.Forrest: "Measurements of Energy Band Offsets Using Capacitance and Current Techniques", in *Heterojunction Band Discontinuities: Physics and Device Applications*, ed. by F. Capasso, G. Margaritondo (North-Holland, Amsterdam 1987) p. 363
10.79 S. Sze: *Physics of Semiconductor Devices* (Wiley, New York 1981) p. 496
10.80 Y. Zohta: Jpn. J. Appl. Phys. **27**, L-906 (1988)
10.81 M.-H. Meynadier, R.E. Nahori, J.M. Worlock, M.C. Tamargo, J.L. de Miguel, M.D. Sturge: Phys. Rev. Lett. **60**, 1338 (1988)
10.82 H.C. Liu: Appl. Phys. Lett. **51**, 1019 (1987)
10.83 J.A. Lott, J.F. Klem, H.T. Weaver: Appl. Phys. Lett. **55**, 1226 (1989)
10.84 D.G. Ong: *Modern MOS Technology: Processes, Devices, and Design* (McGraw-Hill, New York 1984) p. 215

11. Quantum Structural Diagrams

James C. Phillips

With 12 Figures

The structure of matter has fascinated man since early Greek philosophers, e.g., Democritus, first introduced atoms as indivisble components of matter. Today we say that structure is function, but again the Greeks said it first. On mathematical grounds Plato deduced the exact forms of the atoms of the four elements. These were: fire, tetrahedra; air, octahedra; earth, cubes; and water, icosahedra. Dodecahedra (the fifth regular solid) have appeared in modern hydrated clathrate structures $X \cdot n\ H_2O$, where X may be Xe, CH_4, Cl_2 and so on. Today we smile at the simplicity of Plato's ideas without realizing that he introduced them because he realized that modelling atoms as hard spheres was inadequate.

Since the advent in 1910 of Bragg diffraction, supplemented by computer analysis, the structures of millions of compounds and substitutional alloys have been determined with great accuracy. Thus today we are in a position which might have aroused the envy of many earlier philosophers and scientists. Given the available data, we can seek simple sets of rules or principles which can explain these structures in economical ways.

In principle, we have in quantum mechanics the tools for predicting the structure of any crystal at $T = 0$ for any pressure, and with a little statistical mechanics these results can be generalized to describe phase transitions at any temperature T up to the melting temperature T_m. In principle melting itself can be calculated by molecular dynamics simulations of sufficient sophistication. There has been substantial progress in all these areas recently, to which we allude later in this chapter. Nevertheless, all first principles methods at present are restricted to treating cases one at a time, and none of the first principles contain in themselves global descriptions of general trends. As Alice remarked so wisely, she couldn't explain herself.

The procedures of modern quantum and statistical mechanics enable us in principle to solve structural problems one case at a time. What are the possibilities for simplified global surveys of the structures of *all* solids? (Here physicists invariably impose a cutoff by restricting the discussion to inorganic, nonmolecular solids, the conventional ionic, metallic, and covalent materials. It seems that the packing of molecular solids is essentially a geometrical problem which must be solved on a case-by-case basis.) Such surveys can be of enormous benefit to our understanding, if they are successful in identifying the factors based on the quantum structures of atoms which determine the crystal structures and phase diagrams of solids.

Because of the complexity of solid structures (there are more than 200 space group symmetries alone), many scientists have long felt that the most that can be expected in simple terms is a crude guide to structures such as can be obtained by packing hard spheres. These spheres can be charged positively and negatively (cations and anions, as in $Na^{+}Cl^{-}$), and so it has been traditional in metallurgy for decades to speak of the structure of alloys as being determined classically by size and charge differences.

Can materials scientists improve on these classical models? In the early days of quantum mechanics it was hoped that solving the wave equation would routinely lead to the correct prediction of the structures and all physical properties of any properly characterized molecule or solid. With the passage of time these high hopes diminished and were replaced by great pessimism. On the one hand, the numerical complexities of the wave equation were formidable. Differences in technique between different investigators often produced variations in results of order 5 %–10 %. On the other hand, differences in energy between different solid structures are often much less than kT_m per atom, where T_m is the melting temperature. Because atomic energies are of order 10 eV, and kT_m is of order 0.1 eV, an accuracy much better than 1 % is needed for quantum structural predictions.

Within the last several decades reliable methods of computation (notably the pseudopotential method [11.1, 2]) have been developed, which in the hands of expert practitioners have achieved independently reproducible results to a few tenths of a percent. At the same time the development of pseudopotential theory has generated a common language which is just as easy to use as classical hard sphere models. As a result of these developments within the last several decades classical models of the global structure of matter have given way to quantum models. The principles behind both classical and quantum models are not exact, but the statistical success of the quantum models is far higher than that of the classical models. Typically the classical models were 60 %–80 % successful, while the quantum models are 95 %–100 % successful. With the quantum models, errors in the literature, which are sufficiently infrequent to be lost in the noise and which went undetected by classical principles, can easily be located. Thus the quantum models have more than mnemonic value, and the new quantum principles are genuinely useful. They also supplant many familiar and widely accepted classical ideas with quantum ideas that are equally simple and easily applicable, and yet much more accurate. In brief, the new models represent substantial advances in our general understanding.

In the last three years there have been dramatic advances in our understanding of the microscopic origin of exotic properties such as stable quasicrystals, high-T_c superconductivity and high-T_c ferroelectricity. These advances are discussed briefly at the end of this chapter. However, to understand how these advances have come about, we must begin at the beginning and trace the evolution of these abstract ideas which enable us to recognize underlying structural principles and relate them to extreme physical properties.

11.1 Interatomic Forces

We can differentiate ionic, covalent and metallic solids according to the nature of the attractive force responsible for cohesion. (The repulsive force opposing collapse is always the Pauli exclusion principle, which is responsible for valence kinetic energies and core-core repulsion.) In ionic crystals the attractive force is the Coulomb interaction between cations and anions. In covalent crystals the attractive force in principle is the overlap between atomic orbitals, which produces the energy gap between bonding and antibonding states. At long distances this interaction is exponential, but at equilibrium the atomic orbitals are greatly deformed and the interaction has a more nearly power-law behavior. In metals atomic-orbital effects are much confused by high coordination numbers and exchange and correlation among the conduction electrons. For this reason no simple quantum picture of metallic binding, capable of predicting structure and phase diagram of compounds and alloys, has been known until recently.

11.2 Ionic Crystals

Extremely ionic crystals (such as LiF, whose structure is shown in Fig. 11.1) can be treated very well by classical models which idealize the cations and anions (here Li^+ and F^-) as nearly isoelectronic to spherical rare gas atoms (here He and Ne). Cohesive energies and compressibilities are obtained simply in the classical model first discussed around 1910 by Ernst Madelung and Max Born. Later the classical model proved extremely useful in explaining the structures of complex, multicomponent minerals. Linus Pauling developed many rules for these structures, as described in [11.3].

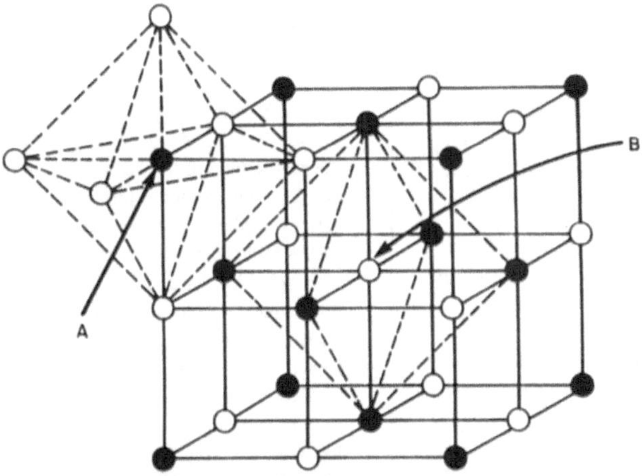

Fig. 11.1. The LiF structure: open (filled) circles, cations (anions). The octahedral prism around atom *A* is shown explicitly

Many of Pauling's rules are geometrical, and the spatial intuition developed in the early days of mineralogical structural analysis by Pauling and other crystallographers has proved of great value in other fields, such as molecular biology. Some of his discussion goes beyond geometry, which would treat cations and anions on an equal footing. Such equality is necessary so far as nearest-neighbor cation–anion interactions are concerned, but second-neighbor interactions are of two types: cation–cation and anion–anion, and classical geometrical models sometimes fail in estimating the relative strength of these second-neighbor interactions.

In geometrical models covering a large number of complex compounds relatively little effort is spent searching for subtle failings of the model which may originate from quantum interactions. On the whole the geometrical models are quite successful, provided that the ionic radii are not regarded as truly rigid. However, even compressible spheres do not predict correctly the relative stabilities of the alkali halides in the NaCl and CsCl structures [11.4], which in the classical model results from anion–anion contacts. This failure is shown in Fig. 11.2 by a classical structureal diagram (CSD).

When one passes from halides to oxides, breakdowns of classical models abound. This is because bonding in oxides contains a large covalent component, and bond angles become almost as important as bond lengths. Another way

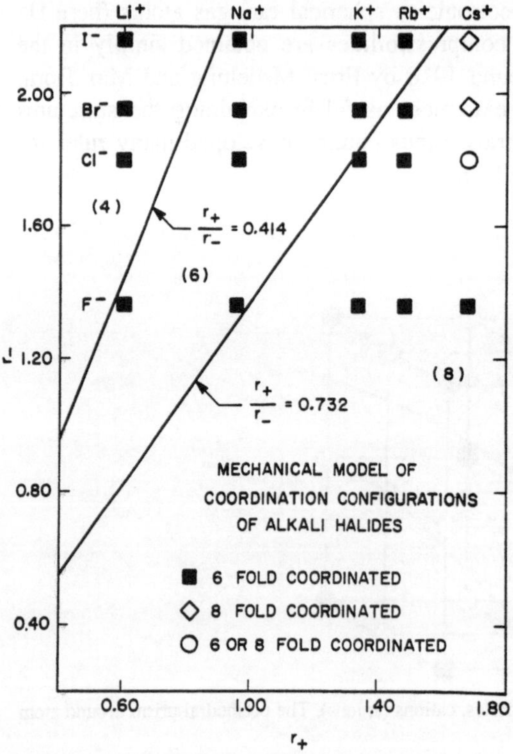

Fig. 11.2. Actual crystal structures of the alkali halides, as contrasted with the predictions of the hard sphere model. The hard sphere radii r_+ (r_-) of cations (anions) are plotted, as well as the radius-ratio lines at which, according to the hard sphere model, anion contact should generate structural transitions to lower coordination numbers (CN, in parentheses). Many contradictions are evident; LiI has CN = 6, but the model predicts CN = 4; CsF has CN = 6, but the model predicts CN = 8

of saying this, which is popular among chemists, utilizes nonbonded (second-neighbor) radii [11.5]. Even the size and formal charge of oxygen anions is highly variable, and the classical polarizability of oxygen anions varies in ionic crystals by a factor of three.

11.3 Covalent Crystals

Whereas ionic crystals represent the most favorable situation for classical models, covalent crystals represent the most favorable situation for quantum models. Crystals with the diamond structure (which includes Si and Ge) have four valence electrons per atom and are tetrahedrally coordinated, as shown in Fig. 11.3. The atomic ground state is s^2p^2, but in the crystal four directed valence orbitals from the sp^3 configuration form bonding and antibonding states. The four valence electrons fill half the s-p valence shell. In the lighter elements from column IV of the periodic table, where the s and p atomic energies in the solid are nearly equal, there is a large energy gap E_g between the bonding and antibonding states. Because the valence shell is half-filled, the bonding states are occupied and the antibonding states are empty. The stabilization energy gained from E_g overcompensates the promotion energy ΔE_p from s^2p^2 to sp^3. The resulting tetrahedral element structural is by far the simplest macroscopic manifestation of quantum mechanics, because if the atoms were merely classical spheres the coordination would be close-packed (12-fold rather than 4-fold coordination).

Crystals such as diamond, Si, and Ge, with four s-p valence electrons per atom, belong to a large family of compounds with eight s-p valence electrons

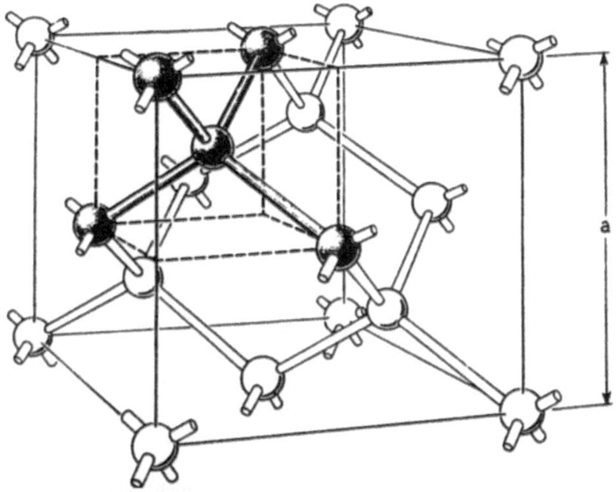

Fig. 11.3. The diamond crystal structure, with overall cubic symmetry and a coordination number of four, corresponding to four linear combinations of s, p_x, p_y and p_z atomic orbitals. These directed valence orbitals are oriented along the interatomic bonds, here represented as rods

per atom pair, represented as $A^N B^{8-N}$ and often called octet compounds. This family includes the alkali halides ($N = 1$), with ionic crystal structures, and many other familiar materials, such as GaAs ($N = 3$), an important material for optical electronics. The properties of these materials can be varied by changing pressure and temperature, but most importantly by chemical substitution to form pseudobinary alloys, such as $Al_xGa_{1-x}As$.

A quantum description of the properties of octet compounds was developed by *Van Vechten* and the present author which focuses on the average spectroscopic energy gap E_g and determines th covalent and ionic contributions (denoted by E_h and C, respectively) to E_g by simple algebraic relations [11.6–8]. This in turn led to the first quantum structure diagram (QSD), which exactly separates the ionic structures (such as Fig. 11.1) from the covalent structures (such as Fig. 11.3). The QSD which is shown in Fig. 11.4 utilizes E_h and C as Cartesian coordinates. This is appropriate because E_h and C both habe the dimensions of energy and are related to E_g by the Cartesian metric $E_g^2 = E_h^2 + C^2$. Note that the boundary dividing the 70 compounds in Fig. 11.4 into ionic and covalent structures is *straight* and corresponds to $(C/E_g)^2 = 0.79 \pm 0.01$. Because $C = 0$ for diamond-type crystals, this line can be defined as the critical ionicity line.

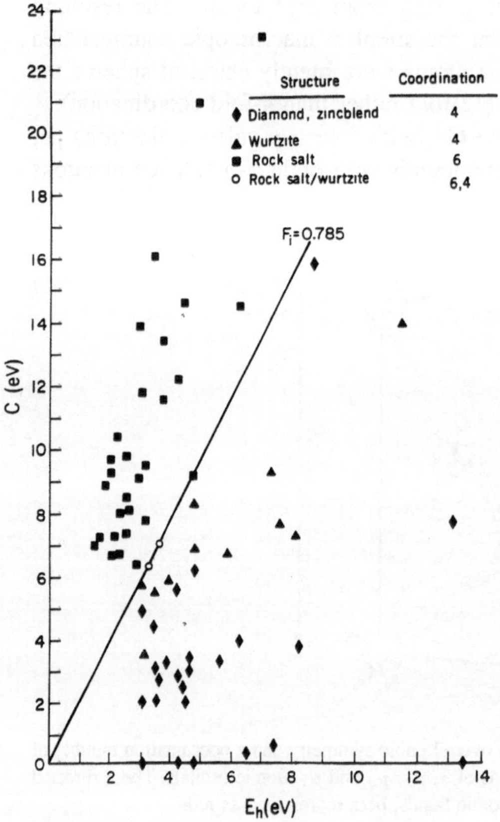

Fig. 11.4. Covalent (E_h) and ionic (C) contributions to the average energy gap E_g are used as Cartesian coordinates to separate the structures of Figs. 11.1 and 11.3 with coordination numbers four and six. The only adjustable parameter in this plot is the slope of the straight line which separates covalent from ionic structures

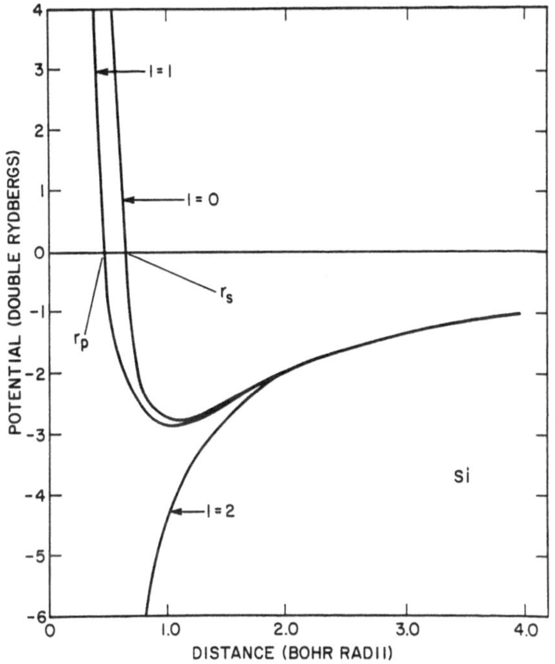

Fig. 11.5. Hard-core pseudopotentials which are used to define the quantum core radii r_s (l = 0) and r_p (l = 1) for Si, from [11.9, 10]. Also shown for comparison only is the potential seen by d states (l = 2)

The remarkable success of Fig. 11.4 as a QSD has led to efforts to extend the theory to other cases. Many of these extensions are to other physical properties of $A^N B^{8-N}$ compounds using the spectroscopic quantum coordinates E_h and C to identify and quantify chemical trends [11.6–8]. The great success of the (E_h, C) theory arises primarily from the connections it establishes between dielectric (or, more generally, optical) properties and chemical bonding through general concepts such as the random-phase approximation and the f-sum rule. These successes can be achieved for crystals which have isotropic dielectric properties and are nonmetallic, so that the covalent and ionic interactions can each be described by only one parameter. However, an aspect of the theory which initially seemed minor, but which has grown steadily in importance and finally has led to the breakthroughs described in this chapter, was the search for alternative quantum coordinates which could be readily generalized to materials other than the $A^N B^{8-N}$ family.

The new quantum coordinates which have enjoyed wide success are based on the very simple idea shown in Fig. 11.5. In an isolated atom we define for each orbital angular momentum l a hard-corre pseudopotential $V_l(r)$ such that for large r the potential is the same as the one-electron potential $V(r)$, whereas for small$r$ a large positive term simulates the effects of valence electron repulsion from the region of the atomic core by the Pauli exclusion principle. The way in which the positive term is chosen has been refined steadily for several decades, and the successes of calculations (mentioned later in this chapter) for special cases, even at very high pressures, suggest that this term is now known quite accurately. However, the success of global structural theories is such that several alternative definitions of core repulsion can be used, provided they are applied consistently throughout the periodic table.

Given a set of s and p pseudopotentials $V_s(r)$ and $V_p(r)$ for neutral atoms, we define the s and p quantum core radii by the condition

$$V_l(r_l) = 0 \tag{11.1}$$

for $l = s$ and p. This condition corresponds to the classical turning point for a probe charge of energy E scattering from an atom in the limit $E \to \mu$ (the chemical potential) = 0. We may say that these radii define the size of the quantum core, the region from which s and p valence electrons are excluded.

The first persuasive demonstration that quantum core radii were valid structural coordinates was obtained by *Bloch* and *St. John* [11.11–13] by showing that a successful QSD for the 80 $A^N B^{8-N}$ octet compounds resulted from using suitable combinations for $r_s(A)$, $r_p(A)$, $r_s(B)$ and $r_p(B)$ as Cartesian coordinates. (Here and later the actual combinations used may not be described in detail. This is because in many cases different combinations have been used by different authors, corresponding to affine transformations of the metric which do not affect the separability of different structures in significant ways.) This is shown in Fig. 11.6. On the one hand some of the simplicity of Fig. 11.4 is lost in the sense that the boundaries between the different crystal structures are not linear. (This was to be expected, because of distortions of the metric due to the fact there is no linear relationship between E_g and the hard-core radii.) However, there is an improvement in another respect. Apart from the broad covalent–ionic structural dichotomy treated in Fig. 11.4, there are subtle structural distinctions on the octet

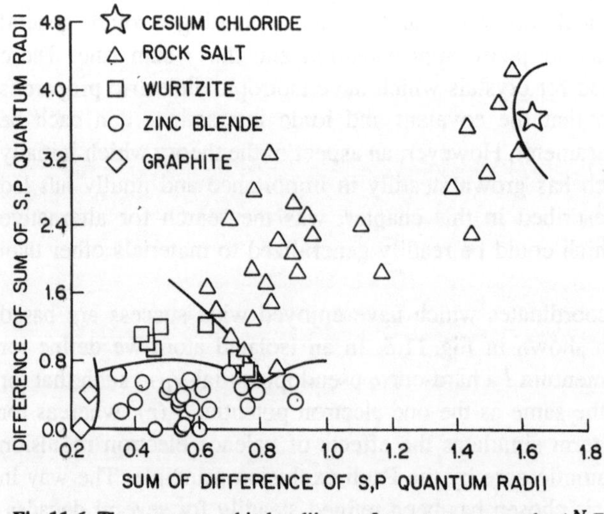

Fig. 11.6. The quantum orbital radii map for octet compounds $A^N B^{8-N}$ constructed by *Bloch* and *St. John* [11.11–13]. Note that five different crystal structures are successfully separated on this diagram

family, corresponding to slightly different crystal structures, which are faithfully resolved. (An example of one of these is the zincblende vs wurtzite pair, which corresponds to cubic-close-packed vs hexagonal-close-packed. The two structures have essentially the same nearest- and next-nearest-neighbor configurations.) Altogether five different crystal structures are distinguished in Fig. 11.6. Moreover, the slightly distorted structures are found in small pockets between the large domains defining more symmetric structures. This is a very surprising success indeed, and it suggests that very small structural differences arise just from small differences in repulsion of electrons from atomic cores, not by attractive interactions near atomic boundaries, where conventional chemical bonding pictures concentrate.

11.4 Metallic Compounds and Alloys

Spurred by the successful application of quantum core radii to octet compounds by *Bloch* and *St. John*, many researchers decided to apply the same coordinates to metals. Here the number of compounds is much larger, while binary alloy phase diagrams fill thousands of pages in the literature. The new quantum coordinates met with almost instant success, and *Chelikowsky, Andreoni, Zunger* and others soon reported QSDs which exhibited successful structural separations involving up to 500 compounds. Our own favorite case [11.14] is shown in Fig. 11.7.

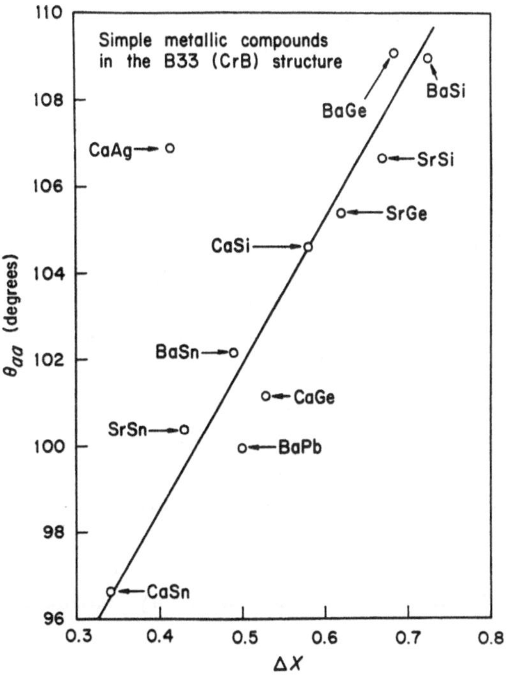

Fig. 11.7. Anion–anion helical chain angle θ as a function of the difference ΔX in average s-p quantum core radii [11.14] in CrB compounds

The pitch angles of helices in a uniaxial crystal structure are described with an accuracy of better than 1°, which is close to the experimental uncertainty.

The next important step came about as a result of a conference on crystal chemistry organized at Castle Hot Springs, Arizona, in January 1980, by *O'Keeffe* and *Navrotsky* [11.15]. This conference brought together scientists not only from many countries, but also from many disciplines. The crucial interaction turned out to be the one between American solid state physicists and physical chemists, on the one hand, and Swiss crystal chemists on the other. The new ideas discussed by the former were seized upon by *Villars* of the ETH at Zürich. What follows is a brief description of his work, which emphasizes mainly its extraordinary scope, which is without parallel in my experience.

Villars began with a healthy skepticism as to the general utility of the new quantum coordinates. True enough, Figs. 11.4 and 11.6 are very successful for octet compounds, but these compounds constitute a very special case not only because the *s-p* valence shell is half full, but also because transition and rare-earth metals are completely excluded. Historically many coordinates (including physical properties such as melting points and boiling points) have been used by chemists and metallurgists as guides in searching for new materials. Many correlations had been found for one or another of these coordinates. Most of these correlations involve only a few examples and are statistically minor compared to those shown in Figs. 11.4 and 11.6, and far less accurate than that shown in Fig. 11.7. However, all the correlations studied by the quantum theorists still spanned only a small fraction of all metallic compounds and alloys.

Given this background, *Villars* decided to search for the best set of configuration coordinates in an unprejudiced way. He reviewed the literature and finally tabulated 182 elemental candidate coordinates derived from atomic or physical properties without introducing adjustable constants. The graphical efficacy of these coordinates was tested systematically with the aid of structural maps of about 1000 simple binary (*AB*) compounds constructed by computer plotting. The details of his reduction procedure for selecting the best coordinates are fascinating and are well presented in his original articles [11.16], and they show good judgment and remarkable insight.

Space does not permit a full discussion of *Villars*'s procedure, but a few comments are appropriate. Both coordinates and properties always exhibit the periodic behavior characteristic of the Mendeleev table. Normal random variable techniques for judging the adequacy of the statistical basis for discriminating among coordinates are not suitable here. Instead, experience has shown that the number of data points needed to identify one coordinate (with the others being fixed or irrelevant) is about 10 or 15, two coordinates require 50 data points and three coordinates about 200 data points. *Villars* identified three coordinates and used a simple binary criterion based on *AB* compounds with the NaCl or CsCl structures, which includes nearly 600 data points. We may then be confident that his choice of coordinates is based on a much more than satisfactory statistical sample.

Villars's analysis of the efficacy of 182 candidate configurational coordinates confirmed some old ideas which were already generally accepted and led to some new results which at first seemed very surprising. The number of valence electrons P per $A^N B^{P-N}$ pair proves to be a good coordinate even for transition and rare-earth metal compounds when the s, p and d electrons are counted as valence electrons. However, to distinguish between s, p effects on one hand and d effects on the other, two more coordinates are necessary. One of these is the average hard-core pseudopotential radius $(r_s + r_p)/2$ already recognized by earlie workers [11.11–14] as a measure of quantum core size and studied most carefully by modern self-consistent field techniques and pseudopotential concepts by *Zunger* and *Cohen* [11.9, 10]. This is truly a quantum coordinate, and its performance is consistently better than that of the classical hard-sphere atomic radius R. Because the latter, based on measured interatomic spacings and corresponding to packing of hard spheres in simple structures (such as close-packed elemental metals), has great intuitive appeal, we pause here for an aside to examine why the pseudopotential radius $(r_s + r_p)/2$ associated with electron repulsion from atomic cores in structurally more significant than the atomic radius R associated with attractive interatomic interactions.

The clue to the origin of the inadequacy of R for structural classification lies in the following relation discussed by *Pauling* [11.1, 2]:

$$R(n) = R(1) - 0.6 \log f \ , \tag{11.2}$$

where R is in angstroms and $f < 1$ measures the fractional bond character. For an atom with v resonating valence bond electrons and with n nearest neighbors, $f = v/n$. Using relation (11.2) *Pauling* studied many intermetallic compounds, and with a relatively small amount of "poetic license" in defining the number of resonating valence bonds v he found a good fit to interatomic spacings in these compounds. However, R as defined by (11.2) is not a property of each element, because it depends on n through f. In different structures with different n, each element will have a different atomic radius. For structure plots it is necessary that each configuration coordinate be fixed for each element. Then one can prepare a periodic table containing all the elemental configuration coordinates, as *Villars* has done [11.16]. Variable configuration coordinates are unsatisfactory because they introduce discontinuities into the metric of structural maps.

Another way of analyzing size effects is to consider the possibility of forming directional s-p bonds; in most solids the d-electron charge density is too concentrated to exert a substantial directional influence. There may be no configuration coordinate which describes s-p directionality successfully on a global scale, but if such a coordinate exists it can only be $(r_s + r_p)/R$, the ratio of s-p core to atomic size. Inspection of periodic trends in $(r_s + r_p)$ and R with valence electron number fixed shows that the former in general varies much more rapidly than the latter. Also qualitatively one expects small values of this ratio to favor directional bonding in the $l \leq 1$ subspace, while large values would "squeeze" the valence bonds and reduce the $l \leq 1$ components of the atomic wave functions while increasing the $l > 1$ components. These qualitative considerations are consistent with the quantitative results of *Villars*'s statistical analysis, which show that $(r_s + r_p)$ is a better structural coordinate than R.

Prior to this long but necessary aside we said a third coordinate was needed to separate d from s and p electrons. For the third coordinate, *Villars* found that atomic ionization potentials are most successful in separating structures. For non-transition metals the ionization potential is closely related to the s-p quantum core radius and was originally used in its definition [11.11–13]. However, for transition metals ionization primarily involves removal of a d electron. We can picture the ionization potential as measuring the total valence (including s, p and d) electron charge transfer, while $(r_s + r_p)/2$ measures the s and p directed valence electron exclusion from the core region. The ionization potentials used by *Villars* are obtained by taking a weighted average of the ionization potentials of all the valence electrons. This weighted average describes cohesive energies of simple elements and compounds with an average accuracy of about 5 %. This accuracy is not high by the standards of first-principles calculations for individual cases, but it seems to be quite adequate for global structural purposes, where the periodic smoothing effects of atomic shell structure are of primary importance.

The structural significance of Villars's coordinates can be understood from the simple example of discrimination between the NaCl (6-fold coordination) and CsCl (8-fold coordination) structures. Small differences between quantum core radii are found to favor higher coordination numbers over lower ones, as do small differences in ionization potential. The quantum core result can be understood from packing considerations, but the ionization potential result may be unexpected, as the Madelung constant is larger for higher coordination numbers. However, in metallic compounds Coulomb interactions are well screened, so that Madelung potentials are of secondary importance. A large difference in ionization potentials implies concentration of most of the valence charge on the anion sublattice, which is close-packed in the NaCl structure. Thus packing considerations also explain the preference for the NaCl structure with large differences in ionization potential.

Of course, the proof of the pudding lies in the eating, and Villars's systematic search did indeed produce very successful structural separations. His map for $P = 8$ (the octet compounds previously described for non-transition, non-rare-earth binaries in Figs. 11.4 and 11.5) is shown in Fig. 11.8. The non-transition, non-rare-earth compounds are indicated, and it can be seen that addition of transition and rare-earth compounds causes no difficulties. Even the (A, B) pairs which form no compounds are themselves grouped near the origin in the structural map.

If the philosophy of using the best configuration coordinates is correct, then the coordinates which are successful for 1000 AB compounds should also succeed for 1000 AB_2 and 1000 $(AB_3 + A_3B_5)$ compounds, as *Villars* indeed found [11.16]. The overall accuracy for more than 3000 compounds is about 97 %. Bearing in mind that *Villars* uses no adjustable parameters (apart from the boundaries between different structural regions), his QSDs are convincing proof of the quantum origin of material structure and properties.

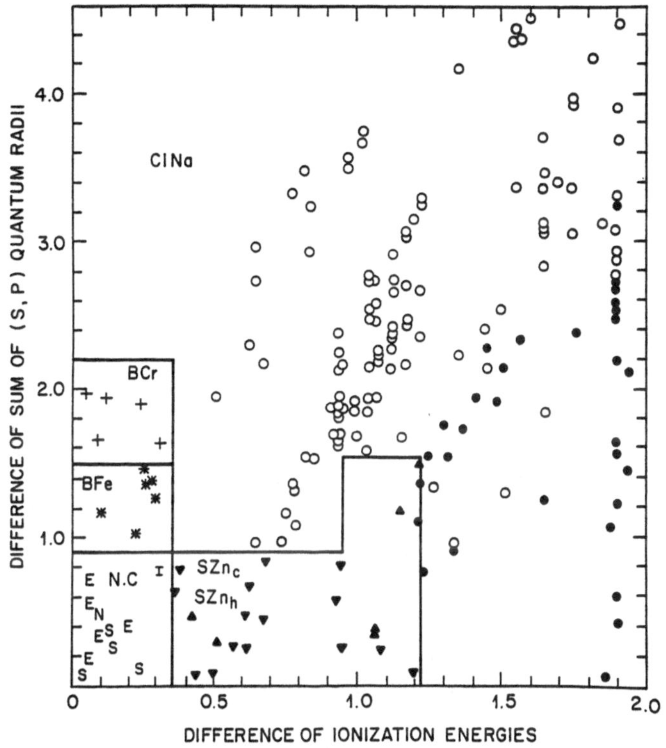

Fig. 11.8. The Villars structural map for octet compounds. Compounds involving only simple metals are indicated by filled symbols, while compounds involving transition or rare-earth metals are represented by open symbols. The region near the origin contains No Compounds (N.C.), which are distinguished in terms of the binary alloy phase diagram as Eutectic (E) or Soluble (S)

There are many possible applications of these ideas, but one of the most demanding has been examined already by *Villars*. Can structural coordinates distinguish compound-forming (A, B, C) ternaries from non-compound-forming ternaries? The three coordinates which best answer this question are the valence electron number difference, the difference in quantum core radii, and the melting point ratio [11.16]. With these coordinates, as shown in Fig. 11.9, the ternary non-compound-forming region is localized close to any two of the three coordinates. Once again quantum core radii play a crucial role.

Purists will notice an interesting feature of the Villars AB coordinates in Fig. 11.8. Both coordinates are difference coordinates which vanish when $A = B$. This means that all elemental structures are located at the origin (the center of the non-compound-forming region), and the Villars coordinates do not discriminate among them. However, the number of elemental structures is only 2 % of the number of simple binary AB structures. Earlier work [11.11–13] showed that r_s and r_p could be used to classify elemental structures, and if these additional coordinates were added to the three used by Villars, higher accuracy would by achieved. However, when three coordinates were 97 % successful, it is doubtful

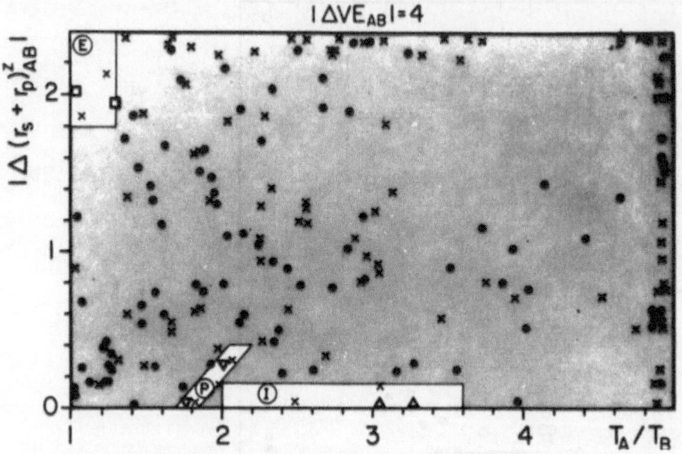

Fig. 11.9. Structural map predicting (non-)compound formation in ternary systems (from [11.16]). Note that in this case the melting temperature ratio T_A/T_B plays an important role, but equally important are the orbital radii coordinates. It is remarkable that the latter (which deal with atomic energies on a scale of 10 eV) can be as structurally significant as the former ($kT_A \sim kT_B \sim 0.1$ eV)

whether achievement of 98 % or 99 % accuracy warrants the use of two additional coordinates. Another way of saying this is that the structures of the elements are really rather exceptional because in the absence of ionic interactions quantum-mechanical resonance effects are abnormally large. Indeed, one of the insights provided by quantum structural diagrams is the realization of just how atypical elemental structures often are compared to compound structures.

Crystalline compounds based on icosahedral building blocks have recently attracted great interest. Small alterations in stoichiometry of known compounds with about 150 atoms/cell generate "quasicrystals" with double periodicities along cubic axes [11.17]. The most characteristic property of these compounds is that their formation as equilibrium phases apparently requires the combination of three elements, for example Si, Mn and Al, or Cu, Li and Al. *Villars*'s QSDs explain this requirement quite simply as the result of having to satisfy *two* conditions [11.18]. The first is a delicate cancellation of size and electronegativity differences (Fig. 11.8) for atoms with average s^2 valence configurations. The second is very close proximity to the compound–non-compound-forming boundary (illustrated in Fig. 11.9). These two conditions are so demanding that no binaries and $< 10^2$ of 10^5 possible ternary combinations meet them [11.18].

11.5 Molecular Structure Diagrams

Many materials scientists may find themselves overwhelmed by the idea of organizing several thousand crystal structures and phase diagrams on three-configurational coordinate maps. What about the simpler case of molecules? Here the number of cases available for study seemed for many years to be statistically insignificant, except for diatomic molecules, which present no structural problem! However, in the last decade the structures of a large number of triatomic molecules $A^N B_2^M$ have been either determined experimentally or calculated by sophisticated quantum-mechanical methods. These include abut 35 "double octet" molecules with $N+2M = 16$, and more than 100 molecules with $N+2M < 16$.

Because the structures of crystalline compounds are so well separated on diagrams with coordinates defined by quantum orbital radii it is natural to construct similar diagrams for triatomic molecules, as recently done by *Andreoni* et al. [11.19]. The simple case is $N + 2M = 16$, which includes familiar triatomic molecules such as CO_2. In this case apex angles different from linear (180°) had previously been calculated with fair accuracy (about ±10°) by *Galli* and *Tosi* using a classical ionic model. The correlation of the apex angle with the difference in quantum orbital radii is shown in Fig. 11.10. Note that the correlation is successful to better than 1°. Such a great improvement between the classical model and the quantum model should not surprise us, because the quantum model for angular geometry was equally successful in a crystalline example (Fig. 11.6) where the interactions are much more complex. Many such examples have been obtained, and this latest success in the molecular case is representative, not exceptional.

Andreoni and coworkers also studied the structures of bent and linear triatomic molecules with $Q = N + 2M < 16$. The values of Q range from 8 (as in K_2O) to 12 (Al_2O) to 15 ($LiCl_2$). Strictly speaking, different values of Q should produce different structures, but the numbers of such molecules which have been studied would not be statistically significant if they were divided into different groups according to each value of Q. Therefore all molecules with $Q < 16$

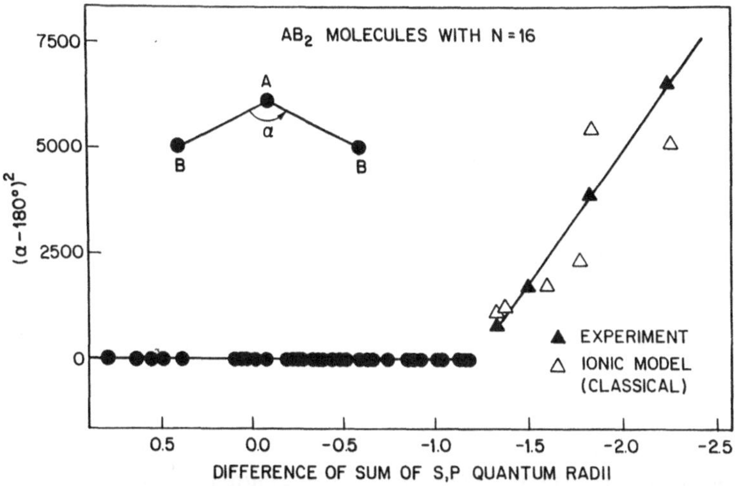

Fig. 11.10. Correlation between triatomic apex angles and quantum orbital s-p radius difference Y [11.9, 10] for double octet molecules. Note the large deviations between theory and experiment for the classical ionic model

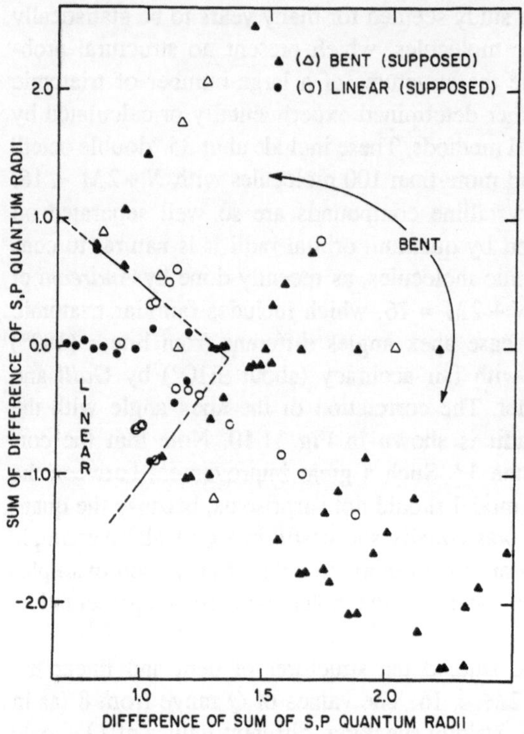

Fig. 11.11. Structural map for electron-deficient triatomic molecules. Here Y is as in Fig. 11.10, while X is the sum of the s-p radii difference [11.9, 10]

were classified as linear or bent, and a structure plot using coordinates derived from quantum orbital radii showed complete separation of all the experimentally known structures (Fig. 11.11). Many structures studied theoretically which were supposed to be linear fall in the bent region of the experimental structure map. This suggests that these theoretical calculations might profitably be re-examined.

The development of quantum structural diagrams reflects a new attitude in our approach to understanding the atomic structure of matter. With classical models one begins with poor configuration coordinates (such as atomic radii) and then carries out either extensive algebraic calculations by hand or even more extensive numerical calculations with supercomupters. (A good example of the latter is numerical simulation of "glass transitions" in hard-sphere systems.) It is then argued that these results have some physical significance because of the quantity of technical effort involved in the computations.

Mindful of the computer acronym GIGO (garbage in, garbage out), the new attitude at the outset focusses on finding good configuration coordinates. With this emphasis on the quality of coordinates, rather than on the quantity of computation, we are led to QSDs, which in a few strokes have already contributed far more to our understanding of solids than many classical computations combined. Together with most of the other sweeping advances which have occurred over the last three decades in the quantum theory of the structure of solids, these new coordinates have emerged from pseudopotential theory [11.1, 2, 9, 10].

11.6 Deductive Calculations

As we mentioned at the beginning, only a case-by-case treatment is possible at present from first principles. In fact, even structurally reliable first-principles calculations for the total energy of solids (other than the alkali metals) have appeared only quite recently. To make an honest test of such calculations, it is not sufficient to show good agreement with the observed structure at $P = T = 0$, since such agreement could be accidental. Fortunately in quantum-mechanical total energy calculations P can be increased to arbitrarily large valuess. Thus if the solid is known to undego structural transformations at $P = P_n$, calculation of P_n constitutes a fair test of the reliability of the total energy calculation. In some respects such a test is more informative than a calculation of the cohesive energy, because the latter involves the total energy of the atom, a few-electron problem which at present is poorly understood in open-shell cases involving several valence electrons.

The total energy of silicon in four crystalline phases has been calculated by *Cohen* and coworkers (see Chap. 5 of this volume) as a function of volume and their results are shown in Fig. 11.12. From these curves transition pressures and volumes are calculated by the Gibbs common tangent (tie line) construction, and the results are in excellent agreement with experiment [11.20]. Similar results

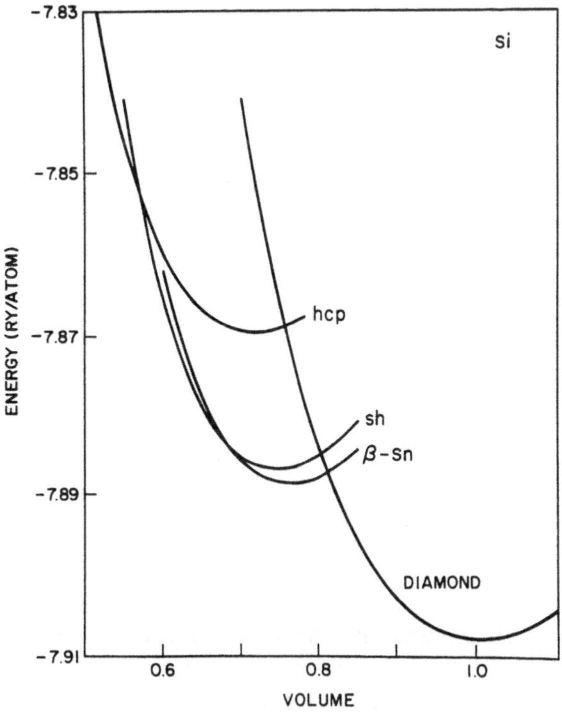

Fig. 11.12. Structural energies for four crystalline phases of silicon as a function of volume as calculated by the pseudopotential method in [11.17]

have been obtained for several $N = 3$ $A^N B^{8-N}$ binary compounds. One of the phases shown in Fig. 11.12 is the simple hexagonal (sh) phase (one atom/cell); this phase, whose existence was predicted theoretically, is observed under pressure for a pure element (rather than a substitutional alloy) only in silicon. One of the advantages of a genuine quantum-mechanical calculation of the high accuracy achieved by *Cohen*'s group is the ability to discuss other properties. In this case the calculation *predicted* that the sh phase would be superconducting with a transition temperature T_c near 10 K. This prdiction has been confirmed [11.21]. Because of the complexity of the microscopic theory of superconductivity, and the relative ease with which the phenomenon can be measured experimentally, this is only the third example known to me of a bona fide theoretical prediction of superconductivity or superfluidity. The first two were He^3 in 1960 by Pitaevski in the USSR and Brueckner, Anderson, and coworkers in the US, and degenerate semiconductors in 1966, also by *Cohen.*

From curves of the kind shown in Fig. 11.12 one can construct simplified classical model force fields [11.22], which can be used to carry out studies of melting by molecular dynamics simulationss, which are beginning to yield interesting results [11.23]. These powerful calculations, which exercise supercomputerss to capacity, are a severe test of the accuracy of the original quantum-mechanical calculations on which they are based. Even so, they still do not provide (nor are they designed to provide) the kind of overview of material trends supplied by QSDs. The success of the latter is a dramatic example of constructive interdisciplinary interactions in materials science [11.1, 2].

11.7 Prospects

The rapidly growing sophistication of materials science, for example in sample preparation contrrolled on an atomic scale by techniques such as molecular beam epitaxy, calls for an equally rapid growth in depth and breadth of theoretical constructs and methods. Much current theory is devoted to abstract mathematical models which rarely, if ever, make contact with current practice. Thus 15 years ago I suggested that global understanding of the structure of solids might have been the most basic unsolved problem in solid state physics [11.24]. The improvements resulting from replacing CSDs by QSDs discussed in this article show that theory, even in the guise of apparently abstract quantum-mechanical calculations, can make contact with larger aspects of science and technology by fostering interdisciplinary interactions in a positive spirit. These advances in our understanding of the structure of solids are part of a recent crystallographic handbook compiled by *Villars* and *Calvert* [11.25]. Their work goes much farther than I had thought possible in an article [11.24] regarded by many at the time as visionary and perhaps even foolish. The handbook includes critically evaluated data on 22 000 individual phases. On the more fundamental side the need for soluble paradigmatic models which can connect QSDs to accurate first-principles calculations is evident. Some progress in this area has recently been made by *Chelikowsky* and *Burdett* who carried out first-principles calculations on a binary potential model of varialbe ionicity [11.26]. Another example of positive interdisciplinary interactions in the quite different field of non-equilibrium statistical mechanics is the new wave of experiments on the physics of glass [11.27], which has stimulated a new area of mathematical physics called vector percolation [11.28]. But most striking of all are recent applications of QSDs to the systematics of quasicrystral formation [11.18] (at present described by mathematicians in terms of tile packing; the tiles have no chemical significance), high-T_c superconductivity [11.29, 30] (where mathematicians use exotic models which do not even identify correctly electron–phonon interactions as the relevant microscopic mechanism), and high-T_c ferroelectricity [11.30] (where even the dependence on valence electron number had not previously been recognized).

References

11.1 M.L. Cohen: Science **179**, 1189 (1983)
11.2 M.L. Cohen, V. Heine, J.C. Phillips: Sci. Am. **246** (6), 82 (1982)
11.3 L. Pauling: *Nature of the Chemical Bond* (Cornell University Press, Ithaca, NY 1960) See also D.M. Adams: *Inorganic Solids* (Wiley, New York 1974)
11.4 J.C. Phillips: In *Treatise on Solid State Chemistry*, Vol. 1, ed. by N.B. Hannay (Plenum, New York 1973) p. 1
11.5 M. O'Keeffe, B.G. Hyde: J. Solid State Chem. **44**, 24 (1982); Nature **309**, 411 (1984)
11.6 J.C. Phillips, J.A. Van Vechten: Phys. Rev. Lett. **22**, 705 (1969)
11.7 J.C. Phillips: Rev. Mod. Phys. **42**, 317 (1970); *Bonds and Bands in Semiconductors* (Academic, New York 1973); Science **169**, 1035 (1970)
11.8 J.C. Phillips: Today **23** (2), 23 (1970)
11.9 A. Zunger: [11.15], p. 73, Phys. Rev. Lett. **47**, 1086 (1981)
11.10 A. Zunger, M.L. cohen: Phys. Rev. B **20**, 4082 (1979)
11.11 J.St. John, A.N. Bloch: Phys. Rev. Lett. **33**, 1095 (1974)
11.12 G. Simons, A.N. bloch: Phys. Rev. B **7**, 2754 (1973)
11.13 G. Simons, R.G. Parr: J. Chem. Phys. **55**, 4197 (1971)
11.14 J.R. Chelikowsky, J.C. Phillips: Phys. Rev. B **17**, 2453 (1978)
11.15 M. O'Keeffe, A. Navrotsky (eds.): *Structure and Bonding in Crystals* (Academic, New York 1981)
11.16 P. Villars: J. Less-Common Met. **92**, 215 (1983); ibid. **99**, 33 (1985); ibid. **109**, 93 (1985); ibid. **119**, 175 (1986)
11.17 V. Elser: Acta Crystallogr. A **42**, 36 (1986)
11.18 P. Villars, J.C. Phillips, H.S. Chen: Phys. Rev. Lett. **57**, 3085 (1986)
11.19 W. Andreoni, G. Galli, M. Tosi: Phys. Rev. Lett. **55**, 1735 (1985)
11.20 K.J. Chang, M.L. Cohen: Phys. Rev. B **30**, 5376 (1984)
11.21 K.J. Chang, M.M. Dacorogna, M.L. Cohen, J.M. Mignot, G. Chouteau, G. Martinez: Phys. Rev. Lett. **54**, 2375 (1985)
11.22 R. Biswas, D.R. Hamann: Phys. Rev. Lett. **55**, 2001 (1985)
11.23 F. Stillinger, T. Weber: Phys. Rev. B **31**, 5262 (1985)
11.24 J.C. Phillips: Comments Solid State Phys. **4**, 91 (1972)
11.25 P. Villars, L.D. Calvert: *Pearson's Handbook of Crystallographic Data for Intermetallic Phases*, Vols. 1–3 (ASM, Metals Park, OH 1985)

11.26 J.R. Chelikowsky, J.K. Burdett: Phys. Rev. Lett. 56, 961 (1986)
11.27 J.C. Phillips: Phys. Today (Feb. 1982)
11.28 J.C. Phillips, M.F. Thorpe: Solid State Commun. 53, 699 (1985)
11.29 P. Villars, J.C. Phillips: Phys. Rev. B 37, 2345 (1988)
11.30 J.C. Phillips: *Physics of High-T_c Superconductivity* (Academic, Boston 1989)

12. Ion and Laser Beam Processing of Semiconductors: Phase Transitions in Silicon

John M. Poate

With 26 Figures

The use of energetic ion and laser beams to analyze and modify surfaces has had a profound impact on the development of many areas of materials science. This scientific field is very large in scope and here we will concentrate on the best understood and most technologically important subject: Si processing science and technology. The emphasis will be on what new properties have been discovered or what new phenomena have been explored using ion or laser beams. Although the physics of the beam/solid interactions are different for ions or lasers, there is a commonality between the two subjects. These techniques have led to a better understanding of many phenomena, from energy transfer to defect reactions. The biggest impact however, on materials science has probably been in the areas of phase transitions and crystal growth and these are the areas we will discuss. The beam techniques have resulted in new findings in the three commonly recognized condensed phases of Si: crystalline (c), amorphous (a) and liquid (l).

The subject of ion implantation will be discussed first with an emphasis on the salient physical parameters. The fact that ion implantation can be used to produce clean amorphous layers with clean a–c interfaces has had several ramifications. Most importantly, the subject of solid phase epitaxy has been developed. We will then digress with the interesting topic of ion-beam-induced epitaxy, as several new phenomena have been discovered with regard to solid phase crystal growth and diffusion. The ability to produce a-Si layers and melt them very rapidly using lasers has opened up the areas of very rapid crystal growth from the liquid phase. Moreover, using these techniques it has been possible to observe and measure the first-order melting transition of a-Si.

The interplay between the science and commercial development of these subjects is fascinating. Figure 12.1 gives some idea of the development of ion implantation. The first report [12.1] came from *Ohl* at Bell Laboratories in 1952 and *Shockley* [12.2] obtained the patent in 1954. The path, however, from invention to commercial acceptance took many years. Groups in the United States and Europe studied, in the 1960s, the basic physics and materials science of the process. Understanding the ranges of the implanted atoms, the damage they produced and annealing mechanisms were key developments. In conjunction with these efforts, the Rutherford backscattering and channeling technique was developed to analyze damage and the lattice-site location of implanted atoms. However, ion implantation did not gain commercial acceptance until the early 1970s, when it was recognized that only implantation doping could be used to shift threshold voltages in field effect transistors in a reproducible and reliable fashion. Since

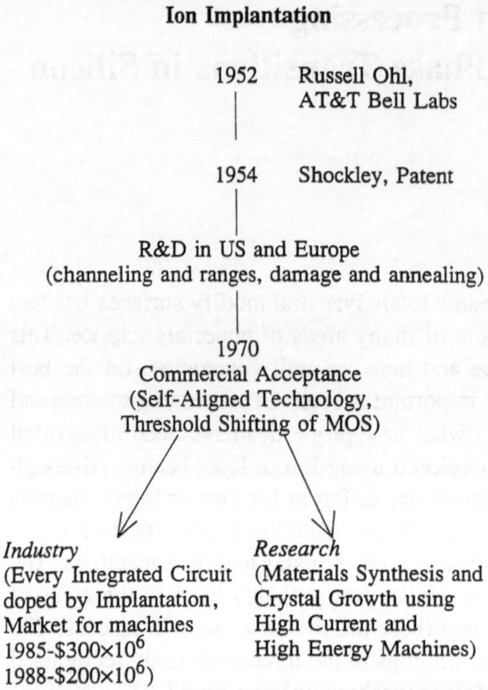

Fig. 12.1. The evolution of ion implantation. As a figure for comparison, a production implanter costs $\sim\$ 1 \times 10^6$

then implantation has become an essential and integral part of the semiconductor industry. Every integrated circuit now made uses ion implantation for doping. As Fig. 12.1 shows, the world-wide market for ion implantation machines alone is quite considerable. The research aspects of implantation continue to flourish. Very high dose implants are now being used to synthesize buried layers such as SiO_2 or silicides. High-energy machines are being used to explore deep implants or unconventional areas of crystal growth.

The subject of laser semiconductor interactions in the context of materials science is somewhat younger. The observation in the latter part of the 1970s that pulsed laser irradiation of ion-implanted and amorphized Si could remove the damage and recrystallize Si gave rise to the field known as laser annealing [12.3]. Soviet scientists pioneered this field and there was an explosion of interest in 1977–1978 with workshops in Albany, New York, and Catania, Italy, followed by a symposium of the Materials Research Society in Boston in 1978. Figure 12.2 outlines the development of the subject since then. The ability to rapidly heat or melt, and quench, well-defined surface layers in novel temperature and time regimes has led to fundamental developments in Si crystal growth, and to an improved understanding of the thermodynamic properties of a-Si.These studies typically take place on nanosecond time scales, but an interesting technological offshoot has been the establishment of rapid thermal annealing (RTA) technology. The fundamental studies in the very short time scale encouraged materials scientist and process engineers to investigate annealing regimes outside

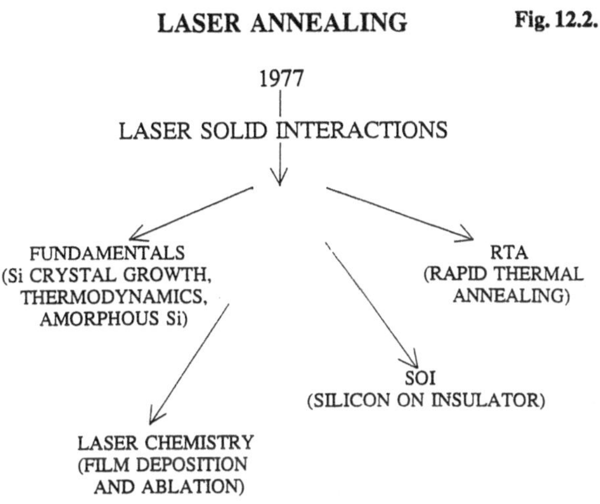

Fig. 12.2. The evolution of laser annealing

those encountered in the usual tube furnaces (e.g., 1000° C for 30 min). Indeed they have demonstrated that Si and GaAs wafers can be processed on the time scale of seconds, and RTA is now a flourishing technology. Other significant offshoots have been the deposition and ablation of films (i.e., laser chemistry) and silicon-on-insulator technology. The laser melting of Si using continuous wave (cw) lasers led to the concept of lateral epitaxy. Polycrystalline Si or a-Si can be melted and the melt puddle scanned over the dielectric SiO_2. If the initial melt takes place over a Si seed very large grains of single crystal Si can be produced on SiO_2. This dielectrically isolated Si has applications in radiation-hard or three-dimensional (3D) integrated circuits.

12.1 Ion Implantation

Ion implantation doping has become one of the dominant processing technologies in Si integrated circuit fabrication. The ability to implant electrical dopants into the surface layers of semiconductors with precise spatial and compositional control is the reason for the success of this technology. Planar Si technology involves the fabrication of many circuit elements in a two-dimensional fashion on a flat Si wafer. This configuration permits exact control over the fabrication steps required to make and connect the many individual circuit elements. The processing steps and materials involved can be exemplified by the field effect transistor (FET) shown in Fig. 12.3. The goal of an electronic device is to control the flow of charge carriers, i.e., electrons or holes. In the FET the flow of carriers from the source to the drain is controlled by means of the gate electrode. The device is fabricated in the following fashion. Perfect Si crystals are cut into wafers that are then polished to give mirror-like and defect-free surfaces. Many different processing steps are required to build the structure and these steps must

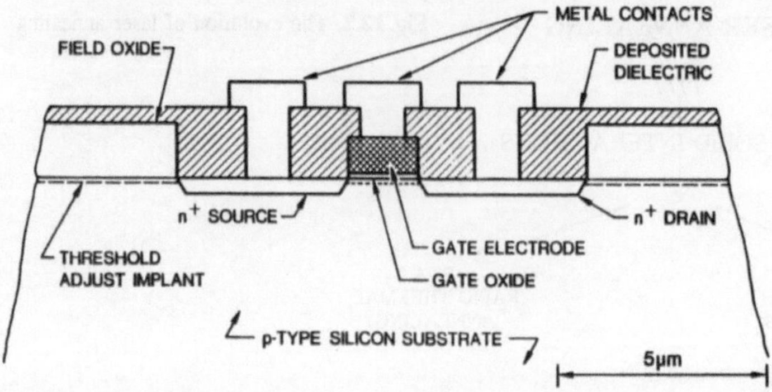

Fig. 12.3. Cross-sectional view of a metal oxide semiconductor field effect transistor (MOSFET) structure showing the patterning and different components required to fabricate such a device structure

be accurately confined to specific regions of the wafer. This control is effected by pattern-transfer techniques that are used to transfer geometric shapes on a mask to the surface of a Si wafer. For example, the Si wafers are heated in oxygen to form insulating SiO_2 layers to form the field oxide. Lithographic techniques are then used to transfer polymeric masks (photoresists) onto the regions that need to be protected in subsequent etching steps. The SiO_2 can be etched chemically or by plasma-assisted etching. In this way the unmasked SiO_2 is removed and source–drain regions are exposed. Pure Si at room temperature has high resistivity, and dopants must be introduced to give the Si its conducting properties. The dopants, such as As, are injected into the Si by means of ion implantation. The whole wafer is exposed to the energetic As beam, but only the exposed or unmasked regions are implanted. This process is known as a self-aligned technology. A complex integrated circuit may have from 6 to 12 separate implantation steps.

The first studies of implantation in semiconductors appear to be due to *Ohl* [12.1]. In 1952 he described the improvement in the electrical characteristics of Si point contact diodes produced by bombardment with H, He, N and Ar. The improvements were, in fact, caused by surface damage and not specific chemical dopings. The concept of chemical doping by implantation is spelled out in the 1954 patent by *Shockley* [12.2]. This patent, entitled "Forming Semiconductive Devices by Ionic Bombardment", has proved remarkably prophetic and still covers the subject as it exists today.

There are now many reviews and books on ion implantation [12.4–6]. We will emphasize here some of the physical concepts and give an idea of the magnitude of the processes involved. We will firstly consider the energy loss and ranges of implanted ions. An ion traversing a solid loses energy by way of the Coulomb interaction with target atoms and electrons of the solid. It is convenient to mathematically separate these two components of energy loss per unit length:

$$\left(\frac{dE}{dx}\right)_{\text{total}} = \left(\frac{dE}{dx}\right)_{\text{nuclear}} = \left(\frac{dE}{dx}\right)_{\text{electronic}} . \tag{12.1}$$

The nuclear term involves substantial energy and momentum in transfer between the incoming ion and the atoms of the solid, so that much energy loss and trajectory deviation can be imparted in these "elastic" collisions. The electronic term refers to energy lost to the electrons of the solid in collisions which do not involve substantial momentum transfer. Such "inelastic" energy loss will not cause important deviations of the particle in its trajectory. The parameter of interest in implantation is the range, which can be obtained by integrating (12.1), i.e.,

$$R = \int_0^E \frac{dE}{(dE/dx)_{\text{total}}} , \tag{12.2}$$

where E is the incident energy. The calculations of the ranges using the Lindhard, Scharff and Schiott (LSS) theory [12.8] are good to within 10%.

Energy loss is a statistical process, thus there will be a spread in the range distribution and, to a first approximation, the range distribution will be Gaussian. The standard deviations in the ranges are large because, for most heavy-ion–target combinations, nuclear stopping predominants, and the path of the ion through the solid will resemble a random walk processs. A more exact fit to range distributions involves a four-moment (Pearson-IV) approach from statistical theory. Figure 12.4 shows such a fit to ranges for Boron [12.9] in fine-grain, polycrystalline Si. Gaussian functions only fit the ranges at low implantation energies or near the maxima of the distributions. The skewness on the shallow part of the profile can be understood in terms of B backscattering off the heavier Si atoms.

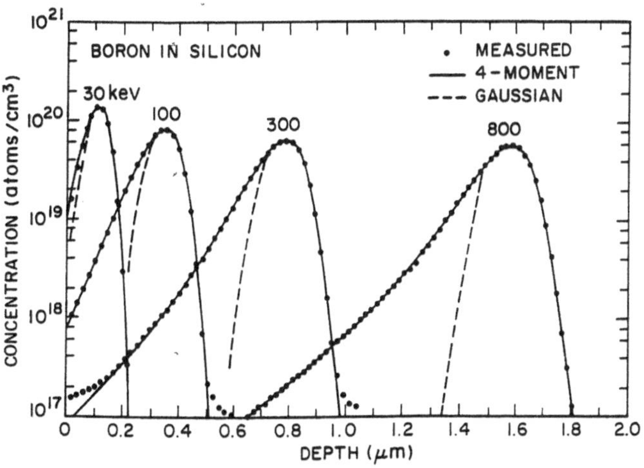

Fig. 12.4. Boron range distribution in Si for 30, 100, 300 and 800 keV respectively. The B was implanted into fine-grain polycrystalline Si to obviate channeling effects. From [12.9]. Calculated ranges are four-moment or Gaussian

Arsenic, which is heavier than Si, shows a skewness toward the deep side of the distribution. The width of the distributions is known as the straggling and can be quite significant, as shown in the figure.

12.2 Amorphization and Solid Phase Epitaxy

An energetic ion penetrating a solid target can impart sufficient kinetic energy to lattice atoms during nuclear collisions to cause atomic displacements. Typical displacement energies are 20 eV. The recoiling lattice atom may itself possess sufficient kinetic energy to displace many other lattice atoms. As a result, a cascade of displaced atoms may originate from a single collision between the implanted ion and a lattice atom. During its path, the implanted ion may initiate many such displacements cascades within a characteristic volume surrounding the ion track. The time scale over which the cascade occurs is typically 10^{-12} s. The effects of these cascades have been captured by Narayan in some beautiful high-resolution transmission electron microscopy. Figure 12.5 shows a lattice image micrograph of an amorphous Si zone produced by a single 100 keV Bi ion during bombardment of Si (100). The ion traversed the solid until it underwent a violent scattering event, which resulted in a collision cascade and amorphous zone formation. It should be noted that each Bi^+ ion has directly amorphized a

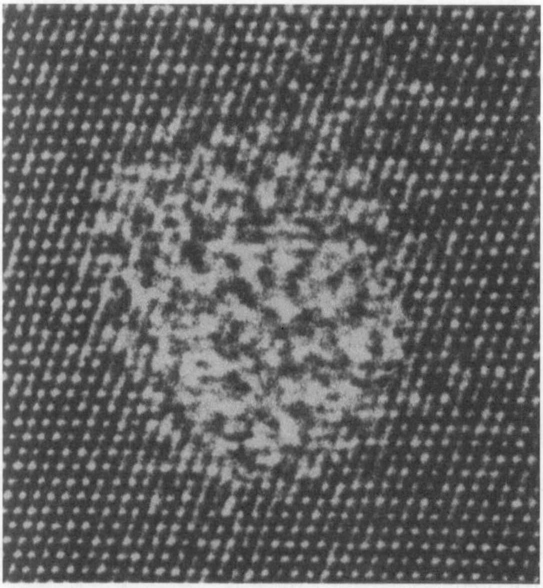

Fig. 12.5. High resolution transmission electron microscope lattice image of an amorphous Si zone produced by a collision cascade from a single 100 keV Bi ion. (From J. Narayan, private communication)

volume containing several thousand Si atoms. This is a very efficient damage event which cannot be modeled using simple collision theory. In contrast to the observations in Fig. 12.5, bombardment of Si with light ions such as B does not result in large amorphous zones about individual ion tracks. Indeed, light ion bombardment produces isolated damage clusters which contain few displaced atoms.

A simple qualitative picture can be used to describe amorphization by heavy ion bombardment. Typical heavy ion damage (as in Fig. 12.5) builds up with ion dose, via an increase in the density of amorphous zones, until zone overlap eventually leads to the formation of a continuous amorphous layer. From the zone dimension for the Bi implant of Fig. 12.5, the minimum dose required to form a continuous amorphous layer would be $\sim 5 \times 10^{12}\,\mathrm{cm}^{-2}$. This number is close to those observed experimentally, which are typically $\gtrsim 10^{13}\,\mathrm{cm}^{-2}$. For light-ion bombardment, during which amorphous layers are produced by the accumulation and overlap of regions of discrete defects, amorphous threshold doses are much higher than those for heavy ions (typically $> 10^{15}$ ions cm^{-2}) and the amorphization process is considerably more complex. The fact that annealing can take place during bombardment severely complicates an assessment of the nature and degree of damage arising from ion bombardment. The observed damage will always be the result of competing disordering and annealing processes.

The ability to form clean, amorphous layers by implantation has contributed significantly to our understanding of solid phase recrystallization processes. For ion-implanted layers, the recrystallization process proceeds epitaxially on the underlying crystalline substrate. This epitaxial growth process is clearly illustrated by the series of cross-section TEM micrographs in Fig. 12.6, from [12.10]. Annealing for various times at 525° C results in regrowth of the amorphous layer produced by the implantation of 200 keV Sb^+ at a dose of $4.4 \times 10^{15}\,\mathrm{cm}^{-2}$ into Si (100). The epitaxial regrowth rates are about 1.5 Å/min for these annealing conditions. The width of the a-Si layer is indicated by the arrows. The a–c inter-

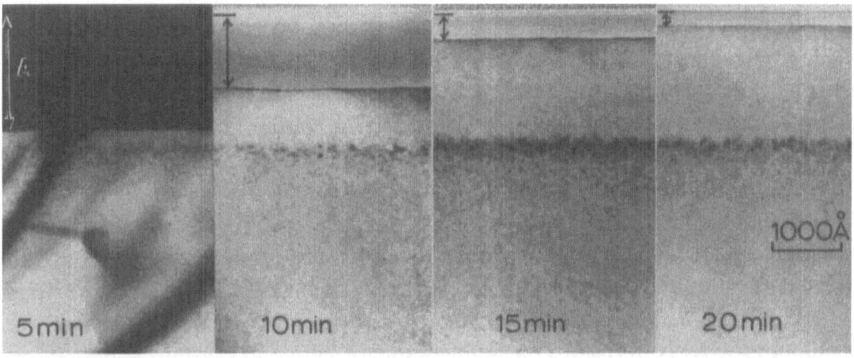

Fig. 12.6. Transmission electron microscope cross-sectional micrographs for 200 keV Sb^+ ($6 \times 10^{15}\,\mathrm{cm}^{-2}$) implants in Si annealed at 525° C. (From [12.10])

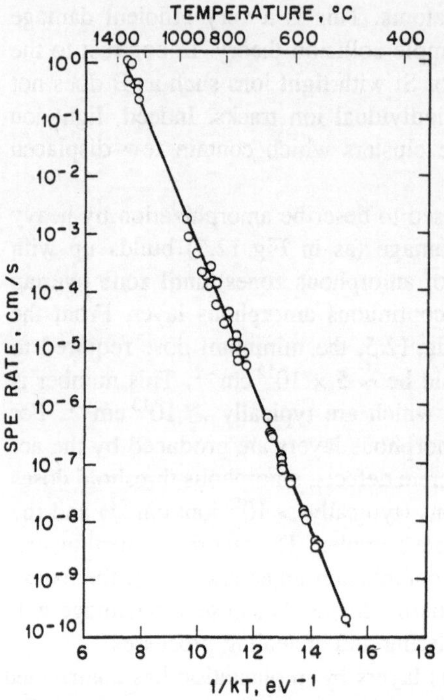

Fig. 12.7. Dependence of solid phase epitaxy rate on temperature in As-implanted Si(100) samples. (From [12.11])

face is atomically sharp with an overall roughness of less than 50 Å. Moreover this planarity is maintained during the epitaxial growth process. The dark band of defects at $\sim$ 1600 Å below the surface corresponds to incompletely annealed end-of-range damage resulting from beam heating. The velocity of the phase boundary as a function of temperature is shown in Fig. 12.7 from [12.11]. These data are remarkable as they show a single activation energy of 2.7 eV for regrowth over a velocity range spanning ten orders of magnitude.

There is still not yet a complete atomistic picture of the crystal growth process at the amorphous crystalline interface. This solid-phase epitaxial process occurs because of the peculiar features of the implantation process. The interfaces are atomically clean and the amorphous Si layers are impurity and void-free (i.e., microstructure-free). It is quite difficult experimentally to achieve the solid-phase epitaxy of deposited amorphous Si films on Si. The smallest amount of interfacial impurities or internal oxidation of microstructures will retard or completely halt interfacial recrystallization, so that polycrystalline films will result.

The successful evolution of implantation as a tool for device fabrication has crucially depended upon the ability to anneal the implantation damage and produce the required electrical activity of the dopants [12.12]. The previous discussion outlined some of the features of damage production. The aim of the annealing process is to restore the semiconductor lattice to crystalline perfection and incorporate the implanted atoms on lattice sites where they can act as electron donors or acceptors. The characteristics of the annealing required will differ for light and heavy ions. The light ions, in general, will not amorphize the surface layers and will produce isolated defect clusters. Dopant activation is more efficient when heavy ions are employed, since annealing results in solid phase epitaxy. Annealing, however, at temperatures greater than 1000° C will produce 100 % electrical activation in Si only as long as the concentration of dopant does not exceed the solubility limit.

12.3 Ion-Beam Induced Epitaxy, Diffusion and Segregation

In the previous section we discussed the formation of planar, a-Si surface layers by ion implantation and the expitaxial recrystallization of these layers by heating. Solid phase epitaxy can be extended to much lower temperatures by means of high energy heavy ion irradiation as shown schematically in Fig. 12.8. Heavy ions with typical energies in the MeV range are shone through the a-Si layer into the underlaying c-Si. The production of defects and cascades at the a–c Si interface enhances the epitaxial recrystallization process. Figure 12.9 shows the a–c interface velocity as a function of temperature during thermal (i.e., furnace) and ion-beam-enhanced recrystallization. The a-Si layers on Si (100) are typically several thousand angstroms thick. In this example, 2.5 MeV Ar ions were used to induce crystal growth at a dose rate of 7×10^{13} ions/cm^{2}s. The interface motion is characterized by an activation energy of 0.3 eV with enormous velocity enhancement over strictly thermal epitaxy for temperatures $< 400°$ C. The ability to induce interface motions at such low temperatures has allowed investigation of several unusual crystal growth and segregation phenomena [12.13]. As in the case of strictly thermal epitaxy, at this time there is not a detailed understanding of the atomic processes which occur at the a–c interface during ion-beam-enhanced epitaxy. *Jackson* [12.14] has recently presented an intracascade model in terms of competition between amorphization and crystallization at the a–c interface. The model fits the data well. It is speculated that amorphization is produced by diffusion and condensation of vacancies and interstitials from c-Si at the interface and that crystallization results from the motion of dangling bonds in the a-Si at the interface.

We will discuss segregation at the a–c Si interface in the thermal and ion-beam-induced cases as several new segregation and diffusion phenomena have

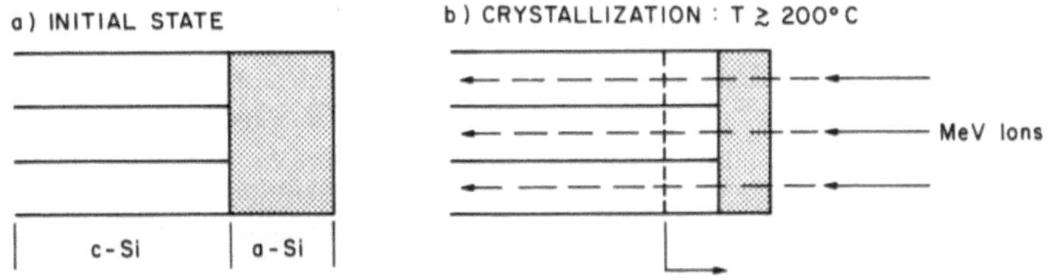

Fig. 12.8a, b. Schematic of the ion-beam-induced crystal growth process

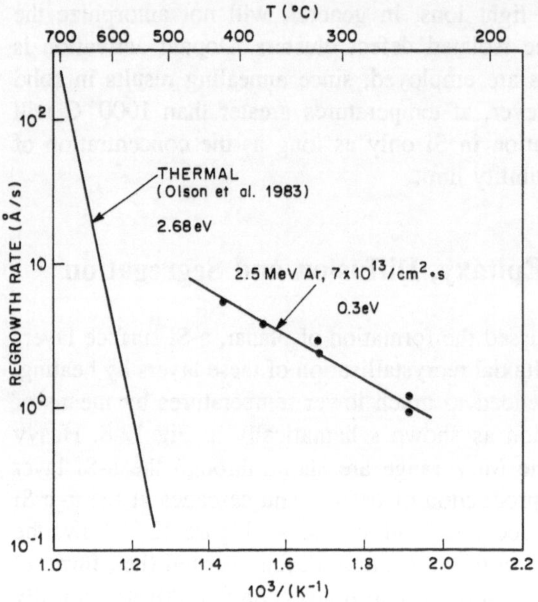

Fig. 12.9. Comparison of Si(100) a–c interface velocities for thermal (i.e., furnace) and ion-beam-enhanced conditions. The 2.5 MeV Ar ion dose rate is 7×10^{13} ions/cm^{2}s

discovered. Solute atoms are segregated at phase boundaries because of differing solubilities in the two phases. Most of our understanding of this process comes from studies of zone-refining during solidification [12.15]. Figure 12.10 shows schematically the segregation of a solute at a phase boundary. We will assume that segregation is occurring at the a–c interface. The ratio of the solute concentration in c-Si to that in a-Si at the interface is given by the interfacial segregation coefficient k'. Assuming a uniform solute concentration n_a in the a-Si, the first a-Si layers to crystallize will have a solute composition of $k'n_a$. As crystallization proceeds, solute accumulates in the a-Si in the near-interface region until a steady-state interface concentration n_a/k' is attained. The width of this segregated spike is given by D/v, where D is the solute diffusivity in a-Si and v is the c–a interface velocity. The fact that solute has to be conserved at steady state requires that the shaded areas be equal, i.e., the solute removed in the initial transient must equal the solute transported in the segregated spike. The width of the initial transient, X_c, is given by $D/k'v$. The condition shown in Fig. 12.10 usually pertains to the case of solidification, when diffusion in the liquid phase is much greater than that in the solid. The solute atoms can therefore diffuse ahead of the interface and build up a segregation spike. This classic behavior is not usually observed during solid phase epitaxy [12.16]. Figure 12.11 shows schematically the observed behavior of impurities in the solid phase. The upper figure shows the behavior of such common dopants as implanted As in a-Si; movement of the ca interface does not measurably perturb the As depth distribution. Such results imply that $k' \approx 1$, i.e., the interfacial solubilities are identical. Moreover, the lack of profile broadening shows that diffusion is negligible in both the c and a phases at typical growth conditions of 500° C.

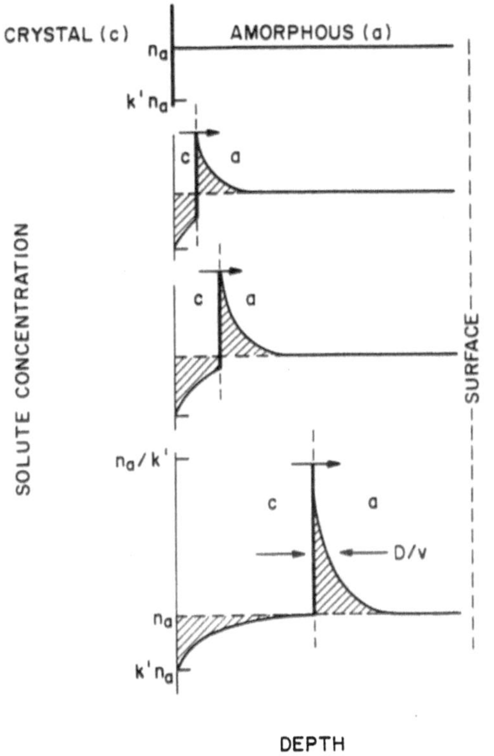

Fig. 12.10. Evolution of the solute spike during interface motion and segregation

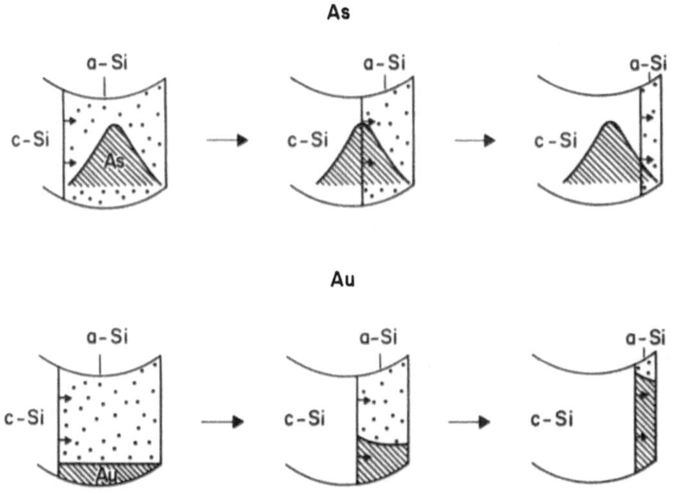

Fig. 12.11. Schematic of c–a Si interface motion and dopant behavior during furnace epitaxy for slow diffusers (*top*, As) and fast diffusers (*bottom*, Au)

While there is much data on diffusion in c-Si, little is known about diffusion in a-Si. We [12.17] and *Kalbitzer* and coworkers [12.18, 19] have pointed out some experimental and phenomenological correlations concerning diffusion in a-Si and c-Si. Firstly, Cu, Ag and Au are relatively fast diffusers in a-Si with well-defined activation energies. Species such as As, In or Sb have immeasurably small diffusion lengths in a-Si for temperatures $< 500°$ C. (By immeasurably small, we mean diffusion lengths smaller than the depth resolution of Rutherford backscattering techniques, i.e. < 100 Å. Measurable diffusion can be observed for very high concentrations of In in a-Si but this is probably a localized melting phenomenon.) This difference correlates with the behavior of these species in c-Si where Cu, Ag and Au are fast (interstitial) diffusers and As, In or Sb are slow (substitutional) diffusers. Figure 12.12 shows diffusion coefficients for Au in a- and c-Si and As in c-Si.

The high temperature diffusion of As in c-Si has been well characterized [12.20]. If As diffusion in a-Si is characterized by the same activation energy and pre-exponential factor as in c-Si, we would expect diffusion lengths of approximately 1 Å for annealing temperatures of 500° C and times of 10^4 s. Figure 12.12 shows our measurements of the diffusivity of Au and a-Si. The diffusion length after 10^4 s at 500° C was found to be 10^3 Å (Fig. 12.12). Recent measurements by *Coffa* et al. [12.21] for the effective diffusion coefficient of Au in c-Si are also shown in Fig. 12.12. A kick-out mechanism was assumed where

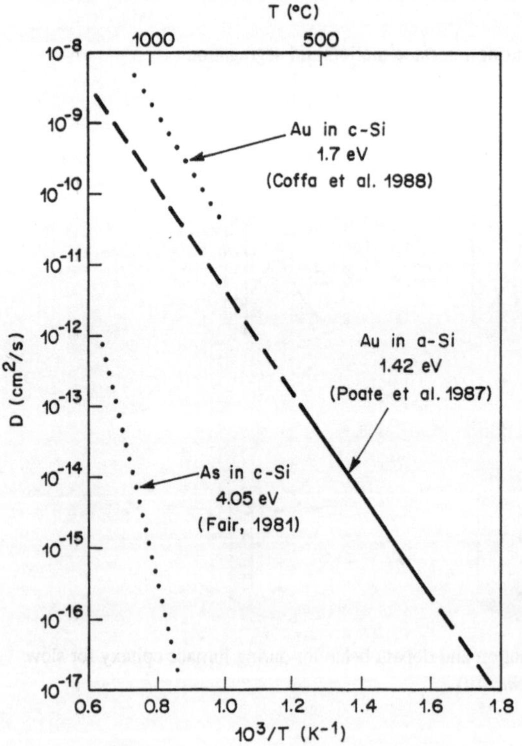

Fig. 12.12. Diffusion coefficients of Au in a- and c-Si and As in c-Si. (From [12.17, 20, 21])

interstitial Au atoms are produced by ejection of Au from substitutional sites by Si interstitials. The diffusion coefficient is then expected to be the product of the interstitial diffusivity multiplied by the ratio of the interstitial and substitutional solubilities. The high-temperature c-Si measurements of Au diffusivity correlate well with our low-temperature a-Si data. This correlation is intriguing, and difficult to understand in terms of the specifics of diffusion in a- and c-Si. It is important to emphasize that fast diffusers in c-Si are also fast diffusers in a-Si. Similarly, slow diffusers in c-Si appear to have small diffusion coefficients in a-Si.

The behavior of the rapid diffusers under solid phase epitaxy conditions is shown schematically for Au [12.22] in the lower part of Fig. 12.11. Not only does Au have high diffusivity, but we estimate Au to be eight orders of magnitude more soluble in a-Si than in c-Si at 515° C. The enhanced solubility and diffusivity is nicely demonstrated by the fact that Au can be found at equilibrium within a-Si layers at temperatures of $\sim$ 500° C without penetrating the underlying c-Si substrate. Solid phase epitaxy at temperatures $\sim$ 500° C results in the Au being retained within the narrowing amorphous layser. The Au diffusivity is so high that the Au outruns the interface motion, thus producing essentially flat segregation profiles. The process has not reached steady state, but one clearly observes $k' \ll 10^{-1}$.

There has been renewed interest in segregation phenomena because of the very rapid interface motion and nonequilibrium crystal growth associated with laser melting of surface layers. As we will show later, the c–a Si interfaces can approach speeds of 15 cm/s during laser melting. Under such conditions, solute can be segregated with k' values exceeding equilibrium values by several orders of magnitude. This solute trapping is a direct consequence of the undercooling (or chemical potential driving force) produced by the high interface velocities, and occurs when the liquid-phase diffusive velocity of the solute is comparable to the interface velocity. In liquid-phase bulk crystal growth, D/v is $\approx 50\,\mu$m whereas in surface laser melting it is $\approx$ 50 Å. As D/v approaches the interatomic spacing, the probability increases for solute atoms to be engulfed or trapped by the advancing interface. For ion-beam-induced crystal growth at 250° C, D/v is less than an interatomic spacing if we use the Au thermal diffusivity presented previously. However, the diffusivity is also enhanced by the ion-beam irradiation to give D/v escape distances of $\approx$ 20 Å. These values result in another intriguing segregation and trapping regime.

We produced the following structures to study ion-beam-induced crystallization and segregation [12.13]. Gold was uniformly diffused through micrometer-thick a-Si layers formed by implantation. Ion-beam-induced crystallization experiments were carried out in the temperature range 250°–420° C using 2.5 MeV Ar ion bombardment. Figure 12.13 shows Rutherford backscattering depth profiles of the Au as a function of Ar dose at 320° C and a dose rate of 7×10^{13} ions/cm^{2}s. The dose rate was kept constant to give an interface velocity of $\sim$ 3 Å/s. The zone-refined profiles display the characteristic features of the segregation process: buildup of segregated solute at the interface and concomitant removal of

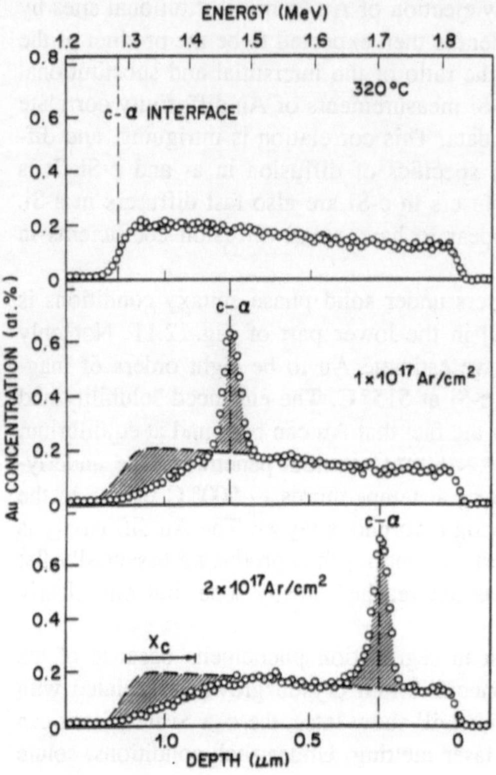

Fig. 12.13. Gold depth profiles in c- and a-Si following 2.5 MeV Ar induced crystallization. The equilibrium distance, X_c, is indicated. (From [12.13])

material during the initial transient. It should be noted that the initial transient, X_c, extends for 0.2 μm. At steady state the amount of material in the segregation spike remains constant and solute is rejected behind the moving interface at the original concentration. It is remarkable that the Au is trapped in c-Si at concentrations some ten orders of magnitude greater than the equilibrium solubility in c-Si. Transmission electron microscopy shows this c-Si to be defect-free without evidence of Au precipitation, although Rutherford backscattering and channeling measurements give no evidence of the Au being substitutional.

The profiles shown in Fig. 12.13 cannot be modeled using the usual thermal diffusion coefficients because the Ar ion irradiation enhances the Au diffusivity at the c–a interface. We have implanted Cu, Ag and Au in a-Si and measured [12.23] the diffusion under MeV Ar ion bombardment. The diffusion coefficients of Cu, Ag and Au in a-Si are shown in Fig. 12.14 versus $1/T$ for irradiation with a 2.5 MeV Ar beam at a dose rate of 7×10^{13} ions/cm^{2}s. The pure thermal diffusion for these three impurities is also shown by heavy solid lines. At the lowest temperatures the diffusion is almost temperature independent (athermal), and due to collisional effects or ion-beam mixing (ballistic regime). Increasing the temperature results in a transition from the ion beam mixing to a thermally activated regime. In this regime the diffusivity is much higher than that encountered during purely thermal diffusion. Finally, at higher temperatures, the purely thermal

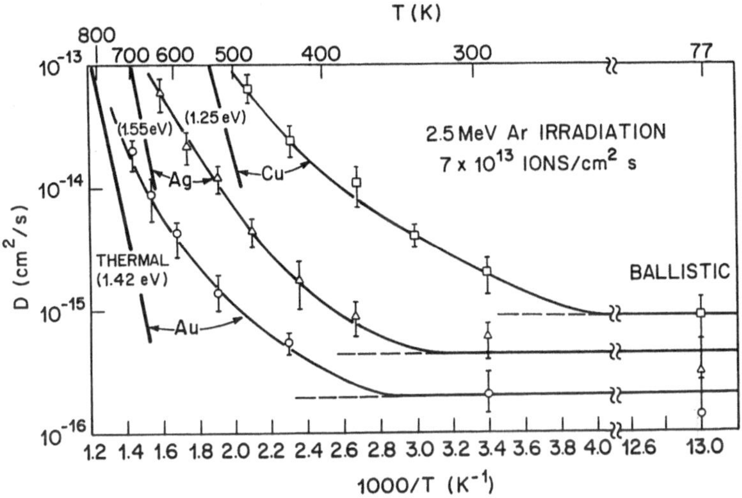

Fig. 12.14. Total diffusion coefficients of Cu, Ag, and Au in a-Si vs $1000/T$ for 2.5 MeV Ar irradiation at dose rates of 7×10^{13} ions/cm^{2}s. The pure thermal components are shown as heavy lines. (From [12.23])

diffusion dominates. We obtain the thermally activated, ion-beam-enhanced coefficient by subtracting both the thermal and ion-beam mixing components from the total diffusivity. An Arrhenius behavior is observed with activation energies of 0.27 ($\pm$0.1), 0.39 ($\pm$0.1) and 0.37 ($\pm$0.1) eV for Cu, Ag and Au, respectively, to be compared with the corresponding thermal values of 1.25 ($\pm$0.04), 1.55 (0.09) and 1.42 ($\pm$0.05) eV.

The measured segregation profiles of Fig. 12.13 can now be combined with the diffusion coefficients of Fig. 12.14 to fit the segregation behavior using only one free parameter, i.e., the interfacial segregation coefficient k'. Figure 12.15a shows a fit to the 320° C data using the measured interface velocity of 2.85 Å/s and radiation-enhanced diffusivity of 4.4×10^{-15} cm^2/s, giving $k' = 0.007 \pm 0.004$. To allow for detector resolution, straggling and the waviness of the interface (as measured by TEM), these fits have been convoluted with a Gaussian function with $\sigma \approx 300$ Å. The width (D/v) of the segregated spike is expected to be ≈ 20 Å, with a peak Au concentration at the interface of 20 at. %. This value is clearly in excess of the equilibrium solubility of Au in a-Si, which, for example, we measure to be 0.7 at. % at 515° C. The solubility of Au in a-Si must therefore be markedly enhanced by ion irradiation over the 20 Å spike width. This is the first time that such segregation phenomena have been observed in the solid phase. All the classic features of segregation and crystal growth are formed, albeit within an intrinsically nonequilibrium process.

The velocity and temperature dependences of k' demonstrate some of the unique characteristics of the ion-beam process. The interface velocity scales with the Ar dose rate. We observe that a one order of magnitude change in velocity (using the same dose but different dose rates) produces identical segregation

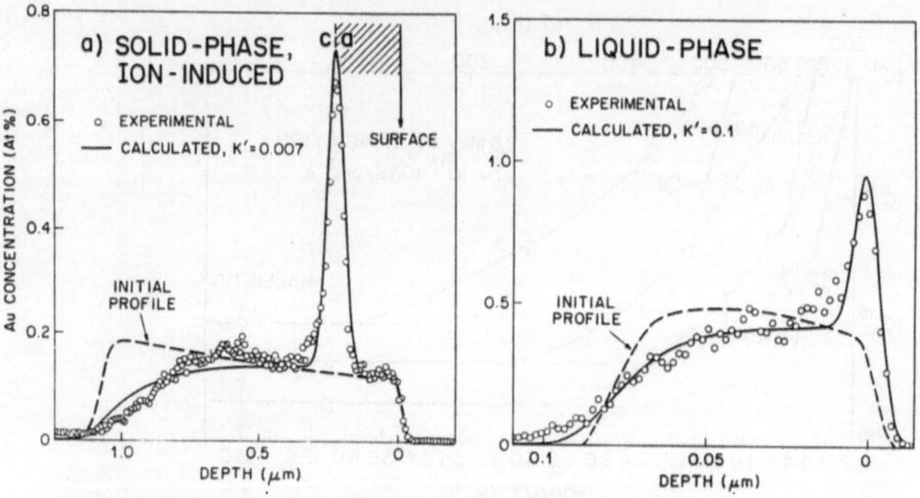

Fig. 12.15. Comparison between ion-beam-induced (a) and laser-induced (b) segregation for the Au-Si system. Note that both the vertical and the horizontal scales are different in the two figures. (a) The Au profile for an ion-beam irradiation with 2.5 MeV Ar at a dose of 2×10^{17} ions/cm^2 and at a temperature of 320° C. (b) The Au profile after 0.55 J/cm^2 laser irradiation. The continuous lines are fits to the data. (From [12.28])

profiles. Specifically, the equilibration distance $D/k'v$ remains constant for all interface velocities. Since both the diffusivity and velocity scale linearly with the dose rate, D/v is independent of interface velocity. Hence the segregation coefficient k' is also found independent of interface velocity.

This behavior is a consequence of the beam-induced crystallization process, and is quite different than that occurring in liquid phase epitaxy, where k' scales with v because the chemical driving force for trapping increases with the velocity-dependent undercooling in the liquid. The segregation coefficient k', however, does vary strongly with temperature, as shown in Fig. 12.16. These data were taken for the same ion beam conditions described previously, with substrate temperature varied in the 250°–420° C range. At high temperatures, k' tends towards the equilibrium value with essentially no trapping of Au in the crystalline phase.

We do not yet have an understanding of this result in terms of the crystal growth processes. The driving force for trapping is related to the chemical potentials of the impurities in a- and c-Si, and hence should depend on temperature but not necessarily on the interface velocity. However, we do not know to what extent the production of defects by the ion beam will change the chemical driving forces. Moreover, ion-beam mixing at the interface may also play a role in the nonequilibrium trapping of Au in the growing crystalline phase.

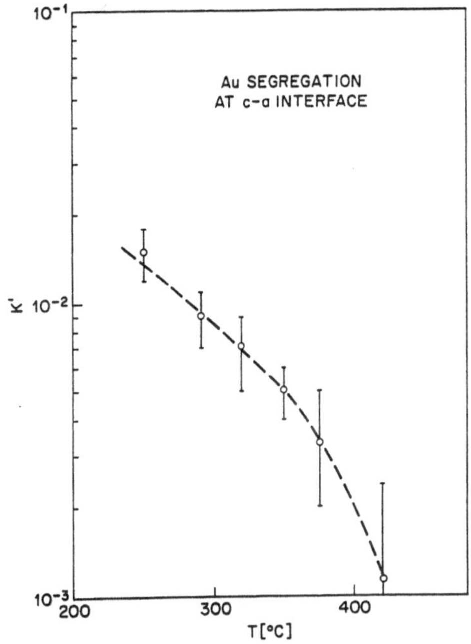

Fig. 12.16. The interfacial segregation coefficients k' of Au at the c–a Si interface as a function of temperature, under Ar-ion-induced crystallization

12.4 Thermodynamic and Kinetic Properties of Amorphous Si

Amorphous Si is in a higher energy state than c-Si and this difference is the basic driving force for crystallization. Although a-Si is thermodynamically unstable in contact with c-Si, it does not recrystallize because of kinetic barriers. In the previous section we discussed some of the ways these barriers can be surpassed. Figure 12.17 shows a plot of the Gibbs free energy of a- and l-Si relative to c-Si. When these calculations were first carried out by *Bagley* and *Chen* [12.24] and *Spaepen* and *Turnbull* [12.25] there were no data on the thermodynamic properties of a-Si, and data from a-Ge were used. We embarked on a program several years ago to measure the thermodynamic properties of a-Si. There were several reasons for this, the principal one being the prediction of the melting temperature of a-Si. The intersection of the a and l free energy curves gives, by definition, the amorphous melting temperature (T_{al}). We see from the curve that $T_{cl} - T_{al} \approx 250°$ C where T_{cl} is the melting temperature of crystalline Si. This is a remarkable difference in melting temperature for two solid phases of the same element.

The most convincing argument to explain this large depression in melting temperature is based on the premise that the phase transition from the amorphous to the liquid phase is discontinuous and first order. The physical reasoning behind this premise is plausible when it is realized that the transition involves a fundamental change in bonding from the covalent, four-fold coordinated amorphous phase to the metallic liquid, with (11–12)-fold coordination. Clearly this

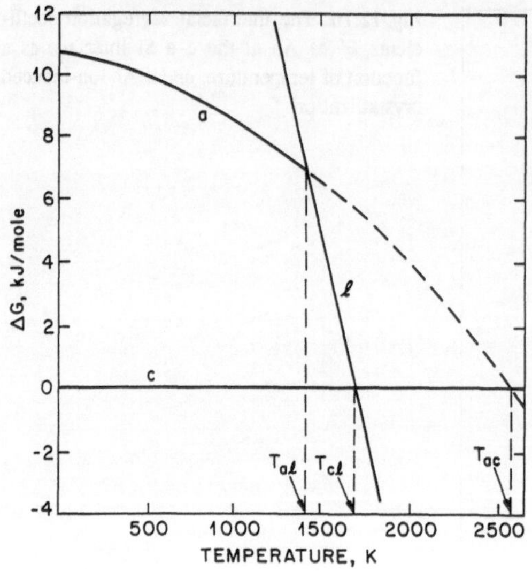

Fig. 12.17. Estimate of the Gibbs free energy of a-Si and l-Si relative to c-Si. (From [12.26])

transition will only occur during very rapid heating, otherwise the amorphous phase will recrystallize directly in the solid phase, as observed during furnace heating, for example.

In this section we will describe experiments carried out to measure the free energy of a-Si (see also Sect. 12.8). In the following sections we will describe rapid melting experiments which demonstrated the first-order nature of the transition. Rather thick a-Si layers have to be used to measure heat release in a calorimeter. *Donovan* et al. [12.26] achieved this using sequential MeV Ar implants which produced continuous a-Si layers some 2 μm thick. These samples were analyzed in a differential scanning calorimeter (DSC). Figure 12.18 shows a DSC scan, 40 K/min, from recent experiments [12.40] using MeV Si ions to produce 2 μm a-Si layers. Silicon energies of 0.5, 1.0 and 2.0 MeV were employed at doses of 5×10^{15} cm^{-2} and a substrate temperature of 110 K. There are several features to this data. The high-temperature heat release starting at 900 K corresponds to heat released at the moving a–c interface. The fall-off at 970 K corresponds to the interface reaching the surface. This heat release is exponential with temperature, and the interface velocity can be determined directly from the DSC curve. These data are in excellent agreement with those of *Olson* and *Roth,* with an activation energy of 2.62 eV. The integrated area under the crystallization peak corresponds to an enthalpy of crystallization of 13.3 ± 0.7 kJ/mol. This value, when used in the Gibbs free energy calculations, gives $T_{\mathrm{cl}} - T_{\mathrm{al}} = 250°$ C.

The long tail in the DSC data with an integrated heat release of 4.85 kJ/mol is not associated with interface motion. It corresponds, instead, to a continuous or homogeneous heat release phenomenon. This heat comes from relaxation of the amorphous network. It is generally believed that the a-Si structure can be adequately described by the tetrahedrally coordinated random network model. The amorphous nature is accounted for by relatively small distortions of the

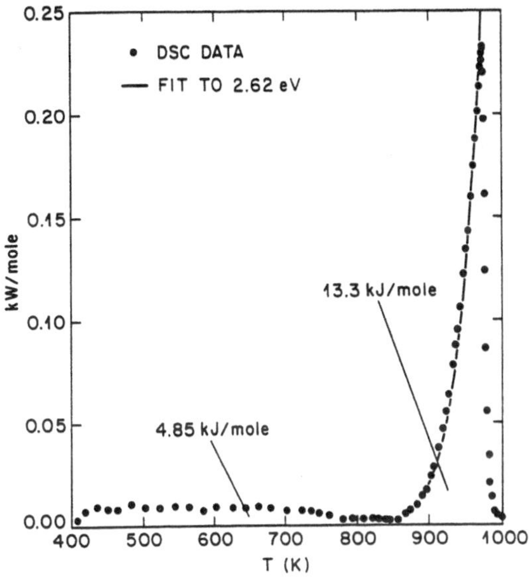

Fig. 12.18. DSC output for a stack of 3 Si wafers implanted on both sides to a depth of 2.1 μm. (From [12.40])

average bond angles. Relaxation could therefore represent changes in the bond angle distortions, or movement and annihilation of defects intrinsic to the amorphous network. We are now trying to distinguish between these different relaxation mechanisms.

12.5 Liquid Phase Crystal Growth and Dopant Segregation

We now change gears and discuss the pulsed laser irradiation of solids to produce heating and melting. It is generally agreed that, for Si, the carriers excited by laser irradiation and the lattice reach a common temperature on a picosecond time scale. Indeed a Si surface layer may melt within such a time scale. This is an interesting regime, as the collision cascades discussed in the previous section have lifetimes $\sim 10^{-1}$ ps. Here we will discuss nanosecond laser irradiation, where solidification velocities can be controlled in the range 1–20 m/s.

Figure 12.19 shows calculations for pulsed ruby laser irradiation of Si and Al. The comparison is made for the same melt depth to show the differences between metals and semiconductors. The melt front propagates at approximately the same speed ($\sim$ 20 m/s). The velocity of the resolidification front is an order of magnitude faster for Al (20 m/s) than Si (2 m/s) because of its higher thermal conductivity. The velocity depends upon the rate at which the latent heat of crystallization can be extracted from the solidifying interface. For these cases the thermal gradients range between 10^6 and 10^7 K/cm with quenching rates of 10^9–10^{10} K/s. Solidification velocities can be varied by changes in laser pulse length or Si substrate temperature. For ⟨100⟩ Si, perfect epitaxy is maintained for l–c interface velocities as fast as 15 m/s.

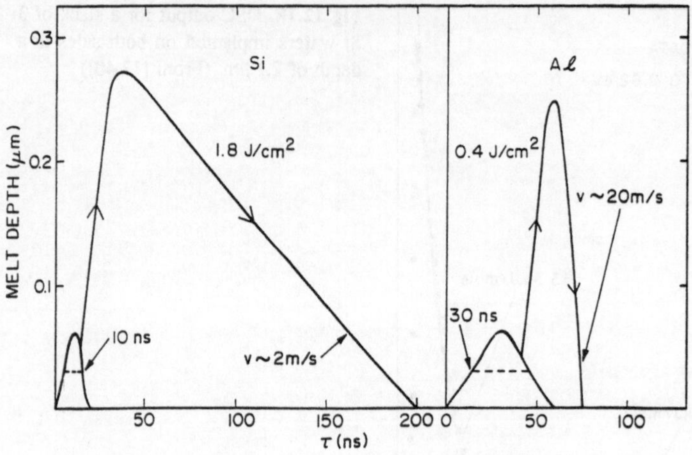

Fig. 12.19. Position of melt-front in Si and Al after irradiation with a pulsed ruby laser

One of the most interesting phenomena observed during laser annealing of Si was impurity atom trapping in c-Si at concentrations far above the equilibrium concentration. Figure 12.20 shows [12.27] depth profiles of Bi trapped in c-Si after pulsed laser irradiation produced solidification velocities of 2 m/s. The triangular points represent the channeling spectra and show that virtually all the Bi is trapped on lattice sites. There are several salient features to these data. The smooth lines show the fit to the data. The segregation coefficients are much lower than the equilibrium value of 7×10^{-4} and there is a strong orientation dependence to the trapping. It should be noted that it is impossible to quench the solidification front and catch the interface in its trajectory, as was done in the case of the ion-beam-induced segregation data (Figs. 12.13 and 12.15).

Recently, laser annealing experiments [12.28] were conducted on the Au-Si systems using a 3 ns Q-switched Nd:YAG laser in the frequency-doubled mode. Gold was implanted into Si, and Fig. 12.15b shows measured and calculated segregation profiles, assuming an interface velocity of 9 m/s and Au liquid phase diffusivity of 10^{-4} cm²/s. Alongside in Fig. 12.15a is the ion-beam-induced segregation profile for the Au-Si system, described earlier. Several noteworthy observations can be made. First, there is a striking similarity between the two Au redistribution profiles despite the fact that one process occurs in the liquid, at an interface velocity of 9 m/s, and the other in the solid phase at an interface velocity of 3 Å/s. The trapping of the Au is greater during laser irradiation at interface velocities of 9 m/s than during ion beam irradiation. However, laser irradiation with a 30 ns pulse (Q-switched ruby laser, 694 nm) at interface velocities of ~ 2 m/s results in substantially lower trapping rates ($k' \sim 10^{-3}$). The equilibrium solid liquid segregation coefficient is estimated to be $\sim 2 \times 10^{-5}$. There is, however, a marked difference between the two segregation processes in Fig. 12.15. During ion-beam-induced segregation Au is not trapped uniquely on lattice sites. In laser-induced-segregation, Au is trapped on near-substitutional sites to concen-

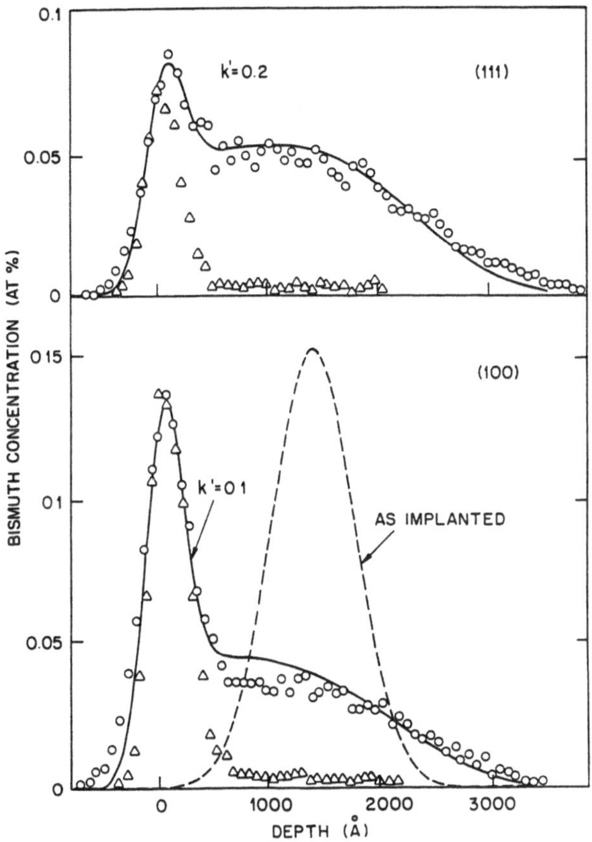

Fig. 12.20. Depth profiles for 240 keV Bi^+ implants in Si(100) and Si(111), annealed using identical laser treatments (ruby, 2.0 J/cm^2, 100 ns). (From [12.27])

trations near 0.5 at. %. Future investigations should determine whether one can model the ion-beam-induced segregation in terms of free-energy or undercooling concepts or if one should consider, instead, kinetic or ballistic ion–beam effects.

In Fig. 12.21 we show data [12.29, 30] for In solute trapping during laser melting. The data for Bi and Sb are similar. These data represent the first determinations of the velocity- and orientation-dependent segregation coefficients. The k' values at 5 m/s are increased by two orders of magnitude above the equilibrium value. This solute trapping is a direct consequence of the high interface velocities. The diffusive velocity of impurities in the liquid phase, away from the interface, is less than the interface velocity. The plot also shows segregation at the (111) interface, where the k''s are considerably higher than the corresponding (100) values. This behavior can be explained in terms of undercooling or in terms of the interface structure. The undercooling is expected to be greater at a (111) interface than a (100) interface for the same velocity. Greater undercooling will produce greater trapping. Structurally, growth occurs at the (111) interface by

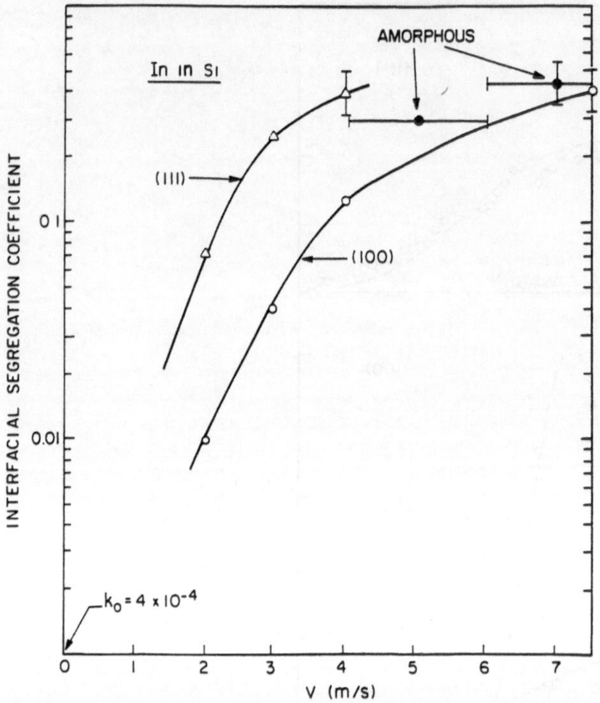

Fig. 12.21. Interfacial segregation coefficients k' for In in crystalline and amorphous Si. The large uncertainties in the velocity of the a–l interface come from uncertainties in the heat-flow calculations. (From [12.29, 30])

lateral motion of ledges. The velocity of the ledges must be increased to maintain the crystallization velocity. Trapping will, therefore, be enhanced by increased velocity of the ledges. Segregation coefficients at the amorphous liquid interface were also measured [12.30]. It was found that low-energy irradiation of a-Si layers containing implanted impurities will produce buried melts which freeze directly into a-Si. Moreover, the solidifying a–l Si interface is well defined, and segregates impurities in a fashion very similar to that of c–l interface. Segregation at a well-defined a–l Si interface demonstrates that the transition is first order and that it is *not* a glass transition.

12.6 Melting of Amorphous Si: A First-Order Phase Transition

As we have discussed, there are cogent arguments to believe that a-Si undergoes a first-order phase transition. These range from the heat of crystallization measurements to segregation phenomena at the a–l interface. An earlier measurements, a peculiar structure was also observed in a-Si layers which had been pulse-heated with electrons at relatively low energies. These structures could only be interpreted in terms of a considerable reduction in the melting temperature. The heat of fusion ΔH_{al} measured from those experiments agreed well with the value estimated using Gibbs free energy concepts. This agreement tended to verify the basic thermodynamics concepts. However, the transition was not dynamically observed until the development of transient measuring techniques.

The development of the transient reflectance [12.31] and conductance [12.32] techniques has greatly aided the understanding of the dynamics of rapid melting and solidification experiments. Time-dependent reflectance $R(t)$ measurements of the Si surface are used to determine when the surface melts and solidifies. The technique is only sensitive to optical property changes within the absorption depth of the probe beam, which is typically 10–20 nm for molten Si. Transient conductance measurements, on the other hand, give the molten layer thicknesses as a function of time, but are insensitive to the location of the melt. The concept behind the transient conductance measurements is that, for most materials, the electrical conductivity changes discontinuously at phase changes. Indeed, many semiconductors become metallic upon melting; for example, molten Si is metallic, and the conductivity increases by a factor ~ 30 upon melting. As a result, if the electrical conductance of a sample with a well-defined geometry is measured in real time, the melt depth $d(t)$ as a function of time can be directly determined from changes in the conductance. The schematic setup which allows transient conductance and reflectance techniques measurements is shown in Fig. 12.22.

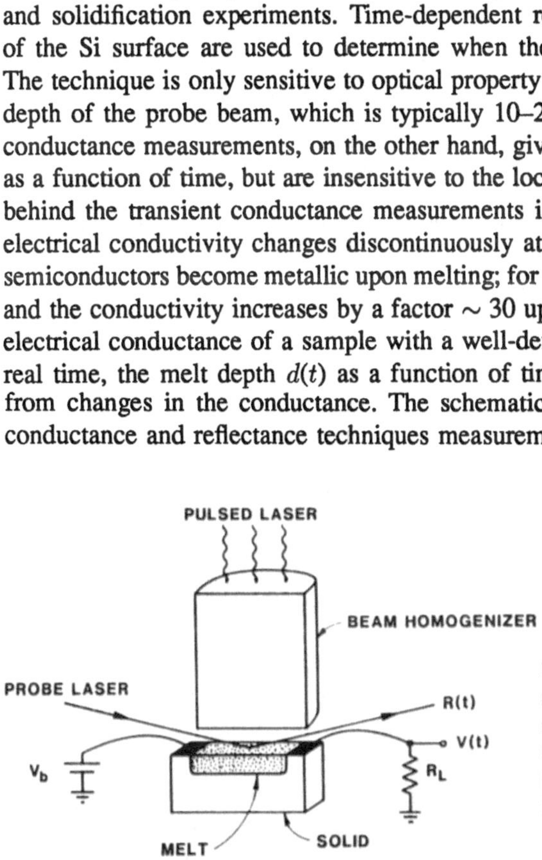

Fig. 12.22. Schematic of the experimental arrangement for simultaneous measurement of the transient conductance and surface reflectance for real-time measurements of melt and solidification dynamics. (From [12.33])

Figure 12.23 shows transient measurements of $R(t)$ and $d(t)$ [12.33] performed on Si on sapphire (SOS). This material is used because its low electron–hole lifetime reduces interference to the transient conductance signal from photoconductivity. The SOS is patterned lithograhically to produce a resistor with a 55.5 length-to-width ratio. Irradiation was achieved with a 30 ns ruby laser pulse at 694 nm, shown by the dot-dash line. The rise in reflectance $R(t)$ indicates surface melting with a melt duration of some 130 ns. The conductance $d(t)$ shows an initial perturbation due to photoconductivity, followed by the effect of

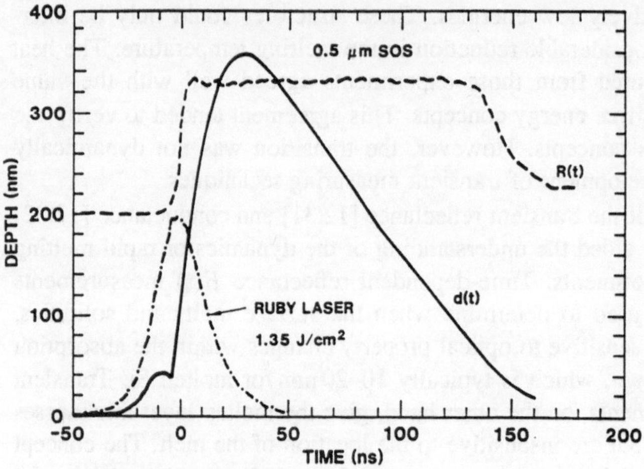

Fig. 12.23. Surface reflectance $R(t)$ and melt depths $d(t)$ versus time measured from transient conductance of SOS. (From [12.33])

a melt-front which propagates into the Si at a velocity greater than 20 m/s until the maximum melt depth of 350 nm is attained. At that point, the liquid/solid interface reverses direction and solidification ensues at a velocity of 3.4 m/s. These measurements in c-Si agree well with heat-flow calculations. The transient conductance technique is a powerful tool for measuring melting and solidification kinetics.

The melting of a-Si can be observed directly [12.34] by irradiating a-Si structures on SOS at low energies. Figure 12.24 shows the melt depth as a function of time for laser irradiation of 300 nm of a-Si on SOS at various incident energies. Transient refelctance measurements demonstrated that the melt originates at the irradiated surface just before the peak of the laser pulse. As the melt reaches th amorphous crystalline interface, a plateau is observed with an incident-energy-density-dependent duration between 2 and 6 ns.

The plateaus shown in Fig. 12.24 are consistent with a first-order phase transition of a-Si to metallic liquid at a melting temperature that is reduced from the melting temperature of c-Si. The data are interpreted as follows. Melt initiates at the irradiated surface when the temperature exceeds the melting temperature T_{al} of a-Si. The melt propagates into the a-Si film at this reduced temperature until it encounters the a–c Si interface. Because the temperature of the molten liquid is below the melting temperature of c-Si, the melt front must pause at this interface until enough energy is absorbed from the laser to raise the liquid temperature to T_{cl}. Once this energy is absorbed, the melt front can propagate into the underlying c-Si. It should be noted that the melt velocity (slope of the melt-depth-versus-time curve) differs in a- and c-Si, reflecting the different latent heats and thermal conductivities of the two phases. The absolute difference between the melting temperature of a-Si and c-Si can be determined from the plateau width and the incident laser intensity. For the duration of the plateau, the

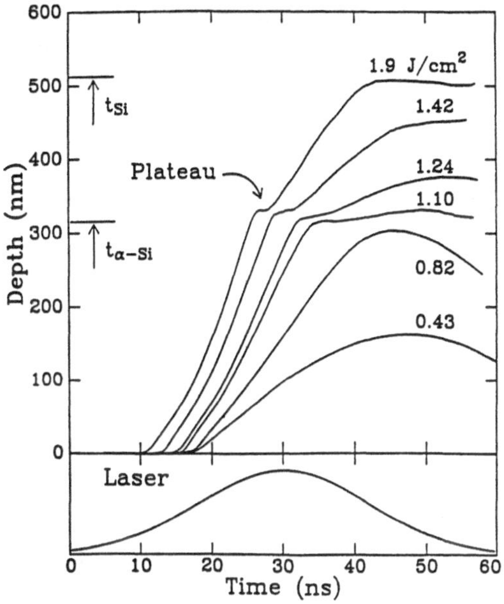

Fig. 12.24. Melt depths versus time (*top*), and laser pulse shape for various irradiations of 300 nm a-Si on SOS. Note the plateau occurring at depths equivalent to a-Si thickness. (From [12.34])

heat necessary to raise the liquid layer by $T_{cl} - T_{al}$ and the heat flow into the substrate must be balanced by the heat supplied by the laser. This analysis yields a melting temperature reduction of 225 ± 50 K, in agreement with the free-energy calculations.

12.7 Conclusion: Undercooling and Explosive Crystallization

We think that a fitting conclusion to this chapter is the illustration of how beam techniques have recently clarified the century-old problem of explosive crystallization.

The amorphous–liquid transition should be reversible so that undercooling the melt beneath T_{al} should lead to growth of a-Si in the melt. When ultrafast laser heating is used, it is possible to quench a-Si from the melt [12.35, 36]. This was the first time that a-Si had been quenched from the melt. By the transient conductance technique, using 2 ns laser pulses in the near UV, it was determined [12.37] that a-Si is formed when the liquid-solid interface velocity exceeds 15 m/s. At such a velocity, therefore, the liquid must have a greater undercooling than 225° C.

The relationship between undercooling and crystal growth, and between superheating and melting, is fundamental to melting–solidification theory. Figure 12.25 illustrates the general trend. The higher the undercooling or superheating, the greater the crystal growth velocity. However, at the highest undercooling, the crystal growth velocity tends asymptotically to zero because of the viscosity

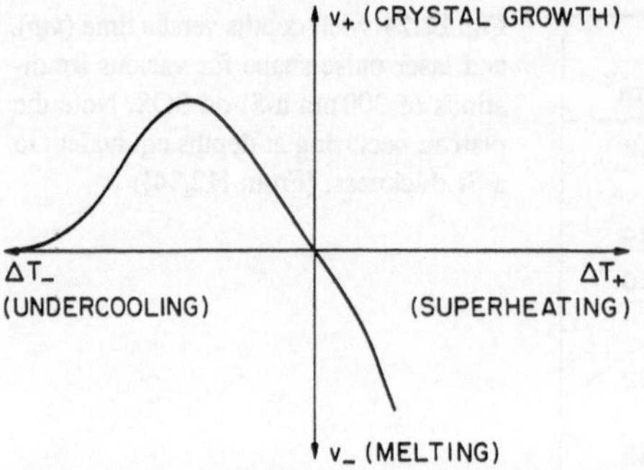

Fig. 12.25. Schematic of crystal growth or melting velocities as a function of interface undercooling or superheating

of the melt. Although Si is such an important material, data on interface velocities during growth or melting are noticeably scarce. In fact the only data we are aware of come from transient conductance experiments [11.33], where the undercooling, at an interface velocity of 6 m/s, was estimated at 90 K. This corresponds to an undercooling rate of 15 K/m s. This value is consistent with laser amorphization experiments, in which the undercooling at 15 m/s must be greater than 225 K (i.e., the freezing temperature depression for a-Si versus c-Si). The various determinations of T_{al} therefore seem somewhat consistent.

The transient conductance and reflectance techniques have allowed an explanation of an unusual crystallization phenomenon, "explosive crystallization", which has a long history. Explosive crystallization involves release of the heat of crystallization from a small region by some initiating disturbance such as a pinprick or laser pulse. This release of energy heats the adjacent material which can then crystallize. *Gore* [12.38] first observed the explosive crystallization of amorphous layers of Sb in 1855. Beautiful crystallization patterns can be generated in amorphous solids. An important feature of the phenomenon is that the crystallizing interfaces can move with velocities of meters per second. If a-Si samples similar to those shown in Fig. 12.24 are irradiated at energies just above the threshold for surface melting, explosive crystallization is observed with the generation of fine-grain polycrystallites. This was believed to be an explosive crystallization phenomenon because the thickness of the polycrystalline layer was much greater than that expected from sample heat-flow calculations.

Figure 12.26 shows transient conductance curves for an incident energy density of 0.20 J/cm^2, which is slightly above the melt threshold energy density of $\sim$ 0.14 J/cm^2. Of particular interest is the double-peaked structure observed in $d(t)$, and the fact that $R(t)$ returns to its solid phase value before the molten layer disappears completely. These data are interpreted as follows. Melt initiates at the irradiated surface, as demonstrated by comparison of $d(t)$ and $R(t)$, and propagates to a depth of $\sim$ 12 nm, whereupon this primary melt begins to solidify. This region is denoted as region *a* and is illustrated schematically in the figure. Associated with the solidification of this primary melt, the latent heat of crystallization ΔH_{lc} is released and heats the underlying a-Si above its reduced melting temperature. The underlying a-Si then melts to create a buried molten layer. While the interface from the primary melt returns to the surface, the buried

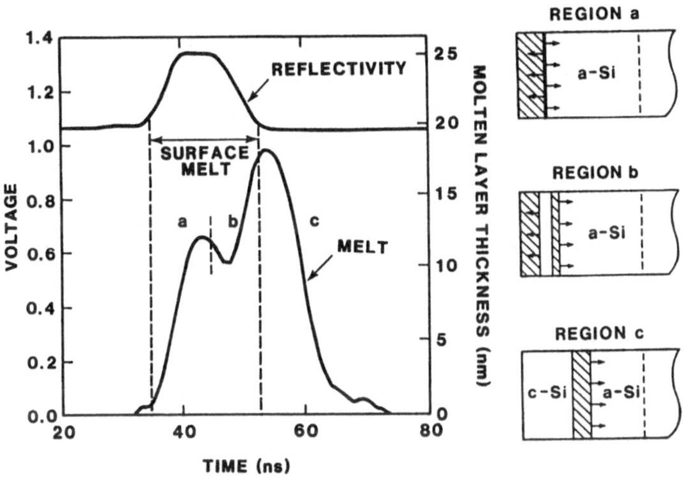

Fig. 12.26. Transient conductance measurements of molten layer thickness and simultaneous surface reflectance for a 0.20 J/cm^2 pulse incident on a 320 nm thick a-Si film. The schematic diagrams at the right illustrate the various regions described in the text for the explosive crystallization process. (From [12.34])

molten layer propagates into the sample, as illustrated schematically as region *b*. At later times, the liquid/solid interface from the original melt has returned to the surface, leaving a buried molten layer in the sample beneath a solid surface. This region is indicated as region *c*. The energy released by crystallization at the rear interface of the buried molten layer is sufficient to raise the temperature of the a-Si at the front interface to the melting temperature of a-Si, and to supply the latent heat ΔH_{al} for the amorphous–liquid phase transition in Si. The buried molten layer therefore propagates into the a-Si. Explosive crystallization thus become a self-sustaining process, mediated by a buried molten layer. Because the melting temperature of a-Si is less than that of c-Si, the melt is severely undercooled. The rapid nucleation at the crystal/liquid interface in this undercooled melt presumably results in the fine-grained polycrystalline Si observed after the explosive crystallization. Using these beam techniques it was therefore finally possible to elucidate the mechanism of a crystallization phenomenon observed over a century ago.

12.8 Update: The State of Amorphous Si

There have been several important developments in our understanding of a-Si formed by implantation since the rest of this chapter was written. The original calorimetry measurements (Sect. 12.4) of *Donovan* et al. [12.26] showed no heat of relaxation, although similar experiments on a-Ge did show such a homogeneous heat release. *Roorda* et al. [12.39] of the FOM, Amsterdam, reasoned that such a heat release should be present in a-Si and repeated the calorimetry measurements. They indeed found a substantial heat release, which we confirmed in subsequent measurements [12.40]. Base line shifts in the calorimetry had caused us to miss the heat release in the early measurements. Having agreed on the existence of the heat release, we collaborated with the FOM group in investigating the basic relaxation mechanisms. These studies have been crucial in throwing light on some of the properties of a-Si.

The as-prepared a-Si can be heated to temperatures approaching 500° C without crystallization occurring but with the heat being released from the homogeneous relaxation process. The a-Si is then in a relaxed state in contrast to the original, unrelaxed state. We have shown [12.41, 42] by calorimetry and Raman techniques that the introduction of defects into the relaxed a-Si by ion bombardment will return it to the unrelaxed state. The a-Si is converted fully to the unrelaxed state by the introduction of 0.02 displacements per target atom. Further bombardment does not introduce any more defects in the a-Si. The heat release from the relaxation of a-Si is therefore due to the annihilation of defects. Moreover, it appears that these defects have characteristics similar to those introduced in c-Si by ion bombardment.

These experiments have established that the as-implanted a-Si contains levels of intrinsic defects of several atomic per cent. Why were such high concentrations of defects not observed in the diffusion experiments discussed in Sect. 12.3? The answer lies in the fact that we did not possess the relevant experimental parameters to home in on the phenomenon. The recent calorimetry experiments, however, give us that information. The defect annealing is a continuous function of temperature. Moreover, the relaxation or defect annihilation process occurs over a time of approximately 100 s. By using fast diffusing Cu as a marker, the defect populations have been clearly identified in relaxed and unrelaxed material [12.43].

Very recent experiments by *Coffa* et al. [12.44, 45] have elucidated the diffusion and solubility mechanisms in a-Si for the fast diffusing species. These species diffuse interstitially in a-Si with essentially the same interstitial diffusivities as in c-Si. The defects, however, can act as trapping sites, thus giving rise to effective diffusivities with activation energies of 1.5 eV. Moreover, the high density of trapping sites explains why these species have such high solubilities in a-Si. The trapping and saturation behavior of the defects has been charted as a function of concentration and annealing times in diffusion experiments. We have, therefore, established a consistent framework to understand the diffusivity and solubility experiments in terms of defects intrinsic to the a-Si network. The next challenge is to identify the nature and structure of these defects within the context of the random network model.

References

12.1 R.S. Ohl: Bell Syst. Tech. J. **31**, 104 (1952)

12.2 W. Shockley: U.S. Patent 2,787,564

12.3 J.M. Poate, J.M.Mayer: *Laser Annealing of Semiconductors* (Academic, New York 1982)

12.4 J.W. Mayer, L. Eriksson, J.A. Davies: *Ion Implantation in Semiconductors* (Academic, New York 1970)

12.5 R.G.Wilson, G.R. Brewer: *Ion Beams with Applications to Ion Implantation* (Wiley, New York 1973)

12.6 G. Dearnaley, J.H. Freeman, R.S. Nelson, J. Stephens: *Ion Implantation* (North-Holland, Amsterdam 1973)

12.7 B. Smith: *Ion Implantation Range Data for Si and Ge Device Technologies* (Research Studies Press, Forest Grove, OR 1977)

12.8 J. Lindhard, M.E. Scharff, H.E. Schiott: K. Dan. Vidensk. Selsk., Mat. Fys. Medd. **13**, No. 14 (1963)

12.9 W.K. Hofker: Philips Res. Rep. Suppl., No. 8 (1975)

12.10 J. Fletcher, J. Narayan, O.N. Holland: In *Microcospy of Semiconducting Materials 1981*, Inst. of Phys. Conf. Ser., Vol. 60, ed. by A.G. Cullis, D.C. Joy (Inst. of Phys., Bristol 1981) p. 295

12.11 G.L. Olson, J.A. Roth: Mater. Sci. Rep. **3**, 1 (1988)

12.12 T.E. Seidel: In *VLSI Technology*, ed. by S.M. Sze (Wiley, New York 1983) Chap. 6

12.13 J.M. Poate, J. Linnros, F. Priolo, D.C. Jacobson, J.L. Batstone, M.O. Thompson: Phys. Rev. Lett. **60**, 1322 (1988)

12.14 K.A. Jackson: J. Mater. Res. **3**, 1218 (1988)

12.15 B. Chalmers: *Principles of Solidification* (Wiley, New York 1964)

12.16 J.S. Williams: In *Surface Modification and Alloying*, ed. by J.M. Poate, G. Foti, D.C. Jacobson (Plenum, New York 1983) Chap. 5

12.17 J.M. Poate, D.C. Jacobson, J.S. Williams, R.G. Elliman, D.O. Boerma: Nucl. Instrum. Methods B **19/20**, 480 (1987)

12.18 M. Reinelt, S. Kalbitzer: J. de Phys. Colloq. C **4**, 843 (1981)

12.19 S. Kalbitzer, M. Reinelt, W. Stolz: In Proc. of the 4th EC Photovoltaic Solar Energy Conference, Stesa, May 1982, ed. by W.H. Bloss, G. Grassi (Reidel, Dordrecht, Holland 1982) p. 163

12.20 R.B. Fair: In *Impurity Doping Processes in Silicon*, ed. by F.F.Y. Wang (North-Holland, New York 1981) Chap. 7

12.21 S. Coffa, L. Calcagno, S.U. Campisano, G. Calleri, G. Ferla: J. Appl. Phys. **64**, 6291 (1988)

12.22 D.C. Jacobson, J.M. Poate, G.L. Olson: Appl. Phys. Lett. **48**, 118 (1986)

12.23 F. Priolo, J.M. Poate, D.C. Jacobson, J. Linnros, J.L. Batstone, S.U. Campisano: Appl. Phys. Lett. **52**, 1213 (1988); Nucl. Instrum. Methods B **39**, 343 (1989)

12.24 B.G. Bagley, H.S. Chen: AIP Conf. Proc. **50**, 97 (1979)

12.25 F. Spaepen, D. Turnbull: AIP Conf. Proc. **50**, 73 (1979)

12.26 E.P. Donovan, F. Spaepen, D. Turnbull, J.M. Poate, D.C. Jacobson: J. Appl. Phys. **57**, 1795 (1985)

12.27 P. Baeri, G. Foti, J.M. Poate, S.U. Campisano, A.G. Cullis: Appl. Phys. Lett. **38**, 800 (1981)

12.28 F. Priolo, J.M. Poate, D.C. Jacobson, J.L. Batstone, J.S. Custer, M.O. Thompson: Appl. Phys. Lett. **53**, 2486 (1988)

12.29 J.M. Poate: In *Laser and Electron-Beam Interactions with Solids*, Mater. Res. Soc. Proc., Vol. 4, ed. by B.R. Appleton, G.K. Celler (North-Holland, New York 1982) p. 121

12.30 S.U. Campisano, D.C. Jacobson, J.M. Poate, A.G. Cullis, N.G.chew: Appl. Phys. Lett. **46**, 846 (1985)

12.31 D.H. Auston, C.M. Surko, T.N.C. Venkatesan, R.E. Slusher, J.A. Golovchenko: Appl. Phys. Lett. **33**, 437 (1978)

12.32 G.J. Galvin, M.O. Thompson, J.W. Mayer, R.B. Hammond, N. Paulter, P.S. Peercy: Phys. Rev. Lett. **48**, 33 (1982)
12.33 P.S. Peercy, M.O. Thompson: In *Energy Beam-Solid Interactions and Transient Thermal Processing*, Mater. Res. Soc. Proc., Vol. 35, ed. by D.K. Biegelsen, G. Rozgonyi, C. Shank (Materials Research Society, Pittsburgh, PA 1985) p. 53
12.34 M.O. Thompson, G.J. Galvin, J.W. Mayer, P.S. Peercy, J.M. Poate, D.C. Jacobson, A.G. Cullis, N.G.Chew: Phys. Rev. Lett. **52**, 2360 (1984)
12.35 P.L. Liu, R. Yen, N. Bloembergen, R.T. Hodgson: Appl. Phys. Lett. **34**, 864 (1979)
12.36 R. Tsu, R.T. Hodgson, T.Y. Tan, J.E.E. Baglin: Phys. Rev. Lett. **42**, 1358 (1979)
12.37 M.O. Thompson, J.W. Mayer, A.G. Cullis, H.C. Webber, N.G. Chew, J.M. Poate, D.C. Jacobson: Phys. Rev. Lett. **50**, 896 (1983)
12.38 G. Gore: Philips. Mag. **9**, 73 (1855)
12.39 S. Roorda, S. Doom, W.C. Sinke, P.M.L.O. Scholte, E. van Loenen: Phys. Rev. Lett. **62**, 1880 (1989)
12.40 E.P. Donovan, F. Spaepan, J.M. Poate, D.C. Jacobson: Appl. Phys. Lett. **55**, 1516 (1989)
12.41 S. Roorda, J.M. Poate, D.C. Jacobson, D.J. Eaglesham, B.S. Dennis, S. Dierker, W.C. Sinke, F. Spaepen: Solid State Commun. **75**, 197 (1990)
12.42 S. Roorda, J.M. Poate, D.C. Jacobson, B.S. Dennis, S. Dierker, W.C. Sinke: Appl. Phys. Lett. **56**, 2097 (1990)
12.43 A. Polman, D.C. Jacobson, S. Coffa, J.M. Poate, S. Roorda, W.C. Sinke: Appl. Phys. Lett. **57**, 1230 (1990)
12.44 S. Coffa, J.M. Poate, D.C. Jacobson, A. Polman: Appl. Phys. Lett., in press
12.45 S. Coffa, J.M. Poate, D.C. Jacobson, W. Frank: To be published

Subject Index